Representations for Genetic
and Evolutionary Algorithms

Studies in Fuzziness and Soft Computing

Editor-in-chief

Prof. Janusz Kacprzyk
Systems Research Institute
Polish Academy of Sciences
ul. Newelska 6
01-447 Warsaw, Poland
E-mail: kacprzyk@ibspan.waw.pl
http://www.springer.de/cgi-bin/search_book.pl?series=2941

Franz Rothlauf

Representations for Genetic and Evolutionary Algorithms

With a Foreword by
David E. Goldberg

With 99 Figures
and 53 Tables

Physica-Verlag

A Springer-Verlag Company

Dr. Franz Rothlauf
University of Bayreuth
Department of Information Systems
Universitätsstraße 30
95440 Bayreuth
Germany
franz@rothlauf.com

ISBN 978-3-642-88096-4   ISBN 978-3-642-88094-0 (eBook)
DOI 10.1007/978-3-642-88094-0

Library of Congress Cataloging-in-Publication Data applied for
Die Deutsche Bibliothek – CIP-Einheitsaufnahme
Rothlauf, Franz: Representations for genetic and evolutionary algorithms: with 53 tables / Franz Rothlauf.
– Heidelberg; New York: Physica-Verl., 2002
  (Studies in fuzziness and soft computing; Vol. 104)
  ISBN-13: 978-3-642-88096-4

Physica-Verlag Heidelberg New York
a member of BertelsmannSpringer Science+Business Media GmbH

© Physica-Verlag Heidelberg 2002
Softcover reprint of the hardcover 1st edition 2002

Hardcover Design: Erich Kirchner, Heidelberg

SPIN 10877108   88/2202-5 4 3 2 1 0 – Printed on acid-free paper

Foreword

It is both personally and intellectually pleasing for me to write a foreword to this work. In January 1999 I received a brief e-mail from a PhD student at the University of Bayreuth asking if he might visit the Illinois Genetic Algorithms Laboratory (IlliGAL). I did not know this student, Franz Rothlauf, but something about the tone of his note suggested a sharp, eager mind connected to a cheerful, positive nature. I checked out Franz's references, invited him to Illinois for a first visit, and my early feelings were soon proven correct. Franz's various visits to the lab brought both smiles to the faces of IlliGAL labbies and important progress to a critical area of genetic algorithm inquiry. It was great fun to work with Franz and it was exciting to watch this work take shape. In the remainder, I briefly highlight the contributions of this work to our state of knowledge.

In the field of genetic and evolutionary algorithms (GEAs), much theory and empirical study has been heaped upon operators and test problems, but problem representation has often been taken as a given. In this book, Franz breaks with this tradition and seriously studies a number of critical elements of a theory of GEA representation and applies them to the careful empirical study of (a) a number of important idealized test functions and (b) problems of commercial import. Not only is Franz creative in *what* he has chosen to study, he also has been innovative in *how* he performs his work.

In GEAs – as elsewhere – there appears sometimes to be a firewall separating theory and practice. This is not new, and even Bacon commented on this phenomenon with his famous metaphor of the spiders (men of dogmas), the ants (men of practice), and the bees (transformers of theory to practice). In this work, Franz is one of Bacon's bees, taking *applicable* theory of representation and carrying it to practice in a manner that (1) illuminates the theory and (2) answers the questions of importance to a practitioner.

This book is original in many respects, so it is difficult to single out any one of its many accomplishments. I do believe five items deserve particular comment:

1. Decomposition of the representation problem
2. Analysis of redundancy
3. Analysis of scaling

4. Time-quality framework for representation
5. Demonstration of the framework in well-chosen test problems and problems of commercial import.

Franz's decomposition of the problem of representation into issues of redundancy, scaling, and correlation is itself a contribution. Individuals have isolated each of these areas previously, but this book is the first to suggest they are core elements of an integrated theory and to show the way toward that integration.

The analyses of redundancy and scaling are examples of *applicable* or *facetwise* modeling at its best. Franz gets at key issues in run duration and population size through bounding analyses, and these permit him to draw definite conclusions in fields where so many other researchers have simply waived their arms.

By themselves, these analyses would be sufficient, but Franz then takes the extra and unprecedented step toward an integrated *quality-time* framework for representations. The importance of quality and time has been recognized previously from the standpoint of operator design, but this work is the first to understand that codings can and should be examined from an efficiency-quality standpoint as well. In my view, this recognition will be understood in the future as a key turning point away from the current voodoo and black magic of GEA representation toward a scientific discussion of the appropriateness of particular representations for different problems.

Finally, Franz has carefully demonstrated his ideas in (1) carefully chosen test functions and (2) problems of commercial import. Too often in the GEA field, researchers perform an exercise in pristine theory without relating it to practice. On the other hand, practitioners too often study the latest wrinkle in problem representation or coding without theoretical backing or support. This dissertation asserts the applicability of its theory by demonstrating its utility in understanding tree representations, both test functions and real-world communications networks. Going from theory to practice in such a sweeping manner is a rare event, and the accomplishment must be regarded as both a difficult and an important one.

All this would be enough for me to recommend this book to GEA aficionados around the globe, but I hasten to add that the book is also remarkably well written and well organized. No doubt this rhetorical craftsmanship will help broaden the appeal of the book beyond the ken of genetic algorithmists and computational evolutionaries. In short, I recommend this important book to anyone interested in a better quantitative and qualitative understanding of the representation problem. Buy this book, read it, and use its important methodological, theoretical, and practical lessons on a daily basis.

David E. Goldberg
University of Illinois at Urbana-Champaign

Preface

This book is about representations for genetic and evolutionary algorithms (GEAs). In writing it, I have tried to demonstrate the important role of representations for an efficient use of genetics and evolutionary optimization methods. Although, experience often shows that the choice of a proper representation is crucial for GEA's success, there are few theoretical models that describe how representations influence GEAs behavior. This book aims to resolve this unsettled situation. It presents theoretical models describing the effect of different types of representations and applies them to binary representations of integers and tree representations.

The book is designed for people who want to learn some theory about how representations influence GEA performance and for those who want to see how this theory can be applied to representations in the real world. The book is based on my dissertation with the title "Towards a Theory of Representations for Genetic and Evolutionary Algorithms: Development of Basic Concepts and their Application to Binary and Tree Representations". To make the book easier to read for a larger audience some chapters are extended and many explanations are more detailed. During the writing of the book many people from various backgrounds (economics, computer science and engineering) had a look at the work and pushed me to present it in a way that is accessible to a diverse audience. Therefore, also people that are not familiar to GEAs should be able to get the basic ideas of the book.

To understand the theoretical models describing the influence of representations on GEA performance I expect college-level mathematics like elementary notions of counting, probability theory and algebra. I tried to minimize the mathematical background required for understanding the core lessons of the book and to give detailed explanations on complex theoretical subjects. Furthermore, I expect the reader to have no particular knowledge of genetics and define all genetic terminology and concepts in the text. The influence of integer and tree representations on GEA performance does not necessarily require a complete understanding of the elements of representation theory but is also accessible for people who do not want to bother too much with theory.

The book is split up into two large parts. The first presents theoretical models describing the effects of representations on GEA performance. The second part uses the theory for the analysis and design of representations. After the first two introductory chapters, theoretical models are presented on how redundant representations, exponentially scaled representations and the locality/distance distortion of a representation influence GEA performance. In chapter 4 the theory is used for formulating a time-quality framework. Consequently, in chapter 5, the theoretical models are used for analyzing the performance differences between binary representations of integers. Finally, the framework is used in chapter 6, chapter 7, and chapter 8 for the analysis of existing tree representations as well as the design of new tree representations. In the appendix common test instances for the optimal communication spanning tree problems are summarized.

Acknowledgments

First of all, I would like to thank my parents for always providing me with a comfortable home environment. I have learned to love the wonders of the world and what the important things in life are.

Furthermore, I would like to say many thanks to my two advisors, Dr. Armin Heinzl and Dr. David E. Goldberg. They did not only help me a lot with my work, but also had a large impact on my private life. Dr. Armin Heinzl helped me to manage my life in Bayreuth and always guided me in the right direction in my research. He was a great teacher and I was able to learn many important things from him. I am grateful to him for creating an environment that allowed me to write this book. Dr. David E. Goldberg had a large influence on my research life. He taught me many things which I needed in my research and I would never have been able to write this thesis without his help and guidance.

During my time here in Bayreuth, my colleagues in the department have always been a great help to overcome the troubles of daily university life. I especially want to thank Michael Zapf, Lars Brehm, Jens Dibbern, Monika Fortmühler, Torsten O. Paulussen, Jürgen Gerstacker, Axel Pürckhauer, Thomas Schoberth, Stefan Hocke, and Frederik Loos. During my time here, Wolfgang Güttler and Tobias Grosche were not only work colleagues, but also good friends. I want to thank them for the good time I had and the interesting discussions.

During the last three years during which I spent time at IlliGAL I met many people who have had a great impact on my life. First of all, I would like to thank David E. Goldberg and the Department of General Engineering for giving me the opportunity to stay there so often. Then, I want to say thank you to the folks at IlliGAL I was able to work together with. It was always a really great pleasure. I especially want to thank Erick Cantú-Paz, Fernando Lobo, Dimitri Knjazew, Clarissa van Hoyweghen, Martin Butz, Martin Pelikan, and Kumara Sastri. It was not only a pleasure working together with them but over time they have become really good friends. My stays at IlliGAL would not have been possible without their help.

Finally, I want to thank the people who were involved in the writing of this book. First of all I want to thank Kumara Sastri and Martin Pelikan again. They helped me a lot and had a large impact on my work. The discussions with Martin were great and Kumara often impressed me with his expansive knowledge about GEAs. Then, I want to say thanks to Fernando Lobo and Torsten O. Paulussen. They gave me great feedback and helped me to clarify my thoughts. Furthermore, Katrin Appel and Kati Sternberg were a great help in writing this dissertation. Last but not least I want to thank Anna Wolf. Anna is a great proofreader and I would not have been able to write a book in readable English without her help.

Finally, I want to say thank you to Kati. Now I will hopefully have more time for you.

Bayreuth, January 2002   *Franz Rothlauf*

Contents

1. Introduction

One of the major duties of researchers in the field of management science, information systems, business informatics, and computer science is to develop methods and tools that should help organizations, such as companies or public institutions, to fulfill their jobs efficiently. However, during the last decade, the dynamics and size of the problems organizations are faced with has changed. Firstly, production and service processes must be reorganized in shorter time intervals and adapted dynamically to the varying demands of markets and customers. Although there is continuous change, organizations must ensure that the efficiency of their processes remains high. Therefore, optimization techniques are necessary that help organizations to reorganize themselves and to stay efficient. Secondly, with increasing organization size the complexity of problems in the context of production or service processes also increases. As a result, standard, traditional, optimization techniques are often not able to solve these problems of increased complexity with justifiable effort in an acceptable time period. Therefore, to overcome these problems, and to develop systems that solve these complex problems, researchers proposed using genetic and evolutionary algorithms (GEAs). Using these nature-inspired search methods it is possible to overcome the limitations of traditional optimization methods, and to increase the number of solvable problems. The application of GEAs to many optimization problems in organizations often results in good performance and high quality solutions.

However, for successful and efficient use of GEAs, it is not enough to simply use efficient genetic operators. It is also necessary to find a proper representation for the problem. The representation must at least be able to encode all possible solutions of an optimization problem, and genetic operators such as crossover and mutation should be applicable to it.

Most of the optimization problems can be encoded by a variety of different representations. In addition to traditional binary and continuous string encodings, a large number of other, often problem-specific representations have been proposed over the last few years. Unfortunately, practitioners often report a significantly different performance of GEAs by simply changing the used representation. These observations were confirmed by empirical and theoretical investigations. The difficulty of a specific problem, and with it the performance of GEAs, can be changed dramatically by using different types of representations. Although it is well known that representations affect the performance of GEAs, no theoretical models exist which describe the effect of representations on the performance of GEAs. Therefore, the design of proper representations for a specific problem depends mainly on the intuition of the GEA designer. Developing new representations is often a result of repeated trial and error. As no theory of representations exists, the current design of proper representations is not based on theory, but more a result of black art.

The lack of existing theory not only hinders a theory-guided design of new representations, but also results in problems when deciding which of the different representations should be used for a specific optimization problem. Currently, comparisons between representations are based mainly on limited empirical evidence, and random or problem-specific test function selection. However, empirical investigations only allow us to judge the performance of representations for the specific test problem, but do not help us in understanding the basic principles behind it. A representation can perform well for many different test functions, but fails for the one problem which one really wants to solve. If it is possible to develop theoretical models which describe the influence of representations on measurements of GEA performance – like time to convergence and solution quality – then representations can be used efficiently and in a theory-guided manner. Choosing and designing proper representations will not remain the black art of GEA research but become a well predictable engineering task.

1.1 Purpose

The purpose of this work is to bring some order into the unsettled situation which exists and to investigate how representations influence the performance of genetic and evolutionary algorithms. This work intends to develop elements of representation theory and to apply them to designing, selecting, using, choosing among, and comparing representations. It is not the purpose of this work to substitute the current black art of choosing representations by developing barely applicable, abstract, theoretical models, but to formulate an applicable representation theory that can help researchers and practitioners to find or design the proper representation for their problem. By providing an applicable theory of representations this work should bring us to a point where the influence of representations on the performance of GEAs can be judged easily and quickly in a theory-guided manner.

The first step in the development of an applicable theory is to identify which properties of representations influence the performance of GEAs and how. Therefore, this work models for different types of representations how solution quality and time to convergence is changed. Using this theory, it is possible to formulate a framework for efficient design of representations for genetic and evolutionary algorithms. The framework describes how the performance of GEAs, measured by run duration and solution quality, is affected by the use of different representations. By using this framework, the influence of different representations on the performance of GEAs can be explained easily. Furthermore, it allows us to compare representations in a theory-based manner, to predict the performance of GEAs using different representations, and to analyze and design representations guided by theory. One does not have to rely on empirical studies to judge the performance of a representation for a specific problem, but can use existing theory for predicting GEA performance. By using this theory, the situation exists where only empirical results are needed to validate theoretical predictions.

However, developing a general theory of how representations affect GEA performance is a demanding and difficult task. To simplify the problem, it must be decomposed, and the different types of encodings must be investigated separately. Three different types of representations are considered: Redundant encodings, exponentially scaled encodings, and encodings that change the distances between corresponding phenotypes and genotypes. For these three types of encodings it is described how GEA performance is affected. Additionally, population sizing and time to convergence models are presented for redundant and non-uniformly scaled encodings. Furthermore, it is shown that representations which change the distances between the individuals can change the difficulty of the problem. For these types of encodings, it can not exactly be modeled how GEA performance is changed, without having complete knowledge regarding the structure of the problem. Although the investigation is limited only to three different types of representations, the understanding of the influence of these three types of encodings on the performance of GEAs brings us a large step forward towards a general theory of representations.

To illustrate the significance and importance of the presented framework on the performance of GEAs, it is used for analyzing the performance of binary representations of integers and tree representations. The investigations show that the framework consisting only of three elements gives us a good understanding of the influence of representations on GEA performance. The theory allows us to predict the performance of genetic and evolutionary algorithms using different types of representations. The results confirm that choosing a proper representation has a large impact on the performance of GEAs, and therefore, a better theoretical understanding of representations is necessary for an efficient use of genetic search.

Finally, it is illustrated how the presented theory of representations can help us in designing new representations more reasonably. It is shown by example for tree representations, that the presented framework allows theory-guided design. Not black art, but a deeper understanding of representations allows us to develop representations which result in a high performance of genetic and evolutionary algorithms.

1.2 Organization

The organization of this work follows its purpose. It is divided into two large parts: After the first two introductory chapters, the first part (chapters 3 and 4) provides the theory regarding representations. The second part (chapter 5, 6, 7, and 8) applies the theory to the analysis and design of representations. Chapter 3 presents theory on how different types of representations affect GEA performance. Consequently, chapter 4 uses the theory for formulating the time-quality framework. Then, in chapter 5, the presented theory of representations is used for analyzing the performance differences between binary representations of integers. Finally, the framework is used in chapter 6, chapter 7, and chapter 8 for the analysis of existing tree representations as well as the design of new tree representations. The following gives a more detailed overview about the contents of each chapter.

Chapter 1 is the current chapter. It sets the stage for the work and describes the benefits that can be gained from a deeper understanding of representations for genetic and evolutionary algorithms.

Chapter 2 provides the background necessary for understanding the main issues of this work about representations for genetic and evolutionary algorithms. Section 2.1 introduces representations which can be described as a mapping that assigns one or more genotypes to every phenotype. The genetic operators selection, crossover, and mutation are applied to the level of genes to the genotypes, whereas the fitness of individuals is calculated from the corresponding phenotypes. Section 2.2 illustrates that selectorecombinative GEAs, where only crossover and selection operators are used, are based on the notion of schemata and building blocks. Using schemata and building blocks explains well why and how GEAs work. This is followed in section 2.3 by a brief review of reasons and measurements for problem difficulty. Measurements of problem difficulty are necessary to be able to compare the influence of different types of representations on the performance of GEAs. The chapter ends with some earlier, mostly qualitative recommendations for design of efficient representations.

Chapter 3 presents three aspects of a theory of genetic and evolutionary algorithms. It investigates how redundant encodings, encodings with exponentially scaled building blocks, and representations that modify the distances between the corresponding genotypes and phenotypes, influence GEA performance. Population sizing models and time to convergence models are presented for redundant and exponentially scaled representations. Section 3.1 illustrates that redundant encodings influence the supply of building blocks in the initial population of GEAs. Based on this observation the population sizing model from Harik et al. (1997) and the time to convergence model from Thierens and Goldberg (1993) can be extended from non-redundant to redundant representations. Because redundancy mainly affects the number of copies in the initial population that are given to the optimal solution, redundant representations increase solution quality and reduce time to convergence if individuals that are similar to the optimal solution are overrepresented. Section 3.2 focuses on exponentially scaled representations. The investigation into the effects of exponentially scaled encodings shows that, in contrast to uniformly scaled representations, the dynamics of genetic search are changed. By combining the results from Harik et al. (1997) and Thierens (1995) a population sizing model for exponentially scaled building blocks with and without considering genetic drift can be presented. Furthermore, the time to convergence when using exponentially scaled representations is calculated. The results show that when using non-uniformly scaled representations, the time to convergence increases, whereas the probability of GEA failure decreases. Finally, section 3.3 investigates the influence of representations that modify the distances between corresponding genotypes and phenotypes on the performance of GEAs. When assigning the genotypes to the phenotypes, representations can change the distances between the individuals. This effect is denoted distance distortion. Investigating its influence shows, that the size and length of the building blocks, and therefore the complexity of the problem is changed if the distances between the individuals are not preserved. Therefore, to assure that an easy problem remains easy, representations which preserve the distances between the individuals are necessary.

Chapter 4 presents the framework for theory-guided analysis and design of representations. The chapter combines the three elements of representation theory from chapter 3 – redundancy, scaling, and distance distortion – to a time-quality framework. It formally describes how the time to convergence and the solution quality of GEAs depend on these three aspects of representations. The chapter ends with implications for the design of representations which can be derived from the framework. In particular, the framework tells us that uniformly scaled representations are robust, that exponentially scaled representations are fast but inaccurate, and that the influence of representations that modify the distances between the individuals on problem complexity is difficult to predict.

Chapter 5 uses the framework for a theory-guided analysis of binary representations of integers. Because the potential number of schemata is higher when using binary instead of integer representations, users often favor the use of binary instead of integer representations, when applying GEAs to integer problems. By using the framework it can be shown that the redundant unary encoding results in low GEA performance if the optimal solution is underrepresented. Both, gray and binary encoding change the distances between the individuals. Therefore, both representations change the complexity of optimization problems. It can be seen that the easy integer one-max problem is easier to solve when using the binary representation, and the difficult integer deceptive trap is easier to solve when using the gray encoding.

Chapter 6 uses the framework for the analysis of tree representations. Representing trees is a difficult task because there is no intuitive "proper" representation. Therefore, researchers have proposed a variety of different, more or less tricky representations. A closer look at the Prüfer number representation in section 6.2 reveals that the encoding does in general modify the distances between the individuals when mapping the phenotypes to the genotypes. As a result, the complexity of the problem is modified, and many easy problems become too difficult to be properly solved using GEAs. The investigation into the redundant link and node biased representation in section 6.3 reveals that the representation overrepresents trees that are either star-like or minimum spanning tree-like. Therefore, GEAs using this type of representation perform very well if the optimal solution is similar to stars or to the minimum spanning tree, whereas they fail when searching for optimal solutions that do not have much in common with stars or the minimum spanning tree. Finally, section 6.4 presents an investigation into the characteristic vector representation. Because invalid solutions are possible when using characteristic vectors, a repair mechanism is necessary which makes the representation redundant. Characteristic vectors are uniformly redundant and GEA performance is independent of the structure of the optimal solution. However, the repair mechanism results in stealth mutation which can be interpreted as a form of background mutation. As a result of stealth mutation the time to convergence increases. With increasing problem size, stealth mutation generates more and more links randomly and offspring trees have not much in common with their parents. Therefore, for larger problems guided search is no longer possible and GEAs behave like random search.

Chapter 7 uses the insights into representation theory for the design of tree representations. To construct a robust and predictable tree representation, it should be non- or uniformly redundant, uniformly scaled, and not change the distances between the corresponding genotypes and phenotypes. When combining the characteristic vector with a weighted representation like the link and node biased representation, the network random key (NetKey) representation can be created (section 7.1). In analogy to random keys, the links of a tree are represented as floating numbers, and a construction algorithm constructs the corresponding tree from the keys. The NetKey representation allows us to distinguish between important and unimportant links, is uniformly redundant, uniformly scaled, and preserves the distances between the individuals well. Section 7.2 presents a direct representation for trees (NetDir). When using direct representations, the phenotype is the same as the corresponding genotype. No explicit genotype-phenotype mapping exists, and the framework for the design of representations can not be used properly. Developing a direct representation illustrates nicely the tradeoff between designing either problem-specific representations or problem-specific operators. For efficient GEAs, it is necessary either to design problem-specific representations and to use standard operators like one-point or uniform crossover, or to develop problem-specific operators and to use direct representations. Consequently, operators for the NetDir representation are developed based on the notion of schemata.

Chapter 8 verifies theoretical predictions concerning GEA performance by empirical verification. It compares the performance of GEAs using different types of representations for the one-max tree problem, the deceptive tree problem, and various instances of the optimal communication spanning tree problem. The instances of the optimal communication spanning trees are presented in the literature (Palmer, 1994; Berry, Murtagh, McMahon, & Sugden, 1997; Raidl, 2001; Rothlauf, Goldberg, & Heinzl, 2002). The results show that with the help of the framework the performance of GEAs using different types of representations can be well predicted.

Chapter 9 summarizes the major contributions of this work, describes how the knowledge about representations has changed, and gives some suggestions for future research.

2. Representations for Genetic and Evolutionary Algorithms

In this second chapter, we present an introduction into the field of representations for genetic and evolutionary algorithms. The chapter provides the basis and definitions which are essential for understanding the content of this work.

Genetic and evolutionary algorithms (GEAs) are nature-inspired optimization methods that can be advantageously used for many optimization problems. GEAs imitate basic principles of life and apply genetic operators like mutation, crossover, or selection to a sequence of alleles. The sequence of alleles is the equivalent of a chromosome in nature and is constructed by a representation which assigns a string of symbols to every possible solution of the optimization problem. Earlier work (Goldberg, 1989c; Liepins & Vose, 1990) has shown that the behavior and performance of GEAs is strongly influenced by the representation used. As a result many recommendations for a proper design of representations were made over the last few years (Goldberg, 1989c; Palmer, 1994; Ronald, 1997). However, many of these design rules are only of a qualitative nature and are not particularly helpful for estimating exactly how different types of representations influence problem difficulty. Consequently, we are in need of a theory of representations which allows us to theoretically predict how different types of representations influence GEA performance. This chapter provides some of the utilities that are necessary for reaching this goal.

The chapter starts with an introduction into genetic representations. We describe the notion of genotypes and phenotypes and illustrate how the fitness function can be decomposed into a genotype-phenotype, and a phenotype-fitness mapping. The section ends with a brief characterization of widely used representations. In section 2.2 we provide the basis for genetic and evolutionary algorithms. After a brief description of the principles of a simple genetic algorithm, we present the underlying theory which explains why and how selectorecombinative GAs using crossover as a main search operator work. The schema theorem tells us that GAs process schemata and the building block hypothesis assumes that many real-world problems are decomposable (or at least quasi-decomposable). Therefore, GAs perform well for these types of problems. Section 2.3 addresses the difficulty of problems. After illustrating that the reasons for problem difficulty depend on the used optimization method, we describe some common measurements of problem complexity. Finally, in section 2.4 we review some former recommendations for the design of efficient representations.

2.1 Genetic Representations

This section introduces representations for genetic and evolutionary algorithms. When using GEAs for optimization purposes, representations are required for encoding potential solutions. Without representations, no use of GEAs is possible.

In subsection 2.1.1 we introduce the notion of genotype and phenotype. We briefly describe how nature creates a phenotype from the corresponding genotype by the use of representations. This more biology-based approach to representations is followed in subsection 2.1.2 by a more formal description of representations. Every fitness function f which assigns a fitness value to a genotype x_g can be decomposed into the genotype-phenotype mapping f_g, and the phenotype-fitness mapping f_p. Finally, in subsection 2.1.3 we briefly review the most important types of representations.

2.1.1 Genotypes and Phenotypes

In the following, we illustrate the basic principles of representations in nature. In 1866, Mendel recognized that nature stores the complete information for each individual in pairwise alleles (Mendel, 1866). The genetic information that determines the properties, appearance, and shape of an individual is stored by a number of strings. Later, it was discovered that the genetic information is formed by a double string of four nucleotides, called DNA.

Mendel realized that nature distinguishes between the genetic code of an individual and its outward appearance. The genotype represents all the information stored in the chromosomes and allows us to describe an individual on the level of genes. The phenotype describes the outward appearance of an individual. A transformation exists – a genotype-phenotype mapping or a representation – that uses the genotypic information to construct the phenotype. To represent the large number of possible phenotypes with only four nucleotides, the genotypic information is not stored in the alleles itself, but in the sequence of alleles. By interpreting the sequence of alleles, nature can encode a large number of different phenotypic expressions using only a few different types of alleles.

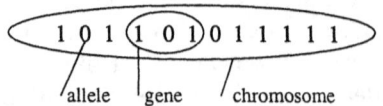

Fig. 2.1. Alleles, genes, and chromosomes

In figure 2.1 we illustrate the differences between chromosome, gene, and allele. A chromosome describes a string of certain length where all the genetic information of an individual is stored. Although nature often uses more than one chromosome, most GEA applications only use one chromosome for encoding the genotypic information. Each chromosome consist of many alleles. Alleles are the smallest information unit in a chromosome. In nature alleles exist pairwise, whereas in most GEA implementations an allele is represented by only one symbol. If for example, we use a binary representation, an allele can have either the value 0 or 1. If a phenotypic property of an individual, like its hair color or eye size is determined by one or more alleles, then these alleles together are denoted to be a gene. A gene is a region on a chromosome that must be interpreted together and which is responsible for a specific phenotypic property.

When talking about individuals in a population, we must carefully distinguish between genotypes and phenotypes. The phenotypic appearance of an individual determines its success in life. Therefore, when comparing the abilities of different individuals we must judge them on the level of the phenotype. However, when it comes to reproduction we must view individuals on the level of the genotype. During sexual reproduction, the offspring does not inherit the phenotypic properties of its parents, but only the genotypic information regarding the phenotypic properties. The offspring inherits genetic material from both parents. Therefore, genetic operators work on the level of the genotype, whereas the evaluation of the individuals is performed on the level of the phenotype.

2.1.2 Decomposition of the Fitness Function

The following subsection provides some basic definitions for our discussion of representations for genetic and evolutionary algorithms. We show how every optimization problem that should be solved by using GEAs can be decomposed into a genotype-phenotype f_g, and a phenotype-fitness mapping f_p.

We define Φ_g as the genotypic search space where the genetic operators such as recombination or mutation are applied to. An optimization problem on Φ_g could be formulated as follows: The search space Φ_g is either discrete or continuous, and the function

$$f(x) : \Phi_g \to \mathbb{R}$$

assigns an element in $\mathbb{R}$ to every element in the genotype space Φ_g. The optimization problem is defined by finding the optimal solution

$$\hat{x} = \max_{x \in \Phi_g} f(x),$$

where x is a vector of decision variables (or alleles), and $f(x)$ is the objective or fitness function. The vector $\hat{x}$ is the global maximum. We have chosen to illustrate a maximization problem, but without loss of generality, we could also model a minimization problem. To be able to apply GEAs to a problem, the inverse function f^{-1} does not need to exist.

In general, the cardinality of Φ_g can be greater than two, but we want to focus for the most part in our investigation on binary search spaces with cardinality two. Thus, GEAs search in the binary space

$$\Phi_g = \{0,1\}^l,$$

with the length of the string x equal to l and the size of the search space $|\Phi_g| = 2^l$.

The introduction of an explicite representation is necessary if the phenotype of a problem can not be depicted as a string or in another way that is accessible for GEAs. Furthermore, the introduction of a representation could be useful if there are constraints or restrictions on the search space that can be advantageously modeled by a specific encoding. Finally, using the same genotypes for different types of problems, and only interpreting them differently by using a different genotype-phenotype mapping, allows us to use standard genetic operators with known properties. Once we have gained some knowledge about specific kinds of genotypes, we can easily reuse that knowledge, and it is not necessary to develop any new operators.

When using a representation for genetic and evolutionary algorithms we have to introduce – in analogy to nature – phenotypes and genotypes (Lewontin, 1974; Liepins & Vose, 1990). The fitness function f is decomposed into two parts. The first maps the genotypic space Φ_g to the phenotypic space Φ_p, and the second maps Φ_p to the fitness space $\mathbb{R}$. Using the phenotypic space Φ_p we get:

$$f_g(x_g) : \Phi_g \rightarrow \Phi_p,$$
$$f_p(x_p) : \Phi_p \rightarrow \mathbb{R},$$

where $f = f_p \circ f_g = f_p(f_g(x_g))$. The genotype-phenotype mapping f_g is determined by the type of representation used. f_p represents the fitness function and assigns a fitness value $f_p(x_p)$ to every individual $x_p \in \Phi_p$. The genetic operators are applied to the individuals in Φ_g (Bagley, 1967; Vose, 1993).

If the genetic operators are applied directly to the phenotype it is not necessary to specify a representation and the phenotype is the same as the genotype:

$$f_g(x_g) : \Phi_g \rightarrow \Phi_g,$$
$$f_p(x_p) : \Phi_g \rightarrow \mathbb{R}.$$

In this case, f_g is the identity function $f_g(x_g) = x_g$. All genotypic properties are transformed to the phenotypic space. The genotypic space is the same as the phenotypic space and we have a *direct representation*. Because there is no longer an additional mapping between Φ_g and Φ_p, a direct representation does not change any aspect of the phenotypic problem such as complexity, distances between the individuals, or location of the optimal solution. However, when using direct representations, we could not in general use standard genetic operators, but have to define problem-specific operators (see section 7.2 regarding the development of a direct representation for trees). Therefore, the key factor for the success of a GEA using a direct representation is not in finding a "good" representation for a specific problem, but in developing proper operators.

We have seen how every optimization problem we want to solve with GEAs can be decomposed into a genotype-phenotype f_g, and a phenotype-fitness mapping f_p. The genetic operators are applied to the genotypes $x_g \in \Phi_g$, and the fitness of the individuals is calculated from the phenotypes $x_p \in \Phi_p$.

2.1.3 Types of Representations

In the following subsection we describe some of the most important and widely used representations, and summarize some of their major characteristics. In this work we do not provide an overview of all representations which appear in the literature because every time a GEA is used, some kind of representation is necessary. This means within the scope of this research it is not possible to cite all representations which have been presented. For a more detailed overview about representations see Bäck, Fogel, and Michalewicz (1997, section C1).

Binary Representations

Binary representations are the most common representations for selectore-combinative GEAs. Selectorecombinative GEAs process schemata and use crossover as the main search operator. Using these types of GEAs, mutation only serves as background noise. The search space Φ_g is denoted by $\Phi_g = \{0, 1\}^l$, where l is the length of a binary vector $x_g = (x_1, \ldots x_l) \in \{0, 1\}^l$ (Goldberg, 1989c).

When using binary representations the genotype-phenotype mapping f_g depends on the specific optimization problem that should be solved. For many combinatorial optimization problems the representation allows a direct and very natural encoding.

When encoding integer problems by using binary representations, specific genotype-phenotype mappings are necessary. Different types of binary representations for integers assign the integers $x_p \in \Phi_p$ (phenotypes) in a different way to the binary vectors $x_g \in \Phi_g$ (genotypes). The most common representations are the binary, gray, and unary encoding. For a more detailed description of these three types of encodings see section 5.2.

When encoding continuous variables by using binary vectors the accuracy of the representation depends on the number of bits that represent a phenotypic continuous variable. By increasing the number of bits that are used for representing one continuous variable the accuracy of the representation can be increased. When encoding a continuous phenotypic variable $x_p \in [0, 1]$ by using a binary vector of length l the maximal accuracy is $1/2^{l+1}$.

Integer Representations

Instead of using binary strings with cardinality $\chi = 2$, where $\chi \in \{\mathbb{N}^+ \setminus \{0, 1\}\}$, higher χ-ary alphabets can also be used for the genotypes. Then, instead of a binary alphabet a χ-ary alphabet is used for the string of length l. Instead of encoding 2^l different individuals with a binary alphabet, we are able to encode χ^l different possibilities. The size of the search space increases from $|\Phi_g| = 2^l$ to $|\Phi_g| = \chi^l$.

For many integer problems, users often prefer to use binary instead of integer representations because schema processing is maximum with binary alphabets when using standard recombination operators (Goldberg, 1990b).

Real-valued Representations

When using real-valued representations, the search space Φ_g is defined as $\Phi_g = \mathbb{R}^l$, where l is the length of the real-valued chromosome. When using real-valued representations, researchers often favor mutation-based GEAs like evolution strategies or evolutionary programming. These types of optimization methods mainly use mutation and search through the search space by adding a multivariate zero-mean Gaussian random variable to each variable. In contrast, when using crossover-based GEAs real-valued problems are often represented by using binary representations (see paragraph about binary representations).

Real-valued representations can not exclusively be used for encoding real-valued problems, but also for other permutation and combinatorial problems. Trees, schedules, tours, or other combinatorial problems can easily be represented by using real-valued vectors and special genotype-phenotype mappings (see also section 7.1 about network random keys).

Messy Representations

In all the previously presented representations the position of each allele is fixed along the chromosome and only the corresponding value is specified. The first gene-independent representation was proposed by Holland (1975). He proposed the inversion operator which changes the relative order of the alleles in the string. The position of an allele and the corresponding value are coded together as a tuple in a string. This type of representation can be used for binary, integer, and real-valued representations and allows an encoding which is independent of the position of the alleles in the chromosome. Later, Goldberg, Korb, and Deb (1989) used this position-independent representation for the messy genetic algorithm.

Direct Representations

In section 2.1.2 we have seen that a representation is direct if $f_g(x_g) = x_g$. Then, a phenotype is the same as the corresponding genotype and the problem-specific genetic operators are applied directly to the phenotype.

As long as $x_p = x_g$ is either a binary, an integer, or a real-valued string, standard recombination and mutation operators can be used. Then, it is often easy to predict GEA performance by using existing theory. The situation is different if direct representations are used for problems whose phenotypes are not binary, integer, or real-valued. Then, standard recombination and mutation operators can not be used any more. Specialized operators are necessary that allow offspring to inherit important properties from their parents (Kargupta, Deb, & Goldberg, 1992; Radcliffe, 1993). In general, these operators depend on the specific structure of the phenotypes x_p and must be developed for every optimization problem separately. For more information about direct representations the reader is referred to section 7.2.

2.2 Genetic and Evolutionary Algorithms

In this section we introduce genetic and evolutionary algorithms. We illustrate the basic principles and outline the basic functionality of GEAs. The schema theorem stated by Holland (1975) explains the performance of selectorecombinative GAs and leads us to the building block hypothesis. The building block hypothesis tells us that short, low-order and highly fit schemata can be recombined to form higher-order schemata and complete strings with very high fitness.

2.2.1 Principles

Genetic and evolutionary algorithms (GEAs) were introduced by Holland (1975) and Rechenberg (1973). By imitating basic principles of nature they created optimization algorithms which have successfully been applied to a wide variety of problems. The basic principles of GEAs are derived from the principles of life which were first described by Darwin (1859):

> "Owing to this struggle for life, variations, however slight and from whatever cause proceeding, if they be in any degree profitable to the individuals of a species, in their infinitely complex relations to other organic beings and to their physical conditions of life, will tend to

the preservation of such individuals, and will generally be inherited by the offspring. The offspring, also, will thus have a better chance of surviving, for, of the many individuals of any species which are periodically born, but a small number can survive. I have called this principle, by which each slight variation, if useful, is preserved, by the term Natural Selection."

Darwin's ideas about the principles of life can be summarized by the following three basic principles:

- There is a population of individuals with different properties and abilities. An upper limit for the number of individuals in a population exists.
- Nature creates new individuals with similar properties to the existing individuals.
- Promising individuals are selected more often for reproduction by natural selection.

In the following section, we briefly illustrate these three principles. We have seen in section 2.1 that the properties and abilities of an individual which are characterized by its' phenotype are encoded in the genotype. Therefore, based on different genotypes, individuals with different properties exist (Mendel, 1866). Because resources are finite, the number of individuals that form a population is limited. If the number of individuals exceeds the existing upper limit, some of the individuals are removed from the population.

The individuals in the population do not remain the same, but change over the generations. New offspring are created which inherit some properties of their parents. These new offspring are not chosen randomly but are somehow similar to their parents. To create the offspring genetic operators like mutation and recombination are used. Mutation operators change the genotype of an individual slightly, whereas recombination operators combine the genetic information of the parents to create new offspring.

When creating offspring, natural selection more often selects promising individuals for reproduction than low-quality solutions. Highly fit individuals are allowed to create more offspring than inferior individuals. Therefore, inferior individuals are removed from the population after a few generations and have no chance of creating offspring with similar properties. As a result the average fitness of a population increases over the generations.

In the following we want to describe how the principles of nature were used for the design of genetic and evolutionary algorithms.

2.2.2 Functionality

Genetic and evolutionary algorithms imitate the principles of life outlined in the previous subsection and use it for optimization purposes.

Researchers have proposed many different variants of GEAs in the literature. For illustrating the basic functionality of GEAs we want to use the traditional standard simple genetic algorithm (GA) illustrated by Goldberg (1989c). These types of GEAs use crossover as the main operator and mutation serves only as background noise. GAs are widely known and well understood. GAs use a constant population of size n, the individuals consist of binary strings with length l, and genetic operators like uniform or n-point crossover are directly applied to the genotypes. The basic functionality of a traditional simple GA is very simple. After randomly creating and evaluating an initial population, the algorithm continuously creates new generations. New generations are created by recombining the selected highly fit individuals and applying mutation to the offspring.

- initialize population
  - create initial population
  - evaluate individuals in initial population
- create new populations
  - select fit individuals for reproduction
  - generate offspring with genetic operator crossover
  - mutate offspring
  - evaluate offspring

One specific type of genetic algorithms are selectorecombinative GAs. These types of GAs use only selection and recombination (crossover). No mutation is used. Using selectorecombinative GAs gives us the advantage of being able to investigate the effects of different representations on crossover alone and to eliminate the effects on mutation. This is useful if we use GAs in a way such that they propagate schemata (compare subsection 2.2.3), and where mutation is only used as additional background noise. When focusing on GEAs where mutation functions as the main search operator, the reader is referred to other work (Rechenberg, 1973; Schwefel, 1975; Schwefel, 1981; Schwefel, 1995; Bäck & Schwefel, 1995).

In the following we briefly explain the basic elements of a GA. For selecting highly fit individuals for reproduction a large number of different selection schemes have been developed. The most popular are proportionate (Holland, 1975) and tournament selection (Goldberg, Korb, & Deb, 1989). When using proportionate selection, the expected number of copies an individual has in the next population is proportional to its fitness. The chance of an individual x_i of being selected for recombination is calculated as

$$\frac{f(x_i)}{\sum_{j=1}^{n} f(x_j)},$$

where n denotes the number of individuals in a population. With increasing fitness an individual is chosen more often for reproduction.

When using tournament selection, a tournament between s randomly chosen different individuals is held and the one with the highest fitness is chosen for recombination and added to the mating pool. After n tournaments of size s the mating pool is filled. We have to distinguish between tournament selection with and without replacement. If we perform tournament selection with replacement we choose for every tournament s individuals from all individuals in the population. After n tournaments the mating pool is filled. If we perform a tournament without replacement there are s rounds. In each round we have n/s tournaments and we choose the individuals for a tournament from those who have not already taken part in a tournament in this round. After all individuals have performed a tournament in one round (after n/s tournaments) the round is over and all individuals are considered again for the next round. Therefore, to completely fill the mating pool s rounds are necessary.

The mating pool consists of all individuals who are chosen for recombination. When using tournament selection, there are no copies of the worst individual, and either an average of s copies (with replacement), or exactly s copies (without replacement) of the best individual in the mating pool. For more information concerning different tournament selection schemes see Bäck, Fogel, and Michalewicz (1997, C2) and Sastry and Goldberg (2001).

Crossover operators imitate the principle of sexual reproduction and are applied to the individuals in the mating pool. In many GA implementations, crossover produces two new offspring from two parents by exchanging substrings. The most common crossover operators are one-point (Holland, 1975), and uniform crossover (Syswerda, 1989). When using one-point crossover, a crossover point $c = \{1, \ldots, l-1\}$ is initially chosen randomly. Two children are then created from the two parents by swapping the substrings. As a result we get for the parents $x^{p1} = x_1^{p1}, x_2^{p1}, \ldots, x_l^{p1}$ and $x^{p2} = x_1^{p2}, x_2^{p2}, \ldots, x_l^{p2}$ the offspring $x^{o1} = x_1^{p1}, x_2^{p1}, \ldots, x_c^{p1}, x_{c+1}^{p2} \ldots x_l^{p2}$ and $x^{o2} = x_1^{p2}, x_2^{p2}, \ldots, x_c^{p2}, x_{c+1}^{p1} \ldots x_l^{p1}$. When using uniform crossover it is decided independently for every single allele of the offspring from which parent it inherits the value of the allele. In most implementations no parent is preferred and the probability of an offspring to inherit the value of an allele from a specific parent is $p = 1/x$, where x denotes the number of parents that are considered for recombination. For example, when two possible offspring are considered with same probability ($p = 1/2$), we could get as offspring $x^{o1} = x_1^{p1}, x_2^{p1}, x_3^{p2}, \ldots, x_{l-1}^{p1}, x_l^{p2}$ and $x^{o2} = x_1^{p2}, x_2^{p2}, x_3^{p1}, \ldots, x_{l-1}^{p2}, x_l^{p1}$. We see that uniform crossover can also be interpreted as $(l-1)$-point crossover.

Mutation operators should slightly change the genotype of an individual. Mutation operators are important for local search, or if some alleles are lost during a GEA run. By randomly modifying some alleles in the population already lost alleles can be reanimated. The probability of mutation p_m must be selected to be at a low level because otherwise mutation would randomly change too many alleles and the new individual would have nothing in common with its parent. Offspring would be generated almost randomly and genetic search would degrade to random search. In contrast to crossover operators, mutation operators focus more on local search because they can only modify properties of individuals but can not recombine properties from different parents.

2.2.3 Schema Theorem and Building Block Hypothesis

In the following, we review the explanations for the performance of genetic algorithms. We start by illustrating the notion of schemata. This is followed by a brief summary of the schema theorem and finally a statement about the building block hypothesis.

Schemata

Schemata were first proposed by Holland (1975) to model the ability of GEAs to process similarities between bitstrings. A schema $h = (h_1, h_2, \ldots, h_l)$ is defined as a ternary string of length l, where $h_i \in \{0, 1, *\}$. $*$ denotes the "don't care" symbol and tells us that the allele at this position is not fixed.

The size or order $o(h)$ of a schema h is defined as the number of fixed positions (0s or 1s) in the string. A position in a schema is fixed if there is either a 0 or a 1 at this position. The defining length $\delta(x)$ of a schema h is defined as the distance between the two outermost fixed bits. The fitness of a schema is defined as the average fitness of all instances of this schema and can be calculated as

$$f(h) = \frac{1}{||h||} \sum_{x \in h} f(x),$$

where $||h||$ is the number of individuals $x \in \Phi_g$ that are an instance of the schema h. The instances of a schema h are all genotypes where $x_g \in h$. For example, $x_g = 01101$ and $x_g = 01100$ are instances of $h = 0*1**$. The number of individuals that are an instance of a schema h can be calculated as $2^{l-o(h)}$.

For a more detailed explanation of schemata in the context of GEAs the reader is referred to Holland (1975), Goldberg (1989c), and Radcliffe (1997).

Schema Theorem

Based on the notion of schemata, Holland (1975) and De Jong (1975) formulated the schema theorem which describes how the number of instances of a schema h changes over the number of generations t:

$$m(h, t+1) \geq m(h, t) \frac{f(h, t)}{\bar{f}(t)} (1 - p_c \frac{\delta(h)}{l-1} - p_m o(h)),$$

where

- $m(h, t)$ is the number of instances of schema h at generation t,
- $f(h, t)$ is the fitness of the schema h at generation t,
- $\bar{f}(t)$ is the average fitness of the population at generation t,
- $\delta(h)$ is the defining length of schema h,
- p_c is the probability of crossover,
- p_m is the probability of mutation,
- l is the string length,
- $o(h)$ is the order of schema h.

The schema theorem describes how the number of copies that are given to a schema h depends on selection, crossover and mutation, when using a standard GA with proportionate selection, one-point crossover, and bit-flipping mutation. Selection favors a schema if the fitness of the schema is above the average fitness of the population $(f(h, t) > \bar{f}(t))$. When using crossover the defining length $\delta(h)$ of a schema must be small because otherwise one-point crossover frequently disrupts long schemata. The bit-flipping mutation operator favors low order schemata because with increasing $o(h)$ the number of schemata which are destroyed increases.

The main contribution of the schema theorem is that schemata, which fitness is above average $(f(h) > \bar{f})$, which have a short defining length $\delta(h)$, and which are of low order $o(h)$, receive exponentially increasing trials in subsequent generations. The theorem describes the hurdle between selection, which preserves highly fit schemata, and crossover and mutation which both destroy schemata of large order or defining-length.

This observation brings us to the concept of *building blocks* (BBs). Goldberg (1989c, page 20 and page 41) defined buildings blocks as "highly fit, short-defining-length schemata" that "are propagated generation to generation by giving exponentially increasing samples to the observed best". The notion of building blocks is frequently used in the literature but rarely defined. In general, a building block can be described as a solution to a subproblem that can be expressed as a schema. A thus-like schema has high fitness and its size is smaller than the length of the string. By combining BBs of lower order, a GA can form high-quality over-all solutions. Using the notion of genes we can interpret BBs as genes. A gene consists of one or more alleles and can be described as a schema with high fitness. The alleles in a chromosome can be separated (decomposed) into genes which do not interact with each other and which determine one specific phenotypic property of an individual like hair or eye color. We see that by using building blocks we can describe with the help of the schema theorem how GAs can solve an optimization problem. If the subsolutions to a problem (the BBs) are short (small $\delta(h)$) and of low order (small $o(h)$), then the number of subsolutions increases over the generations and the problem can easily be solved by a GA.

Building Block Hypothesis

Using the definition of the building blocks as being highly fit solutions to sub-problems, the *building block hypothesis* can be formulated. It describes the processing of building blocks and is based on the quasi-decomposability of a problem (Goldberg, 1989c, page 41):

> Short, low order, and highly fit schemata are sampled, recombined, and resampled to form strings of potentially higher fitness.

The building block hypothesis basically states that GEAs mainly work due to their ability to propagate building blocks. By combining schemata of lower order which are highly fit, a GEA can construct overall good solutions.

We can use the building block hypothesis for explaining the high performance of GEAs in many real-world applications. A schemata processing GEA performs well, if we assume that the problems it is applied to are quasi-decomposable, that means the overall problem can be separated into smaller subproblems. If the juxtaposition of smaller, highly fit, partial solutions (building blocks) does not result in good solutions, GEAs would fail in many real-world problems. Only by decomposing the overall problem into many smaller subproblems, solving these subproblems separately, and combining the good solutions, can a GEA find good solutions to the overall optimization problem.

This observation raises the question of why the approach of separating complex problems into smaller ones and solving the smaller problems to optimality is so successful. The answer can be found in the structure of the problems themselves. Many of the problems in the real world are somehow decomposable, because otherwise all our design and optimization methods which try to decompose complex problems could not work properly. A look in the past reveals that approaching real world problems in the outlined way has resulted in quite interesting results. Not only do human designers or engineers use the property of many complex real world problems to be decomposable, but nature itself. Most living organisms are not just one complex system where each part interacts with all others, but they consist of various separable subsystems for sensing, movement, reproduction, or communication. By optimizing the subsystems separately, and combining efficient subsystems, nature is able to create complex organisms with surprising abilities.

Therefore, if we assume that many of the problems in the real world can be solved by decomposing them into smaller subproblems, we can explain the good results obtained by using genetic and evolutionary algorithms (GEAs) for real world problems. These types of optimization algorithms perform well because they try, in analogy to human intuition, to decompose the overall problem into smaller parts (the so called building blocks), solve the smaller subproblems, and combine the good solutions. By identifying the interdependencies between the different alleles, a problem can be properly decomposed.

The purpose of the genetic operators is to decompose the problem by detecting which alleles in the chromosome influence each other, to solve the smaller problems efficiently, and to combine the sub-solutions (compare Harik and Goldberg (1996) and Harik and Goldberg (1996)).

2.3 Problem Difficulty

Previous work has shown that representations influence the behavior and performance of GEAs (Goldberg, 1989c; Liepins & Vose, 1990). The results revealed that when using specific representations some problems become easier, whereas other problems become more difficult to solve for GEAs. To be able to systematically investigate how representations influence GEA performance, a measurement of problem difficulty is necessary. With the help of a difficulty measurement, it can be determined how representations change the complexity and difficulty of a problem. However, a problem does not have the same difficulty for all types of optimization algorithms, but difficulty always depends on the optimization method used. Therefore, focusing on selectorecombinative GEAs also determines the reasons of problem difficulty: building blocks.

Consequently, in subsection 2.3.1 we discuss reasons of problem difficulty and illustrate for different types of optimization methods that different reasons for problem difficulty exist. As we focus on selectorecombinative GEAs and assume that these types of GEAs process building blocks, we decompose problem difficulty with respect to BBs. This is followed in subsection 2.3.2 by an illustration of different measurements of problem difficulty. The measurements of problem difficulty are based on the used optimization method. Because we focus in this work on schemata and BBs, we later use the schemata analysis as a measurement of problem difficulty.

2.3.1 Reasons for Problem Difficulty

One of the first approaches to the question of what makes problems difficult for GEAs, was the study of deceptive problems by Goldberg (1987). His studies were mainly based on the work of Bethke (1981). These early statements about deceptive problems were the origin of a discussion about the reasons of problem difficulty in the context of genetic and evolutionary algorithms. Searching for reasons of problem difficulty means investigating what makes problems difficult for GEAs.

Researchers recognized that there are other possible reasons of problem difficulty besides deception. Based on the structure of the fitness landscape (Weinberger, 1990; Manderick, de Weger, & Spiessens, 1991), the correlation between the fitness of individuals describes how difficult a specific problem is to solve for GEAs. By the modality of a problem, which is more popular for mutation-based search methods, problems can be classified into easy unimodal problems (there is only one local optimum), and difficult multimodal problems (there are many local optima). Another reason for difficulty is found to be epistasis, which is also known as the linear separability of a problem. This describes the interference between the alleles in a string and measures how well a problem can be decomposed into smaller subproblems (Holland, 1975; Davidor, 1989; Davidor, 1991; Naudts, Suys, & Verschoren, 1997). A final reason for problem difficulty is additional noise which makes most problems more difficult to solve for GEAs.

Many of the earlier approaches use only one aspect of problem difficulty and are not focused on schemata-processing selectorecombinative GEAs. Goldberg (2002) presented a more general approach of understanding problem difficulty based on the schemas theorem and the building block hypothesis. He viewed problem difficulty for selectorecombinative GEAs as a matter of building blocks and decomposed it into

- difficulty within a building block (intra-BB difficulty),
- difficulty between building blocks (inter-BB difficulty), and
- difficulty outside of building blocks (extra-BB difficulty).

This decomposition of problem difficulty is based on the schema theorem and assumes that difficult problems are building block challenging. In the following we briefly discuss these three aspects of BB-complexity.

If we count the number of schemata of order $o(h) = k$ that have the same fixed positions, there are 2^k different competing schemata. Based on their fitness, the different schemata compete against each other and GEAs should increase the number of the high-quality schemata. Identifying the high quality schemata and propagating them properly is the main difficulty of intra-BB difficulty. Goldberg measures intra-BB difficulty with the deceptiveness of a problem. Deceptive problems (Goldberg, 1987) are most difficult to solve for GEAs because GEAs are led by the fitness landscape to a deceptive attractor which has maximum distance to the optimum. To reliably solve difficult, for example deceptive problems, GEAs must increase the number of copies of the best BB by giving enough copies to them.

One basic assumption of the schema theorem is that a problem can be decomposed into smaller subproblems. GEAs solve these smaller sub-problems in parallel and try to identify the correct BBs. In general, the contribution of different BBs to the fitness function is not uniform and there can be interdependencies between the different BBs. Because the different building blocks have different contributions to the fitness of an individual, the loss of low salient BBs during a GEA run is one of the major problems of inter-BB difficulty. Furthermore, a problem can often not be decomposed into completely separate and independent subproblems, but there are still interdependencies between the different BBs which are an additional source of inter-BB difficulty.

Even for selectorecombinative GEAs there is a world outside the world of schemata and BBs. Sources of extra-BB difficulty like noise have an additional influence on the performance of GEAs because selection is based on the fitness of the individuals. Additional, non-deterministic noise randomly modifies the fitness values of the individuals. Therefore, selection decisions are no longer based on the quality of the solutions (and of course the BBs) but on stochastic variance. A similar problem occurs if the evaluation of the individuals is non-stationary. Evaluating fitness in a non-stationary way means that individuals have a different fitness at different moments in time.

In the remainder of the subsection we discuss that reasons of problem difficulty must be seen in the context of a specific optimization method. If different optimization methods are used for the same problem then there are different reasons of problem difficulty. As a result, there is no general problem difficulty for all types of optimization methods but we must independently identify for each optimization method the reasons of problem difficulty. We illustrate how problem difficulty depends on the used optimization method with two small examples.

When using random search, the discussion of problem complexity is obsolete. During random search new individuals are chosen randomly and no prior information about the structure of the problem is used. As a result all possible types of problems have the same difficulty. Although measurements of problem complexity, like correlation analysis or the analysis of intra-, inter-, or extra-BB difficulty, lead us to believe that some problems are easier to solve than others, there are no easy or difficult problems. Independently of the complexity of a problem, random search always needs on average the same number x of fitness evaluations for finding the optimal solution. All problems have the same difficulty regarding random search, and to search for reasons of problem difficulty makes no sense.

When comparing crossover- and mutation-based evolutionary search methods, different reasons of problem complexity exist. From the schema theorem we know that selectorecombinative GEAs propagate schemata and BBs. Therefore, BBs are the main source of complexity for these types of GEAs. Problems are easy for selectorecombinative GEAs if the problem can be properly decomposed into smaller sub-solutions (the building blocks) and the intra-, and inter-BB difficulty is low. However, when using mutation-based approaches like evolution strategies (Rechenberg, 1973; Schwefel, 1975), these reasons for problem difficulty are not relevant any more. Evolution strategies perform well if the structure of the solution space guides the population to the optimum (compare the good performance of evolution strategies on uni-modal optimization problems). Problem complexity is not based on BBs, but more on the structure of the fitness landscape. To use the notion of BBs for mutation-based optimization methods makes no sense because they propagate no schemata.

We have illustrated how the difficulty of a problem for selectorecombinative GEAs can be decomposed into intra-BB, inter-BB, and extra BB-difficulty. The decomposition is based on the assumption that GEAs decompose problems and work with schemata and BBs. The proposed BB-based reasons of problem difficulty can be used for selectorecombinative GEAs but can not be applied to other optimization methods like evolution strategies or random search.

2.3.2 Measurements of Problem Difficulty

In the previous section we discussed the reasons for problem difficulty. In the following subsection we describe some measurements of problem difficulty.

To investigate how different representations influence the performance of GEAs, a measurement of problem difficulty is necessary. Based on the different reasons for problem difficulty which exist for different types of optimization methods, we discuss some common measurements of problem difficulty:

- Correlation analysis,
- polynomial decomposition,
- walsh coefficients, and
- schemata analysis.

These four measurements of problem difficulty are widely used in the GEA literature for measuring different types of problem difficulty (Goldberg, 1989b; Goldberg, 1992; Radcliffe, 1993; Horn, 1995; Jones & Forrest, 1995). For an overview see Bäck, Fogel, and Michalewicz (1997, chapter B2.7). The specific properties of the four measurements are briefly discussed in the following.

Correlation analysis is based on the assumption that the high and low quality solutions are grouped together and that GEAs can use information about individuals whose genotypes are very similar for generating new offspring. Therefore, problems are easy if the structure of the search space guides the search to the high quality solutions. Consequently, correlation analysis is a proper measurement for the difficulty of a problem when using mutation-based search approaches. Correlation analysis exploits the fitness between neighboring individuals of the search space as well as the correlation of the fitness between parents and their offspring (For a summary see Deb, Altenberg, Manderick, Bäck, Michalewicz, Mitchell, and Forrest (1997)). The most common measurements for distance correlation are the autocorrelation function of the fitness landscape (Weinberger, 1990), the fitness correlation coefficients of genetic operators (Manderick, de Weger, & Spiessens, 1991), and the fitness-distance correlation (Jones, 1995; Jones & Forrest, 1995; Altenberg, 1997).

The linearity of an optimization problem can be measured by the polynomial decomposition of the problem. Each function f defined on $\Phi_g = \{0,1\}^l$ can be decomposed in the form

$$f(x) = \sum_{N \subset \{1,...,l\}} \alpha_N \prod_{n \in N} e_n^T x,$$

where the vector e_n contains 1 in the nth column and 0 elsewhere, T denotes transpose, and the α_N are the coefficients (Liepins & Vose, 1991). Regarding the vector x having components $x_1, \ldots, x_l$, we may view f as a polynomial in the variables $x_1, \ldots, x_l$. The coefficients α_i describe the non-linearity of the problem. If there are high order α_N in the decomposition of the problem, the function is highly nonlinear. If the decomposition of a problem only has order 1 coefficients, then the problem is linear and easy for genetic and evolutionary algorithms. It is possible to determine the maximum non-linearity of $f(x)$ for a GA by its highest polynomial coefficients. The higher the order of the α_i, the more nonlinear the problem is.

There is some correlation between the non-linearity of a problem and the difficulty of a problem for selectorecombinative GEAs (Mason, 1995), but the order of non-linearity can only give an upper limit of the problem difficulty. As illustrated in the following example there could be high order α_i although the problem still remains easy for a GA. The function

$$f(x) = \begin{cases} 1 & \text{if } x = 00, \\ 2 & \text{if } x = 01, \\ 4 & \text{if } x = 10, \\ 10 & \text{if } x = 11, \end{cases} \tag{2.1}$$

could be decomposed in $f(x) = \alpha_0 + \alpha_1 x_0 + \alpha_2 x_1 + \alpha_3 x_0 x_1 = 1 + x_0 + 3x_1 + 5x_0 x_1$. The problem is easy for selectorecombinative GEAs (all BBs are of order $k = 1$), but nonlinear.

Instead of decomposing a problem into its polynomial coefficients, it can also be decomposed into the corresponding Walsh coefficients. The Walsh transformation is analogous to the discrete Fourier transformation but for functions whose domain is a bitstring. Every real valued function $f : \Phi_g \to \mathbb{R}$ over an bitstring of length l, can be expressed as a weighted sum of a set of 2^l orthogonal functions called Walsh functions:

$$f(x) = \sum_{j=0}^{2^l - 1} w_j \, \psi_j(x),$$

where the Walsh Functions are denoted $\psi_j : \Phi_g \to \{-1, 1\}$. The weights $w_j \in \mathbb{R}$ are called Walsh coefficients. The indices of both Walsh functions and coefficients may be expressed as the numerical equivalent of the binary string j. The jth Walsh function is defined as:

$$\psi_j(x) = (-1)^{bc(j \wedge x)},$$

with x, j are binary strings and elements of Φ_g, $\wedge$ denotes the bitwise logical and, and $bc(x)$ is the number of one bits in x. For a more detailed explanation the reader is referred elsewhere (Goldberg, 1989a; Goldberg, 1989b; Vose

& Wright, 1998a; Vose & Wright, 1998b). The Walsh coefficients can be computed by the Walsh transformation:

$$w_j = \frac{1}{2^l} \sum_{i=0}^{2^l-1} f(i)\psi_j(i).$$

The coefficients w_j measure the contribution to the fitness function by the interaction of the bits indicated by the positions of the 1's in j. The larger the value of j, the higher the order of the interactions between the bits in j is. For example, w_{001} measures the linear contribution to f associated with bit position 0. w_{111} measures the nonlinear interaction between all three bits. Any function f over an l bit space can be represented as a weighted sum of all possible 2^l bit interaction functions ψ_j.

Walsh coefficients are an important feature in measuring the problem difficulty for GEAs (Goldberg, 1989a; Oei, 1992; Goldberg, 1992; Reeves & Wright, 1994). It was shown that problems are easy for GEAs if the Walsh coefficients are of order 1. Furthermore, difficult problems tend to have high order Walsh coefficients, but nevertheless the Walsh coefficients do not give us an exact measurement of problem complexity. The highest order of the coefficient w_i can only give an upper limit of the problem complexity. Therefore, Walsh coefficients show the same behavior as polynomials. This behavior is expected as it has already been shown that Walsh functions are polynomials (Goldberg, 1989a; Goldberg, 1989b; Liepins & Vose, 1991). The insufficient measurement of problem difficulty for selectorecombinative GEAs can be illustrated with the earlier example (equation 2.1). The Walsh coefficients for the former example are $w = \{4.25, -1.75, -2.75, 1.25\}$. Although the problem is quite simple, there are high order Walsh terms.

If we assume that selectorecombinative GEAs process schemata and BBs, then the most natural and direct way to measure problem complexity is to analyze the size and length of the building blocks in the problem. If we assume that GEAs process building blocks, the intra-BB difficulty of a problem can be measured by the maximum length $\delta(h)$ and size $k = o(h)$ of the BBs h (Goldberg, 1989c).

A problem is denoted to be deceptive of order k_{max} if for $k < k_{max}$ all schemata that contain parts of the best solution have lower fitness than their competitors (Deb & Goldberg, 1994). Schemata are competitors if they have the same fixed positions. An example for competing schemata of size $k = 2$ for a bitstring of length $l = 4$ are $h = 0 * 0*$, $h = 0 * 1*$, $h = 1 * 0*$, and $h = 1 * 1*$. Therefore, the highest order k_{max} of the schemata that are not misleading determines the complexity of a problem for selectorecombinative GEAs. The higher the maximum order k_{max} of the schemata, the more difficult the problem is to solve for GEAs.

The average fitness of the schemata for the brief example illustrated in equation 2.1 is shown in Table 2.1. All schemata that contain a part of the

Table 2.1. Average schema fitness for example described by equation 2.1.

order	2	1		0
schema	11	1*	*1	**
fitness	10	7	6	4.25
schema		0*	*0	
fitness		1.5	2.5	

optimal solution are above average and better than their competitors. Calculating the deceptiveness of the problem based on the fitness of the schemata correctly classifies this problem to be very easy. Since we analyze the influence of representations on selectorecombinative GEAs, and we assume that these types of GEAs process schemata, schema fitness averages are used to measure problem difficulty in the remainder of this work. Problems of length l are defined to be fully easy if $k_{max} = 1$, and to be fully difficult (compare Goldberg (1992)) if $k_{max} = l$. Therefore, when using selectorecombinative GEAs fully easy problems are the most easy problems, whereas fully difficult problems are the most difficult problems.

2.4 Existing Recommendations for the Design of Efficient Representations for Genetic and Evolutionary Algorithms

Although the application of GEAs to optimization problems is not possible without using representations, mainly intuitive knowledge exists about how to choose proper representations. Up till now, there is no proven theory regarding the influence of representations on the performance of genetic and evolutionary algorithms. To help users with the difficult task of finding good representations, some researchers have made recommendations for the design of efficient representations over the last few years. In the following we review some recommendations which are important from the authors point of view. For a more detailed overview the reader is referred to Ronald (1997).

We start in subsection 2.4.1 with the principle of meaningful building blocks and minimal alphabets which were proposed by Goldberg (1989c). From the authors point of view, this is the most important early work in this field. Some years later, Palmer (1994) presented more concrete guidelines about proper design of representations. We briefly discuss his guidelines in subsection 2.4.2. Finally, we illustrate in subsection 2.4.3 the recommendations made by Ronald (1997).

2.4.1 Goldberg's Meaningful Building Blocks and Minimal Alphabets

Some of the first recommendations for the construction of representations were made by Goldberg (1989c). He proposed the principle of minimal alphabets and of meaningful building blocks.

It is known that the design of an encoding has a strong impact on the performance of a genetic algorithm and should be chosen carefully (compare Coli and Palazzari (1995a), Ronald (1997), and Albuquerque, Chopard, Mazza, and Tomassini (2000)). Therefore, Goldberg (1989c, p. 80) proposed two basic design principles for encodings:

- Principle of meaningful building blocks: The schemata should be short, of low order, and relatively unrelated to schemata over other fixed positions.
- Principle of minimal alphabets: The alphabet of the encoding should be as small as possible while still allowing a natural representation of solutions.

The principle of meaningful building blocks is directly motivated by the schema theorem (see subsection 2.2.3). If schemata are highly fit, short, and of low order, then their number exponentially increase over the generations. If the high-quality schemata are long or of high order, they are disrupted by crossover and mutation and they can not be propagated properly by GEAs. Consequently, representations should modify the complexity of a problem in a way that phenotypically long or high order BBs become genotypically short, and of low order. Then, the problem becomes easier for selectorecombinative GEAs.

The principle of minimal alphabets tells us to increase the potential number of schemata by reducing the cardinality of the alphabet. When using minimal alphabets the number of possible schemata is maximal. This is the reason why Goldberg advises us to use bitstring representations, because high quality schemata are more difficult to find when using alphabets of higher cardinality (Goldberg, 1989c, pp. 80ff). But of course we have a tradeoff between the low cardinality of an alphabet and the natural expression of the problem. Therefore, sometimes a higher cardinality of the alphabet could be helpful for GEAs (Goldberg, 1991b).

Goldberg's two design principles of representations are based on the assumption that GEAs process schemata and BBs, and aid us in a better understanding of how to design efficient representations for selectorecombinative GEAs. However, both principles are very abstract and general, and do not provide the user with exact and applicable guidelines.

2.4.2 Palmer's Tree Encoding Issues

Several years later Palmer analyzed properties of tree representations (Palmer, 1994). The recommendations he gave for the design of tree representations can also be applied to other types of representations. Palmer proposed the following representation issues:

- An encoding should be able to represent all possible phenotypes.
- An encoding should be unbiased in the sense that all possible individuals are equally represented in the set of all possible genotypic individuals.
- An encoding should encode no infeasible solutions.
- The decoding of the phenotype from the genotype should be easy.
- An encoding should possess locality. Small changes in the genotype should result in small changes in the phenotype.

Although Palmer formulated the design issues based mainly on intuition rather than on theoretical investigation, the guidelines can advantageously be used for the design of proper representations. For a more detailed description and discussion of these representation issues in the context of tree network representations, the reader is referred to subsection 6.1.6. We will also see later in this work that the encoding of infeasible solutions (subsection 6.4.2), a bias of the individuals (subsection 6.3.3), or low locality of an encoding (subsection 8.1.3) is not always necessarily disadvantageous for the performance of GEAs.

2.4.3 Ronald's Representational Redundancy

A few years ago, Ronald presented a survey of encoding issues (Ronald, 1997; Ronald, 1995). Representations should be chosen according to the following guidelines:

- Encodings should be adjusted to a set of genetic operators in a way that the building blocks are preserved from the parents to the offspring (Fox & McMahon, 1991).
- Encodings should minimize epistasis (Beasley, Bull, & Martin, 1993).
- Feasible solutions should be preferred.
- The problem should be represented at the correct level of abstraction.
- Encodings should exploit an appropriate genotype-phenotype mapping process if a simple mapping to the phenotype is not possible.
- Isomorphic forms, where the phenotype of an individual is encoded with more than one genotype, should not be used.

Many of the representation issues can be put down to the principles of representations illustrated in subsection 2.4.1. The design issue concerning isomorphic forms will be discussed in section 3.1. The results will show that by using isomorphic or redundant representations the performance of GEAs can easily be increased.

3. Three Elements of a Theory of Genetic and Evolutionary Representations

In this chapter, we study an often ignored aspect of genetic optimization, namely the theory of representations for genetic and evolutionary algorithms. Although the importance of choosing proper representations for the performance of genetic and evolutionary algorithms (GEAs) is already recognized (Caruana & Schaffer, 1988; Goldberg, 1989c; Liepins & Vose, 1990; Ronald et al., 1995; Coli & Palazzari, 1995b; Ronald, 1997; Albuquerque et al., 2000; Kargupta, 2000a; Schnier & Yao, 2000; Hinterding, 2000), we are still far from a complete theory of representations.

Due to the fact that developing a general theory of representations is a formidable challenge, we decompose this task into smaller parts. In the following, we start by presenting three elements of representation theory which are the basis of the time-quality framework of representations for genetic and evolutionary algorithms which we present in chapter 4. Namely, we focus on redundancy, the scaling of building blocks (BBs), and the modification of distances between individuals when mapping the phenotypes on the corresponding genotypes. We present theoretical models for these three aspects of representation theory and show how they affect the performance of genetic and evolutionary algorithms.

The following paragraphs discuss the three aspects of representation theory. A representation is denoted to be redundant if the number of genotypes is higher than the number of phenotypes. In redundant representations, a phenotype is represented on average by more than one genotype. Investigating redundant representations more closely, we recognize that linear redundancy can be addressed as a matter of BB-supply. Representations that give more copies to high quality solutions in the initial population result in a higher performance of GEAs, whereas encodings where high quality solutions are underrepresented make a problem more difficult to solve. Uniform redundancy, however, has no influence on the performance of GEAs. Based on the Gambler's ruin population sizing model by Harik, Cantú-Paz, Goldberg, and Miller (1997) we are able to present a theoretical model describing the effects of redundant representations on the performance of GEAs.

The order of scaling of a representation describes how different the contribution of the BBs to the fitness of an individual is. It is well known that if the BBs are uniformly scaled, GEAs solve all BBs implicitly in parallel.

In contrast, for non-uniformly scaled BBs, domino convergence occurs and the BBs are solved sequentially starting with the most salient BB (Thierens, 1995). As a result, the convergence time increases and the search is affected more strongly by genetic drift. Lower salient alleles are not opposed to selection pressure unless they are reached by the solving process. Therefore, some of the lower salient alleles can loose their diversity and are randomly fixed. To model more exactly the effects of non-uniformly scaled representations on GEA performance, we extend the work of Thierens and Goldberg (1993) and present a more general theory of non-uniformly scaled encodings. It allows us to predict more accurately the performance of GEAs using non-uniformly scaled representations under the influence of genetic drift.

In general, the representation (the genotype-phenotype mapping) used by GEAs should have no influence on the ability of GEAs to solve easy problems. However, previous work has shown that representations can easily change the difficulty of a problem (Liepins & Vose, 1990). By the use of representations, fully easy problems can become fully difficult, and vice versa. Section 3.3 reveals that the distance distortion of a representation determines whether the difficulty of a problem is changed by the representation. The distance distortion of a representation describes how great the distances between the genotypes are compared to the distances between the corresponding phenotypes. We illustrate that the high locality of a representation, that means neighboring phenotypes correspond to neighboring genotypes, is a necessary condition for a representation to have low distance distortion.

Furthermore, we are able to show both theoretically and empirically, that by using representations where the distance distortion is not equal to zero, fully easy problems become more difficult, whereas fully difficult problems become easier. Therefore, if our aim is to reliably solve problems of bounded difficulty, we demand high quality representations to have low distance distortion and to preserve the complexity of the BBs. We discuss why it is in general not possible to create representations that preserve complexity for easy problems, and reduce complexity for difficult problems.

In the following, we give a brief overview about the structure of the chapter. Section 3.1 shows how the usage of redundant encodings affects genetic search. We use redundancy on the level of bitstrings and investigate the effects of encoding the information content of one Bit with more than one bit in a string. Based on the Gambler's ruin model (Harik, Cantú-Paz, Goldberg, & Miller, 1997), we develop a quantitative model of redundancy and verify it empirically for one-max and deceptive problems.

In section 3.2, we show how the behavior of GEAs changes for exponentially scaled encodings by using the existing models of genetic drift and population sizing. We use two different drift models and develop a population sizing and convergence time model that allows us to predict the solution quality more accurately than the previous models. To verify the theoretical models, we present an empirical investigation into the performance of GEAs using exponentially scaled representations.

Section 3.3 shows that only representations where the distance distortion is zero guarantee that phenotypicly easy problems remain genotypicly easy and can still be solved using GEAs. If the distance distortion of a representation is unequal to zero, the size and length of the BBs can be different for the genotypes and phenotypes and the complexity of the problem is changed. Fully easy problems can only become more difficult, and fully difficult problems can only become easier to solve for GEAs. The chapter ends with concluding remarks.

3.1 Redundancy

This section provides the first of three elements of a theory of genetic and evolutionary representations. It identifies redundancy to be important for the design of representations, investigates the effects of redundancy for bitstrings, and uses existing complexity models to characterize the effect of redundancy on encodings. It presents a theoretical model on how the population size, run duration and overall problem difficulty is changed. The model is verified using empirical results.

3.1.1 Definitions and Background

Information theory provides us with a measurement of information. The information content[1] (measured in Bits) of a sequence is defined as the number of bits required to represent a given number of s possibilities using an optimal encoding (Shannon, 1948; Shannon & Weaver, 1949). It is calculated as $\log_2(s)$. Redundant encodings are less efficient codings that require more bits to represent the information but do not increase the amount of information represented.

For encoding one Bit of information content (for example the possibilities 0 and 1) a binary string of at least length one is necessary (one bit). However, it is also possible to encode one Bit (either 0 or 1) using a bitstring of length $l > 1$. Then, more than one bit of the bitstring encodes one Bit of information, and the representation becomes redundant. We want to emphasize that it is important to distinguish between the amount of information (**Bit**) that should be represented and the number of bits in a string that are used to represent the information.

k_r is defined as the order of redundancy and measures the amount of redundant information in the encoding (in bit). There are k_r bits and 2^{k_r} different possibilities (individuals) to encode 1 Bit of information. Using no redundancy in an encoding results in $k_r = 1$. r is defined as the number of

---

[1] other notations are information, self-information, entropy, or Shannon entropy.

genotypic BBs of length kk_r that represent the optimal phenotypic BB of length k. If we encode 1 Bit of information content (two possibilities) with a bitstring of length $l = 1$ the encoding is not redundant. The optimal solution (one of the two possibilities) is represented by exactly one genotype, and $r = 1$. In general, when using uniformly redundant encodings

$$r_{uniform} = 2^{k(k_r - 1)}, \tag{3.1}$$

where k denotes the order of the phenotypic BB. An encoding is uniformly redundant if each phenotype is represented by the same number of genotypes. For redundant encodings

$$r \in \{1, 2, \ldots, 2^{kk_r} - 2^k + 1\}. \tag{3.2}$$

In general, when focusing on BBs of length k there are 2^k different phenotypes and they are represented by 2^{kk_r} different genotypes. We want to denote a representation to be linear redundant if one phenotypic bit is represented by k_r genotypic bits and k_r is an integer. Then we get

$$r = i^k, \text{ with } i \in [1, 2, \ldots, 2^{k_r} - 1]. \tag{3.3}$$

Each phenotype bit is determined by k_r genotype bits. Furthermore, linear redundant representations assign the genotypes to the phenotypes in such a way that the distances between genotypes which are assigned to the same phenotype are minimal. The distance d_{x_i, x_j} between two bitstrings x_i and x_j is calculated according to the Hamming distance (Hamming, 1980) and counts the number of different bits in the two individuals. For example, when using for $k = 1$ a redundant representation with $k_r = 2$ and $r = 2$, we can assign $x_g^1 = 00$ and $x_g^2 = 01$ to $x_p = 0$ and $x_g^3 = 10$ and $x_g^4 = 11$ to $x_p = 1$. The distance $d_{x_g^1, x_g^2} = 1$ and $d_{x_g^3, x_g^4} = 1$ and the encoding is linear redundant. If we would assign x_g^1 and x_g^4 to $x_p = 0$ and x_g^2 and x_g^3 to $x_p = 1$ the representation would be not linear redundant any more because $d_{x_g^1, x_g^4} = 2$ and $d_{x_g^2, x_g^3} = 2$. The distances between genotypes that represent the same phenotype are not minimal any more in comparison to other redundant representations.

For example, if $k_r = 2$ and $k = 3$ the genotypic BBs have size $kk_r = 6$ and r can either be $1^3 = 1$ (optimum is underrepresented), $2^3 = 8$ (uniform redundancy), or $3^3 = 27$ (optimum is overrepresented). Figure 3.1 gives another example for a redundant encoding. There are always two bits in the phenotypic space Φ_p represented by 3 bits in the genotypic space Φ_g ($k_r = 1.5$). With $|\Phi_p| = 2^k = 2^2$ the size of the genotypic space is $|\Phi_g| = 2^{kk_r} = 2^3$.

By introducing redundancy the search space for a GA using phenotypes of string length $l = l_p$ is increased from $|\Phi_p| = 2^l$ to $|\Phi_g| = 2^{lk_r}$. The length of the individuals increases from $l = l_p$ in the phenotypic space to $l_g = k_r * l$ in the genotypic space. To represent all phenotypes, each individual $x_p \in \Phi_p$ must be represented by at least one genotype $x_g \in \Phi_g$. If $|\Phi_g| = |\Phi_p|$, and each phenotype is represented by at least one genotype, we have a redundant-free, one-to-one mapping.

genotypes   phenotypes

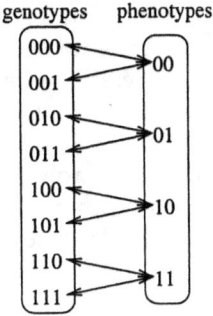

Fig. 3.1. A redundant encoding ($k_r = 1.5$, $r = 2$)

Similarly to our notion of uniformly redundant representations, other work (Shipman, 1999; Shipman, Shackleton, & Harvey, 2000; Shackleton, Shipman, & Ebner, 2000; Shipman, Shackleton, Ebner, & Watson, 2000; Ebner, Langguth, Albert, Shackleton, & Shipman, 2001) introduced the so called Trivial Voting mappings. For these types of encodings a phenotype bit is represented by a number of genotype bits. The value of the phenotype bit follows the majority of the corresponding genotype bits. If the majority of the genotype bits have value 1 then the phenotype bit is set to 1. If the majority of the genotype bits have value 0 then the phenotype bit is set to 0. The purpose of these publications was to combine redundant encodings with encodings that do not preserve the distances between the individuals (compare section 3.3). Then GAs could more easily find good solutions which have a large phenotypic distance by performing random neutral walks in the genotype space. As a result, GAs would be able to perform better on fully difficult problems. Some researchers investigated in this context the role of neutral search spaces (Huynen, Stadler, & Fontana, 1996; Schuster, 1997; Smith, Husbands, & M., 2001c; Reidys & Stadler, 2001; Smith, Husbands, & M., 2001a; Smith, Husbands, & M., 2001b). If the order of redundancy is high, a lot of phenotypes have the same fitness value, and GEAs start to drift on this neutral search space. If this drift behavior can be used advantageously then difficult problems can become easier.

There are differing opinions regarding the effects of redundancy on the performance of GEAs. Some work has addressed redundancy as a problem either of diversity loss (see Ronald, Asenstorfer, and Vincent (1995)), or of cross-competition where useful schemata compete against each other. In both cases, using redundancy leads to a reduction in performance for genetic and evolutionary search (Davis, 1989; Eshelman & Schaffer, 1991; Ronald, Asenstorfer, & Vincent, 1995). Other work however reports higher performance with additional redundancy (Gerrits & Hogeweg, 1991; Cohoon, Hegde, Martin, & Richards, 1988; Julstrom, 1999), which some researchers ascribe to be due to an increase in diversity that hinders premature convergence. At the moment there exists no quantitative modeling for the effects of redundancy.

One interesting approach to redundancy was presented by Kargupta (2000b). This work considers the computational implications of genetic code-like representations in gene expression. Interpreting gene expression as a genotype-phenotype mapping, the work shows that adding redundant information that favors optimal solutions affects the energy spectrum of a code. The genome representation becomes more efficient, and the overall problem easier to solve if more copies are given to solutions with higher fitness. In the following, we present a similar approach and interpret redundancy as a matter of building block supply.

3.1.2 Decomposing Redundancy

For modeling the effects of redundant representations, we use the complexity model from Goldberg (1991a) and Goldberg, Deb, and Clark (1992). If we want to understand the effects of redundancy, we must decompose the problem into smaller sub-problems and try to solve these separately. We could decompose the problem step by step and subsequently collate all the sub-problems. The decomposition (Goldberg, 1998) takes place as follows:

- GAs process building blocks.
- Problems are tractable by BBs.
- GAs must ensure proper supply of BBs in the initial generation.
- GAs must grow the high quality BBs.
- GAs must mix the BBs well.
- GAs must decide well among competing BBs.

This decomposition gives us a framework for investigating the effects of redundancy on the performance of GAs. We want to check how redundancy affects the problem decomposition point by point:

Using redundant encodings does not change the principal behavior of GAs. After adding redundancy GAs still process building blocks.

A problem is still tractable by building blocks when using a redundant encoding. However, the question arises as to whether the size of building blocks is changed by redundant encodings. On one hand, redundant representations increase the number of bits that are part of a building block in the genotype from k to kk_r. On the other hand, the number of building blocks in the genotype that represent the same BB in the phenotype increases from 1 to on average $2^{k(k_r-1)}$. Furthermore, the number of fitness values that can be assigned to the genotypes stays constant. Taking these effects into account, we assume that the effective size of the different building blocks remains unchanged by redundancy.

How do redundant encodings change the initial supply of BBs in the initial population? For uniform redundancy (every phenotype $x_p \in \Phi_p$ is represented by the same number of genotypes $x_g \in \Phi_g$), the initial supply of BBs $x_0/n = 1/2^k$ is the same as for non-redundant representations. If the number of genotypes x_g that represent a phenotype x_p is above average, then the phenotype x_p, and the containing schemata h_p, are overrepresented in the initial population. GEAs are shifted more towards solutions that are similar to these x_p. Analogously, the performance of GEAs decreases if the proportion of lower fit phenotypes in the initial population is increased by redundancy. As a result, the supply of BBs could be modified by redundant encodings.

For a proper growth and mixing of BBs, above average BBs must be preferred by selection, and BBs should not be disrupted by the crossover operator. Despite the fact that the effective size of BBs stays constant, the actual defining length of a genotypic BB $\delta(h_g)$ increases with redundancy. Therefore, GEA operators that do not obey the linkage disrupt BBs more frequently. To overcome this problem, competent GAs (Goldberg, 1999) could be used. These kind of GEAs obey the linkage and genotypic building blocks h_g are not disrupted by recombination. Thus, no reduction of performance should occur, and redundancy should have no negative effect on the proper mixing of BBs. However, when using redundant representations there are different genotypes x_g and y_g that represent the same phenotype x_p. Recombining x_g and y_g could result in offspring that do not represent x_p. This effect is also known as cross-competition among isomorphic identical BBs. As an example for a one-bit problem, the genotype $\{00\}$ represents the phenotype $x_0 = 0$, and the genotypes $\{01, 10, 11\}$ represent the phenotype $x_1 = 1$. If the genotypes 01 and 10 are recombined the possibilities for the offspring are 00 and 11. Although, both parental genotypes represent the same phenotype x_1, one of the offspring represents x_0.

Finally, the decision making between competing BBs is not affected by adding redundancy. With redundancy there are different genotypic BBs h_g that represent the same phenotypic BB h_p, but the selection process does not decide between the different h_g because they all have the same fitness. The fitness evaluation is based on the fitness of the phenotypic BBs h_p, and not their genotypic representation.

After recognizing that redundancy could have a major effect on the supply, and a minor effect on the mixing of building blocks, we want to quantify its effect on BBs supply in the following subsection.

3.1.3 Population Sizing

We use the insights into the effects of redundancy on the performance of GEAs from the previous section and quantify them using existing theory. For a problem with string length l, and order of BBs k, we know from Feller (1957) the probability that a particle is captured in a random walk by the absorbing boundary at $x = n$:

$$P_n = \frac{1 - (q/p)^{x_0}}{1 - (q/p)^n},$$

where p is the probability of making the right choice between a single sample of each BB

$$p = N\left(\frac{d}{\sqrt{2m'}\sigma_{BB}}\right). \tag{3.4}$$

d is the signal difference between the best BB and its strongest competitor, $m' = m - 1$ with m is the number of BBs in the problem, σ_{BB}^2 is the variance of a BB, and x_0 is the expected number of copies of the best BB in the randomly initialized population, $q = 1 - p$ is the probability of making the wrong decision between two competing BBs, and n is the population size. It has been shown in Harik, Cantú-Paz, Goldberg, and Miller (1997) that this random walk or Gambler's ruin model can be used for describing the behavior of selectorecombinative GAs propagating schemata and BBs. We also want to use this model for describing the influence of redundant representations on the behavior of GAs using only crossover and selection and no mutation.

For a randomly initialized population with no redundancy $x_0 = n/2^k$. When using redundant encodings the initial supply of BBs is changed. With r the number of genotypic BBs h_g of length kk_r that represent the best phenotypic BB h_p of length k, we get

$$x_0 = n\frac{r}{2^{kk_r}}, \tag{3.5}$$

where k_r is the order of redundancy. If each phenotypic bit is represented by k_r genotypic bits and we use linear redundancy, we get $r = i^k$, where $i \in [1, 2^{k_r} - 1]$. For uniform redundancy, $r = 2^{k(k_r-1)}$ and $x_0 = n/2^k$. The variance σ_{BB}^2 and the number of BBs, m, is not affected by using redundancy. So the probability of GA failure $\alpha = 1 - P_n$ can be calculated as

$$\alpha = 1 - \frac{1 - (q/p)^{x_0}}{1 - (q/p)^n}. \tag{3.6}$$

If we assume that x_0 is small and $q < p$ we can assume that $1 - (q/p)^n$ converges faster to 1 than $1 - (q/p)^{x_0}$. Using these approximations (see also Harik et al. (1997)) the equation can be simplified to

$$\alpha \approx \left(\frac{1-p}{p}\right)^{x_0}.$$

Therefore, we get for the population size

$$n \approx \frac{2^{kk_r}}{r}\left(\frac{\ln(\alpha)}{\ln\left(\frac{1-p}{p}\right)}\right). \tag{3.7}$$

The normal distribution in equation 3.4 can be approximated using the first two terms of the power series expansion (see Abramowitz and Stegun (1972)) as $\mathbb{N}(x) \approx 1/2 + x/2$. Substituting p from equation 3.4 into equation 3.7 we get:

$$n \approx \frac{2^{kk_r}}{r} \ln(\alpha)/\ln\left(\frac{1-x}{1+x}\right),$$

where $x = d/\sqrt{2m'}\sigma_{BB}$. Since x is a small number, $\ln(1-x)$ can be approximated with $-x$ and $\ln(1+x)$ with x. Using these approximations we finally get for the population size n:

$$n \approx -\frac{2^{k_r k - 1}}{r} \ln(\alpha) \frac{\sigma_{BB}\sqrt{\pi m'}}{d}. \tag{3.8}$$

The population size n goes with $O\left(\frac{2^{k_r}}{r}\right)$. With increasing r the number of individuals that are necessary to solve a problem decreases. Using uniformly redundant representations $r = 2^{k(k_r - 1)}$ does not change the population size n in comparison to non-redundant representations.

3.1.4 Run Duration and Overall Problem Complexity

For describing the performance of GAs, not only the number of individuals that are necessary for solving a problem, but also the expected number of generations until convergence must be calculated.

Based on Thierens and Goldberg (1993), Miller and Goldberg developed a convergence model for selectorecombinative GAs (Miller & Goldberg, 1996b; Miller & Goldberg, 1996a). The convergence time t_{conv} depends on the length of the phenotypes $l = l_p$ and the used selection scheme. Using the selection intensity I the convergence model is

$$p(t) = 0.5\left(1 + \sin\left(\frac{It}{\sqrt{l}} + \arcsin(2p(0) - 1)\right)\right),$$

where $p(0) = x_0/n$ is the proportion of best building blocks in the initial population. I depends only on the used selection scheme and we get for example $I = \frac{3}{2\sqrt{\pi}}$ for tournament selection with tournament size 3 (Bäck, Fogel, & Michalewicz, 1997, C 2.3). The number of generations t_{conv} it takes to fully converge the population can be calculated by putting $p(t_{conv}) = 1$:

$$t_{conv} = \frac{\sqrt{l}}{I}\left(\frac{\pi}{2} - \arcsin(2p(0) - 1)\right). \tag{3.9}$$

If we assume $k = 1$ and equal proportion of 1s and 0s in the initial population we get $p(0) = 0.5$. Then the number of generations until convergence simplifies to

$$t_{conv} = \frac{\pi}{2}\frac{\sqrt{l}}{I}.$$

With redundancy the initial proportion of building blocks is $p(0) = \frac{r}{2^{kk_r}}$ (see equation 3.5). Using $\arcsin(x) = x + o(x^3)$ the time until convergence could be approximated by

$$t_{conv} = \frac{\sqrt{l}}{I}\left(1 + \frac{\pi}{2} - \frac{r}{2^{k_r k - 1}}\right). \tag{3.10}$$

With increasing $r/2^{k_r}$ the time to convergence t_{conv} is reduced. For uniform redundancy $r = 2^{k(k_r-1)}$, we get

$$t_{conv} = \frac{\sqrt{l}}{I}\left(1 + \frac{\pi}{2} - \frac{1}{2^{k-1}}\right).$$

The time until convergence when using uniformly redundant representations is the same as without redundancy.

After we have calculated the number of individuals that are necessary for solving a problem (see equation 3.8), and the number of generations that GEAs using only crossover need to converge (see equation 3.10), we can calculate the absolute number of fitness calls that are necessary for solving a problem:

$$n * t_{conv} \approx -\frac{2^{k_r k - 1}}{r} \ln(\alpha) \frac{\sigma_{BB}\sqrt{\pi m'}}{d} * \frac{\sqrt{l}}{I}\left(1 + \frac{\pi}{2} - \frac{r}{2^{k_r k - 1}}\right) =$$

$$= \frac{\sqrt{\pi l m'}}{I} \ln(\alpha) \frac{\sigma_{BB}}{d}\left(1 - \frac{2^{kk_r}}{4r}(2 + \pi)\right)$$

The overall number of fitness calls goes with $O(2^{k_r}/r)$. In comparison to non-redundant representations the number of fitness calls stays constant for redundant representations if $r = 2^{k(k_r-1)}$. Then $x_0/n = 1/2^k$ and the representation is uniformly redundant.

3.1.5 Empirical Results

In this subsection we provide empirical verification for the results we derived in the previous subsection.

We want to use redundant representations where the genotypes and phenotypes are defined on bitstrings. For our empirical investigation we use linear redundancy with $k_r = 2$. This means each bit of $x_p \in \Phi_p$ is represented by two bits in $x_g \in \Phi_g$. There are four different genotypes to represent two different phenotypes. The string length of a genotype x_g increases from $l_p = l$ to $l_g = 2l$. Therefore, the size of the genotypic search space is $|\Phi_g| = 2^{2l}$. In Table 3.1 the three possibilities of assigning at least one genotype $\{00, 01, 10, 11\}$ to one of the phenotypes $\{0, 1\}$ are illustrated. If we assume that the size of BBs k equals 1, then the number of genotypic BBs that represent the optimal phenotypic BB $h_p = 1$ is either $r = 1$, $r = 2$, or $r = 3$. With denoting x_i^p

Table 3.1. Interpretation of redundant bits

x_i^p	$x_{2i}^g x_{2i+1}^g$ (with $k_r = 2$)		
	$r = 1$	$r = 2$	$r = 3$
0	00, 01, 10	00, 01	00
1	11	10, 11	01, 10, 11

the value of the ith bit in the phenotype x^p, the $2i$th and $(2i + 1)$th bit of a genotype determines x_i^p.

The first test example for our empirical investigation is the one-max problem, where $k = 1$. It uses the number of ones in the phenotype x_p as the fitness of the individual. We want to ensure that recombination results in a proper mixing of the building blocks and therefore, we use uniform crossover for all our one-max experiments. In all our runs we use tournament selection without replacement and a tournament size of 2. For the one-max function the signal difference d equals 1, the size k of the building blocks is 1, and the variance of a building block $\sigma_{BB}^2 = 0.25$. This problem is easy to solve for GEAs.

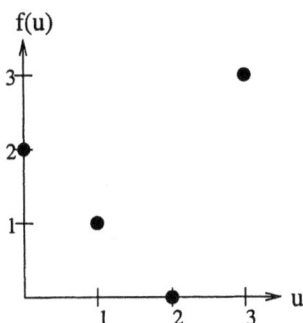

f(u)

Fig. 3.2. A 3-bit deceptive trap function

The second test example uses deceptive trap functions. These functions were first used by Ackley (1987). Figure 3.2 depicts a 3-bit trap-function where the size of a BB is $k = 3$. The fitness value of a phenotype x_p depends on the number of ones u in the string and there is a deceptive attractor that misleads GEAs. The optimal BB (fitness l) has three ones in the string, whereas the misleading attractor has none (fitness $l - 1$). The signal difference d in this function is 1, and the fitness variance equals $\sigma_{BB}^2 = 0.75$. We construct a test function by concatenating $m = 10$ of the 3-bit traps so we get a 30-bit problem. The fitness of an individual x is calculated as $f(x) = \sum_{i=0}^{m-1} f_i(u)$, where $f_i(u)$ is the fitness of the ith 3-bit trap function from Figure 3.2. Although this function is difficult for GEAs, it can be solved with proper population size n.

In Figure 3.3, the proportion of correct BBs at the end of a run for a 150 bit one-max problem using a linear redundant representation with $k_r = 2$

is shown. We know from equation 3.3 that $r = i^k$, with $i \in [0, 2^{k_r}]$. With $k_r = 2$ and $k = 1$ we get either $r = 1^1 = 1$ (each $x_i^p = 1$ is represented by one genotype), $r = 2^1 = 2$ (each $x_i^p = 1$ is represented by two genotypes), or $r = 3^1 = 3$ (each $x_i^p = 1$ is represented by three genotypes).[2] The lines without line points show the theoretical predictions from equation 3.6, and the lines with line points show the empirical results which are averaged over 250 runs. For uniform redundancy, $r = 2$, we get the same performance as for without redundancy. From the figures it is evident that the developed models accurately predict the necessary population size n. As discussed previously, redundancy only influences the supply of BBs in the initial generation. The higher the value of r, the lower is the required population size n. The effect of improper mixing and cross-competition can be neglected as the size of BBs in Φ_p is 1.

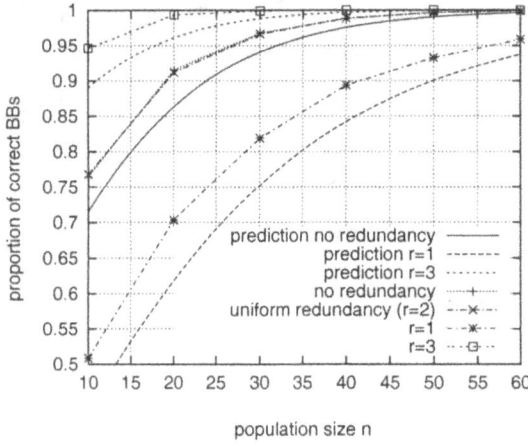

proportion of correct BBs

population size n

prediction no redundancy
prediction r=1
prediction r=3
no redundancy
uniform redundancy (r=2)
r=1
r=3

Fig. 3.3. Experimental and theoretical results of the proportion of correct BBs on a 150-bit one-max function with redundancy. The lines without line points are the predictions from the Gambler's ruin model. Adding redundancy only influences the initial BB supply. Plots are shown for $r = 1$, $r = 2$ (uniform redundancy), and $r = 3$. The order of redundancy is $k_r = 2$.

In Figure 3.4 we show the proportion of correct BBs at the end of a run over different population sizes for the 3-bit deceptive function with 10 BBs using a linear redundant representation with $k_r = 2$. We also use tournament selection without replacement of size 2. In contrast to the one-max problem, two-point crossover was chosen for recombination. Uniform crossover would result in an improper mixing of the BBs because the genotypic BBs are of length $l_g = k_r l_p = 6$ which would result in excessive BB disruption. Again, the lines without line points show the predictions of the model for different r (see equation 3.6). Furthermore, empirical results which are averaged over 250 runs, are shown for $r = 1^3 = 1$, $r = 2^3 = 8$ and $r = 3^3 = 27$ $(r_k = 2)$. For $r = 2^{k(k_r-1)} = 8$ (uniform redundancy) we get the same performance with and without redundancy. As in the previous case the model approximates well to the experimental results (lines with line points).

---

[2] We do not consider the cases $i = 0$ or $i = 4$ because then either $x_i^p = 1$, or $x_i^p = 0$ can not be represented.

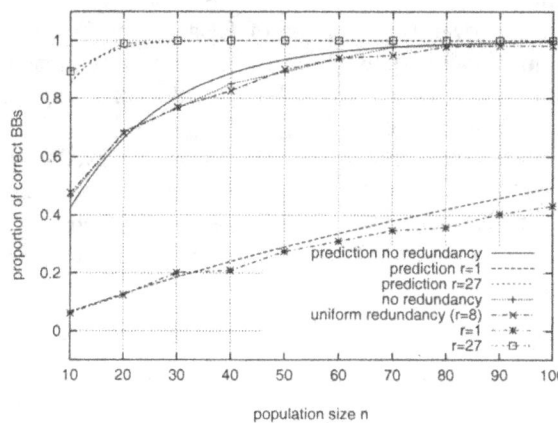

Fig. 3.4. Experimental and theoretical results of the proportion of correct BBs for a 3-bit deceptive function ($k = 3$) with 10 BBs and redundancy ($l_p = 30$, $l_g = 60$). The solid lines are the predictions from the Gambler's ruin model. Adding redundancy only influences the initial BB supply. Plots are shown for $r = 1^3 = 1$, $r = 2^3 = 8$ (uniform redundancy), and $r = 3^3 = 27$. The order of redundancy $k_r = 2$.

The results show that the effect of redundant representations on the performance of GEAs can be explained well with different initial supplies of high quality BBs in the population. If redundancy favors high quality BBs then the performance of GEAs is increased. If BBs of lower fitness are favored the performance is reduced. When using uniformly redundant encodings GEAs show the same performance as when using non-redundant encodings.

In the remainder of the section we perform an empirical investigation into the effect of redundancy on the number of generations until the population is converged. We use the one-max problem from above, uniform crossover and set the order of redundancy k_r to 2. The population size in all runs is $n = 2l$. This allows reliable decision making for the one-max problem (Goldberg, Deb, & Clark, 1991). Results are shown for tournament selection without replacement of size 2. Therefore, the selection intensity $I = 1/\sqrt{\pi}$.

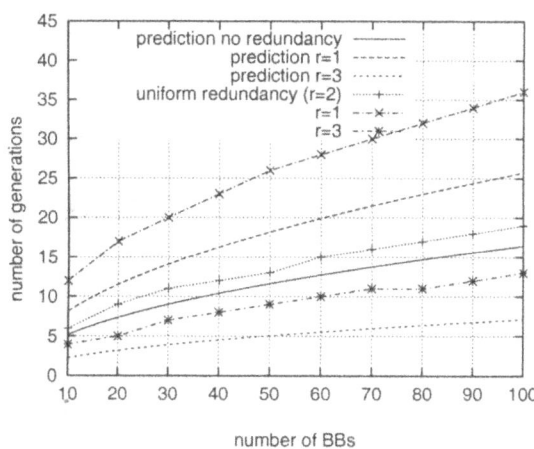

Fig. 3.5. Convergence model and experimental results of the number of generations that are necessary until 90% of all phenotypic BBs are found when optimizing a one-max function using tournament selection and uniform crossover. Plots are shown for $r = 1$, $r = 2$ (uniform redundancy), and $r = 3$. Each bit of a phenotype is represented by two bits ($k_r = 2$) in the genotype.

Figure 3.5 shows the number of generations that are necessary until 90% of all phenotypic BBs are found over the number of BBs, which is equal to $l = l_p$. The lines without line points show the predictions from equation 3.9 and the results for $r = 1$, $r = 2$, and $r = 3$. The results for using no redundancy are the same as for uniform redundancy.

In agreement with the results from Thierens and Goldberg (1993), we report a small underestimation of the expected number of generations when using non-redundant, or uniformly redundant, representations. When using non-uniformly redundant encodings the underestimation is larger but nevertheless the model gives a good approximation for the expected number of generations. We believe that the difference is due to the effects of imperfect mixing and disruption of BBs.

When adding redundancy, the expected number of generations shows the same behavior as predicted. The necessary number of generations increases by about $O(\sqrt{l})$. As proposed, the run duration decreases with larger r. For uniform redundancy, the expected time to convergence remains unchanged. Although we do not consider the effect of BB disruption, the proposed model gives a good approximation for the expected run duration.

3.1.6 Conclusions, Restrictions and Further Research

We have developed a model describing the influence of linear bitstring redundancy on the performance of GEAs. We modeled redundancy as a matter of building block supply in the initial population. Results show that uniform redundancy has no influence on the performance of GEAs. For non-uniform redundancy the performance of GEAs is increased if redundancy favors high quality solutions.

Although the presented approach to redundant representations explains the behavior of GEAs accurately, researchers should be aware of the following assumptions for the model:

- We defined redundancy on bitstrings.
- We used linear redundancy.
- We neglected the effect of improper mixing.

For the proposed model we added redundancy at the level of bitstrings. Furthermore, we used linear redundancy and the redundant genotypic alleles are tight together. Therefore, the effective order of the BBs is not changed. We did not consider any forms of higher order redundancy. We must be aware that the proposed form of linear bitstring redundancy is not the only possibility for constructing redundant encodings, and that there has to be more work done in this field to obtain the complete picture.

Using non-linear redundant representations can be a source of additional problem complexity. An example for non-linear redundancy is that $x_p = 0$ is represented by the genotypes $\{00, 11\}$, and $x_p = 1$ is represented by $\{01, 10\}$. Genotypes that represent the same phenotype are no longer neighbors which can have a negative influence on GEA performance (compare the results about locality in section 3.3).

During a GEA run, recombination operators often use redundant information for constructing the offspring. Lower quality BBs can be created even if the parents only consist of high quality BBs. Therefore, improper mixing of BBs can result in larger problems for GEAs when using redundant representations.

Although some open questions remain that are interesting directions for further research, the developed model of redundant representations allows accurate predictions of the performance of GEAs using redundant encodings. GEA performance, measured by probability of GEA failure and time to convergence, remains unchanged if representations are uniformly redundant. If representations are non-uniformly redundant the performance of GEAs goes with $O(r/2^{k_r})$, where r denotes the number of genotypic BBs that represent the best phenotypic BB and k_r denotes the order of redundancy. Therefore, if the high-quality solutions are overrepresented ($r > 2^{k(k_r-1)}$) GEA performance increases, and if the high-quality solutions are underrepresented ($r < 2^{k(k_r-1)}$) GEA performance is reduced.

3.2 Building Block-Scaling

This section provides the second of three elements towards a theory of representations. Common representations for genetic and evolutionary algorithms often encode the phenotypes by using a sequence of alleles. We know from previous work that the BBs are solved in parallel if the BBs are uniformly scaled (Goldberg, 1989c). However, for non-uniformly scaled BBs, the BBs are solved sequentially and domino convergence occurs. Therefore, the time to convergence increases and the genetic search is affected by genetic drift. This means lower salient BBs are fixed before they can be reached by the search process.

Based on previous work (Rudnick, 1992; Thierens, 1995; Thierens, Goldberg, & Pereira, 1998; Harik, Cantú-Paz, Goldberg, & Miller, 1997), we describe how the performance of GEAs is influenced by the use of representations with non-uniformly scaled BBs. We develop a population-sizing model with, and without, considering genetic drift. The theoretical models are verified with empirical results.

In the following subsection, we review the effects of domino convergence and genetic drift. In subsection 3.2.2 and 3.2.3 we develop population sizing models for domino convergence with, and without, considering drift. We present empirical verification of the proposed models in subsection 3.2.4 and end with concluding remarks.

3.2.1 Background

The term "domino convergence" was introduced by Rudnick (1992). He defined domino convergence as the sequential convergence of the alleles in a bitstring. Domino convergence occurs if the alleles are scaled non-uniformly and solved sequentially. Rudnick showed that there is a convergence window of size λ_c. The convergence window is a set of λ_c contiguous genes that have started to converge but are not yet fully converged. More salient genes have already converged completely, whereas lower salient genes are not yet touched by convergence. The existence of the convergence window means that not all parts of a problem are solved at the same speed. The higher the contribution of one allele to the overall fitness of an individual, the earlier this allele is solved by GEAs.

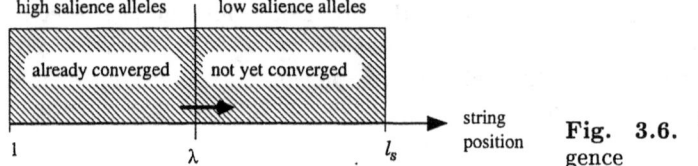

Fig. 3.6. Domino convergence

The λ model is a formal approach for modeling domino convergence (Thierens, Goldberg, & Pereira, 1998). $\lambda \in [1, l_s]$ defines the dividing line between the buildings blocks that have already converged and those that are still not touched by the selection pressure of GEAs. l_s is the length of the exponentially scaled BB, and λ moves on from $\lambda = 1$, where all bits are assumed to be in the initial random state, to $\lambda = l_s$ where all bits are converged (see Figure 3.6). For intermediate states, all lower salient bits have never been exposed to directed selection pressure and remain in their initial state as long as there is no genetic drift. In the approach used by Thierens, Goldberg, and Pereira (1998) the convergence window has size $\lambda_c = 1$.

If the population size is not large enough, some of the lower salient alleles are randomly fixed due to genetic drift. This results in a degradation of GEAs. The existence of genetic drift is widely known and has been addressed in the field of population genetics (Kimura, 1962; Kimura, 1964; Gale, 1990; Nagylaki, 1992; Hartl & Clark, 1997), and also in the field of genetic algorithms (Goldberg & Segrest, 1987; Asoh & Mühlenbein, 1994; Thierens, Goldberg, & Pereira, 1998; Lobo, Goldberg, & Pelikan, 2000). Lower salient bits drift randomly because selection does not take these bits into account when deciding between solutions. As two absorbing states exist for binary alleles (all individuals have at their ith position either a zero or a one), lower salient bits could be fixed because of random genetic drift before they are directly exposed to selection pressure and solved by GEAs.

Thierens (1995) developed a convergence time model for non-uniformly scaled problems. This work modeled the domino convergence for tournament selection by a wave equation, and showed that the overall convergence time complexity for an exponentially scaled fitness function is approximately of order $O(l)$. This is much slower than for uniformly scaled problems where ranked based selection mechanisms have a convergence time of order $O(\sqrt{l})$ where l is the length of the bitstring.

In the following, we want to use some existing models for developing a population sizing model for binary encodings, with, and without considering the effects of genetic drift.

3.2.2 Domino Model without Genetic Drift

In the following, we develop a population sizing model for exponentially scaled encodings neglecting the effect of genetic drift. Representations are exponentially scaled if different alleles have a different, exponentially scaled, contribution to the fitness of an individual. The binary representation is the most common representative of exponentially scaled representations.

We assume domino convergence and there should be no overlapping between the solving process of the current and the next allele. The next lower salient bit is only solved after the current allele is completely converged. Thus, we expect our models to only give us a lower bound on the solution quality and run duration. Although this is a very conservative assumption, this model (Thierens, Goldberg, & Pereira, 1998) predicts the behavior of a GA well (Lobo, Goldberg, & Pelikan, 2000). If we want to overcome the limitation of strict sequential solving, a convergence window of size λ_c can be introduced. This would result in an overlapping solving process for the different alleles in the convergence window. The convergence window would move through the string from the most salient to the least salient alleles and would allow us to create a more exact model for domino convergence.

Assuming strictly sequential solving of alleles ($\lambda_c = 1$), the fitness variance of an exponentially scaled string when λ bits are converged was calculated in Thierens, Goldberg, and Pereira (1998) as

$$\sigma_N^2(\lambda) = \frac{x_0}{n}(1 - \frac{x_0}{n})\frac{2^{2(l_s - \lambda)} - 1}{3} \approx \frac{x_0}{n}(1 - \frac{x_0}{n})\frac{2^{2(l_s - \lambda)}}{3},$$

where l_s is the length of the exponentially scaled string, and x_0 is the average number of best solutions in the initial population. For $x_0 = n/2$ (we assume $k = 1$) the variance simplifies to $\sigma_N^2(\lambda) \approx \frac{1}{12}2^{2(l_s - \lambda)}$. This means that the fitness variance is determined only by the non-converged region. The more alleles that are solved during the run, the less noise we get. The fitness variance of the allele that is currently solved is

$$\sigma_{BB}^2(\lambda) = \frac{x_0}{n}(1 - \frac{x_0}{n})2^{2(l_s - \lambda)}.$$

For $x_0 = n/2$ we get $\sigma_{BB}^2 = 2^{2(l_s - \lambda - 1)}$. As the contribution of the alleles to the fitness of an individual becomes less with lower salience, σ_{BB}^2 becomes smaller with increasing time. The fitness distance d between the best individual and its strongest competitor could be calculated as

$$d(\lambda) = 2^{l_s - \lambda}.$$

If we concatenate m exponentially scaled BBs of size l_s, we get competing BBs and an overall string length of $l = l_s m$. When solving the λth bit of a BB there is noise σ_{BB}^2 from the λth bit of the competing $m' = m - 1$, other BBs, and noise σ_N^2 from the yet unfixed bits in each BB. We know from past work that the probability of making the right choice between a single sample of each bit is (Miller, 1997):

$$p = \mathbb{N}\left(\frac{d}{\sqrt{2(m'\sigma_{BB}^2 + m\sigma_N^2)}}\right).$$

Using the equations from above results in

$$p = \mathbb{N}\left(\frac{1}{\sqrt{2\frac{x_0}{n}\left(1 - \frac{x_0}{n}\right)\left(\frac{4}{3}m - 1\right)}}\right) \tag{3.11}$$

For $x_0 = n/2$ we finally get:

$$p = \mathbb{N}\left(\sqrt{\frac{2}{\frac{4}{3}m - 1}}\right). \tag{3.12}$$

The probability of deciding well is independent of the position λ in the string as long as there is no genetic drift, and the proportion of zeros and ones remains constant for the yet unfixed bits. This means that for m exponentially scaled BBs the probability p of deciding well is independent of the length l_s of the exponentially scaled BB and that it stays constant over all alleles. Thus, the proportion of correct bits in the string at the end of a run depends only on the number of BBs m. For $m = 1$ (there is only one exponentially scaled BB) there is no noise from competing BBs and we get $p = \mathbb{N}(\sqrt{6})$. Using p we can calculate the proportion of incorrect bits at the end of the run according to the Gambler's ruin model as (Feller, 1957; Harik, Cantú-Paz, Goldberg, & Miller, 1997)

$$\alpha = 1 - \frac{1 - (1/p - 1)^{x_0}}{1 - (1/p - 1)^n}. \tag{3.13}$$

For $x_0 = n/2$ and $z = \sqrt{\frac{2}{\frac{4}{3}m - 1}}$ we get:

$$\alpha = 1 - \frac{1 - \left(\frac{1}{\mathbb{N}(z)} - 1\right)^{n/2}}{1 - \left(\frac{1}{\mathbb{N}(z)} - 1\right)^{n}} = \frac{\left(\frac{1}{\mathbb{N}(z)} - 1\right)^{n/2}}{1 + \left(\frac{1}{\mathbb{N}(z)} - 1\right)^{n/2}} = \frac{1}{1 + \left(\frac{1}{\mathbb{N}(z)} - 1\right)^{-n/2}}. \tag{3.14}$$

The probability α of GA failure for exponentially scaled problems only depends on the population size n and the number of competing BBs m as long as the population size is large enough, and no genetic drift occurs. Notice that in contrast to the proportion of correct bits $1 - \alpha$, the number of correctly exponentially scaled BBs is $(1 - \alpha)^{l_s}$. When using the first two terms of the power series expansion as an approximation for the normal distribution (Abramowitz & Stegun, 1972) from equation 3.12 we get

$$p = \frac{1}{2} + \frac{1}{\sqrt{2\pi}} z,$$

where $z = \sqrt{\frac{2}{\frac{4}{3}m-1}}$. Substituting this approximation in equation 3.13 results in

$$n = 2\ln\left(\frac{\alpha}{1-\alpha}\right) / \ln\left(\frac{1 - \sqrt{\frac{2}{\pi}}z}{1 + \sqrt{\frac{2}{\pi}}z}\right).$$

Since z tends to be a small number, $\ln(1 \pm z\sqrt{\frac{2}{\pi}})$ may be approximated as $\pm z\sqrt{\frac{2}{\pi}}$. Using these approximations and substituting the value z into the equation finally gives

$$n \approx -\frac{1}{2}\ln\left(\frac{\alpha}{1-\alpha}\right)\sqrt{\pi\left(\frac{4}{3}m - 1\right)}. \tag{3.15}$$

This rough approximation determines more clearly the variables the population size n depends on. We see that for exponentially scaled representations, the necessary population size n grows with the square root of the size m of the problem. In contrast to the more general population sizing equation from Harik, Cantú-Paz, Goldberg, and Miller (1999) n does not depend on the distance d and the variance of an allele σ_{BB} if genetic drift is neglected.

Finally, we give an estimation for the convergence time t_{conv} for exponentially scaled BBs of length l_s. The time until a uniformly scaled string of length m is converged can be calculated (Thierens & Goldberg, 1994) as

$$t = \frac{\pi}{2}\frac{\sqrt{m}}{I},$$

where I denotes the selection intensity. For tournament selection without replacement of size 2, $I = 1/\sqrt{\pi}$. As there are m exponentially scaled BBs, and therefore m alleles of the same salience, the GEAs solve m bits in parallel. The next m lower salient bits are solved when all m currently solved bits are fully converged. Thus, the solving process for exponentially scaled problems is strictly serial and goes with $O(l_s)$ (Thierens, 1995, pp. 66ff). The overall time to convergence can be calculated as:

$$t_{conv} = l_s \frac{\pi}{2} \frac{\sqrt{m}}{I} = \frac{l}{\sqrt{m}} \frac{\pi}{2I}, \tag{3.16}$$

where $l = l_s m$. In contrast to the uniformly scaled one-max problem ($t_{conv} = O(\sqrt{l_s m})$) the time to convergence goes when using a representation with m exponentially scaled BBs of size l_s with $O(l_s \sqrt{m})$. We see clearly that GEAs need more time to converge if the problem is not uniformly scaled.

3.2.3 Population Sizing for Domino Model and Genetic Drift

In the previous subsection we have developed a population sizing and convergence time model for non-uniformly scaled representations. However, the model does not consider the effects of genetic drift. In the following, we want to lift this restriction and investigate the modifications necessary for considering genetic drift.

Drift affects genetic search if the drift time is lower than the time to convergence $t_{drift} < t_{conv}$. Low salient bits are fixed due to drift before they can be reached by the solution process. The drift time has been studied in the context of genetic and evolutionary algorithms (Goldberg & Segrest, 1987; Asoh & Mühlenbein, 1994). These studies show that the expected time for a gene to converge due to genetic drift is proportional to the population size n. For an equal proportion of ones and zeros in the start population, using tournament selection of size $s = 2$, the size of BBs $k = 1$ (Lobo, Goldberg, & Pelikan, 2000), and random sampling with replacement, we get for the drift time

$$t_{drift} \approx 1.4n. \tag{3.17}$$

For $t_{conv} > t_{drift}$ genetic drift fixes some low salient bits before they can be solved by the search process. Using equation 3.16, we can calculate the population size n_{drift} for which GEAs using tournament selection of size 2 (selection intensity $I = 1/\sqrt{\pi}$) are affected by genetic drift as

$$n_{drift} < \frac{5\pi}{14} \sqrt{\frac{\pi}{m}} l. \tag{3.18}$$

If the population size n of GEAs using an exponentially scaled representation is lower than n_{drift} then domino convergence does not reach the lowest salient alleles and these alleles are randomly fixed at one of the two absorbing states 0 or 1. In the following, we want to propose two approaches modeling the influence of drift on the performance of GEAs using exponentially scaled representations. At first, we need a model that describes the drift process itself. An approximation for the probability $s(t)$ that an allele is fully converged due to genetic drift at time t was given by Kimura (1964):

$$s(t) \approx 1 - 6\frac{x_0}{n}(1 - \frac{x_0}{n}) \exp(-t/n).$$

Using this approximation we can calculate how the probability of randomly fixing an allele depends on the population size n and the number of generations t. With $x_0 = n/2$ is the expected number of 1s in the randomly initialized population (we assume $k = 1$) we get

$$s(t) = 1 - \frac{3}{2}\exp(-t/n).$$

As $s(t)$ is only an approximation ($s(t) < 0$ for small t) we define the convergence probability

$$s'(t) = \begin{cases} 0 & \text{for } t < -n\ln(2/3), \\ 1 - \frac{3}{2}\exp(-t/n) & \text{for } t > -n\ln(2/3). \end{cases}$$

For $t < -n\ln(2/3)$ the probability that an allele converges due to genetic drift is zero. In Figure 3.7 we plot the probability that an allele is fully converged at generation t for $n = 10$.

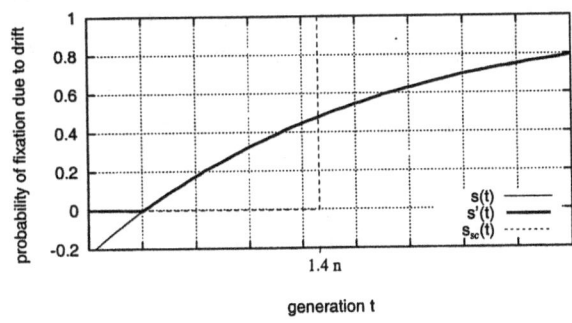

Fig. 3.7. Genetic drift models

We know from equation 3.16 that GEAs need $t = \frac{\pi}{2}\frac{\sqrt{m}}{l}$ generations for solving one allele in the bitstring ($l_s = 1$). When considering genetic drift, the bits in the string are either solved correctly with probability $1 - \alpha$ or they are randomly fixed due to genetic drift. If domino convergence reaches after $t = \lambda\frac{\pi}{2}\frac{\sqrt{m}}{l}$ generations an allele, this allele is not converged due to genetic drift with probability $1 - s'(t)$. Therefore, we want to assume that with probability $1 - s'(t)$, this allele remains in its initial state and is solved correctly with probability $1 - \alpha$. In contrast, with probability $s'(t)$, this allele is converged due to genetic drift. If the allele is converged due to genetic drift, it converges to the correct solution with probability 0.5. Therefore, using the probability of GA failure α from equation 3.14 we can calculate the probability of GA failure when considering genetic drift as

$$\alpha_{drift}(\lambda) = \left(1 - s'(\lambda\frac{\pi}{2}\sqrt{\pi m})\right)\alpha + \frac{1}{2}s'(\lambda\frac{\pi}{2}\sqrt{\pi m}),$$

when using tournament selection of size 2 ($I = 1/\sqrt{\pi}$). $1 - s'(\lambda\frac{\pi}{2}\sqrt{\pi m})$ is the probability that the allele at the λth position is not converged due to genetic drift at time $t = \lambda\frac{\pi}{2}\sqrt{\pi m}$. The probability of error for this allele is α. Furthermore, the alleles are affected by genetic drift with probability $s'(\lambda\frac{\pi}{2}\sqrt{\pi m})$, and converge to the wrong solution with probability 0.5. Consequently, the overall average percentage $\bar{\alpha}_{drift}$ of incorrect alleles in one BB can be calculated as

$$\bar{\alpha}_{drift} = \frac{1}{l_s} \sum_{\lambda=0}^{l_s-1} \alpha_{drift}(\lambda) \tag{3.19}$$

and is simply the sum over all l_s bits of one BB. In the following, we refer to this model as the approximated drift model.

We assume in this model for the unconverged alleles that the proportion of the best building block in the population is still 0.5, and that α does not depend on λ. We also assume that the solving time for one allele stays constant over the whole solving process, and that it is independent of λ.

Instead of using the approximation from Kimura (1964), we could also use a simple drift/no-drift approach for the population sizing model. We assume that all instances of an allele are either fixed due to genetic drift, or remain in the initial state. Thus, we get a linear, stair-case model for the estimation of α_{drift}, if we assume that for $t_{drift} > \lambda\frac{\pi}{2}\sqrt{\pi m}$ all bits can be solved with probability $1 - \alpha$, and for $t_{drift} < \lambda\frac{\pi}{2}\sqrt{\pi m}$ the remaining low salient $l_s - \lambda$ bits of each BB are fixed randomly at the correct solution with probability 0.5. As previously, the drift time can be calculated for GAs using tournament selection of size $s = 2$ as $t_{drift} \approx 1.4n$. Therefore, the probability s_{sc} that an allele is fully converged due to genetic drift at time t is

$$s_{sc}(t) = \begin{cases} 0 & \text{for } t < t_{drift}, \\ 1 & \text{for } t > t_{drift}. \end{cases}$$

We illustrate this in Figure 3.7. For $t < t_{drift}$ we assume no genetic drift, and for $t > t_{drift}$ all the remaining bits are randomly fixed. Therefore, the probability of GA failure can be calculated as:

$$\alpha'_{drift}(\lambda) = \begin{cases} \alpha & \text{for } \lambda < \frac{2.8n}{\pi\sqrt{\pi m}}, \\ 0.5 & \text{for } \lambda \geq \frac{2.8n}{\pi\sqrt{\pi m}}. \end{cases}$$

By using equation 3.17 and 3.18 the average percentage of incorrect alleles is:

$$\bar{\alpha}'_{drift}(\lambda) = \begin{cases} \frac{1}{l_s}\left(\sum_{\lambda=0}^{\lfloor\frac{2.8n}{\pi\sqrt{\pi m}}\rfloor} \alpha + \sum_{\lambda=\lceil\frac{2.8n}{\pi\sqrt{\pi m}}\rceil}^{l_s-1} \frac{1}{2}\right) & \text{for } n < \frac{5\pi}{14}\sqrt{\pi m l_s}, \\ \alpha & \text{for } n \geq \frac{5\pi}{14}\sqrt{\pi m l_s}. \end{cases} \tag{3.20}$$

For large n, no genetic drift occurs and we get the same failure probability as for the non-drift case (equation 3.14). For small n the most salient $\lfloor\frac{2.8n}{\pi\sqrt{\pi m}}\rfloor$

bits are solved correctly with probability $1 - \alpha$ and the rest of the alleles are fixed randomly due to genetic drift.

As the drift time has a standard deviation of approximately the same order (Gale, 1990, pp. 82ff.) as its mean ($\approx 1.4n$), the model underestimates the solution quality for $t < 1.4n$, and overestimates it for $t > 1.4n$. The probability of converging to the correct solution has a stair-cased slope regarding n as long as $t_{drift} < t_{conv}$. Thus, we refer to this model in the following as the stair-case drift model.

3.2.4 Empirical Results

In this subsection, we illustrate that the proposed models considering genetic drift predict the behavior of GEAs well for exponentially scaled problems and small populations. We show that with decreasing number of competing exponentially scaled BBs, and increasing length of the BBs, genetic drift leads to a stronger decline of GEAs.

For our empirical investigation we use the Bin-Int problem (Rudnick, 1992). There are l_s alleles and the contribution of an allele to the fitness is exponentially scaled. Therefore, the overall fitness of an individual x can be calculated as:

$$f(x) = \sum_{i=0}^{l_s-1} x_i 2^i$$

Thus, the optimal solution is a string with only ones. If the population size is large enough, this problem is easily solved by GEAs according to the domino convergence model. We present no results for exponentially scaled deceptive problems because the population size n that is necessary to solve these types of problems is in general large enough to ensure that no genetic drift occurs, and the available population sizing models from subsection 3.2.2 can be used.

For all problems we use uniform crossover, no mutation, and tournament selection of size 2 without replacement. The initial state of the population has an equal proportion of zeros and ones ($x_0 = n/2$). To gain statistical evidence we performed 250 runs for each problem. We present results for three test cases:

- one exponentially scaled BB ($m = 1$),
- 10 concatenated exponentially scaled BBs ($m = 10$),
- 50 concatenated exponentially scaled BBs ($m = 50$).

The fitness of an individual is the sum over all m exponentially scaled BBs. For each of these test cases we present results for the length of an exponentially scaled BBs $l_s = 5$ and $l_s = 10$. In Table 3.2 we present the overall string length l, the probability p of making the right choice between a single sample of each BB, and the overall convergence time t_{conv} when assuming no genetic drift. If drift occurs some lower salient alleles are randomly fixed at 0

Table 3.2. Some properties of the three problem instances.

	$m = 1$		$m = 10$		$m = 50$	
l_s	5	10	5	10	5	10
l	5	10	50	100	250	500
p	$N(\sqrt{6})$		$N(\sqrt{6/37})$		$N(\sqrt{6/197})$	
t_{conv}	$5\frac{\pi}{2}\sqrt{\pi}$	$10\frac{\pi}{2}\sqrt{\pi}$	$5\frac{\pi}{2}\sqrt{10\pi}$	$10\frac{\pi}{2}\sqrt{10\pi}$	$5\frac{\pi}{2}\sqrt{50\pi}$	$10\frac{\pi}{2}\sqrt{50\pi}$

or 1 before they can be reached by the search process and t_{conv} is an upper bound for the overall convergence time.

In Figure 3.8 ($m = 1$), 3.9 ($m = 10$), and 3.10 ($m = 50$) we present results for the BinInt-problem of length $l_s = 5$ (left) and $l_s = 10$ (right). The solid lines with line points show the empirical results. We show predictions for considering no drift (dotted line), for the stair-case drift model (dashed line with line points), and for the approximated drift model (dashed line). All predictions take domino convergence into consideration.

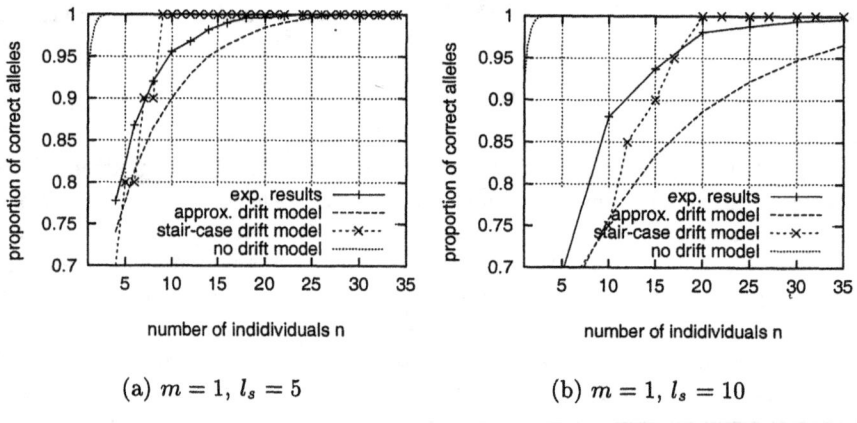

(a) $m = 1$, $l_s = 5$ (b) $m = 1$, $l_s = 10$

Fig. 3.8. Experimental and theoretical results of the proportion of correct BBs for a BinInt problem of length $l_s = 5$ (left) and $l_s = 10$ (right). The solid lines with line points show the empirical results. All shown predictions consider domino convergence. We show predictions for considering no drift (dotted line), for the stair-case drift model (dashed line with line points), and for the approximated drift model (dashed line). With increasing length of the exponentially scaled BBs l_s, genetic drift has a major impact on the GA, and the GA performance declines.

For all three cases, genetic drift has a large impact on the GA performance especially with increasing string length l_s, and decreasing number of competing BBs m. In contrast to the no-drift model which could not predict the behavior of the GA well, both the stair-case drift model, as well as the approximated drift model, are able to accurately describe the behavior of GEAs

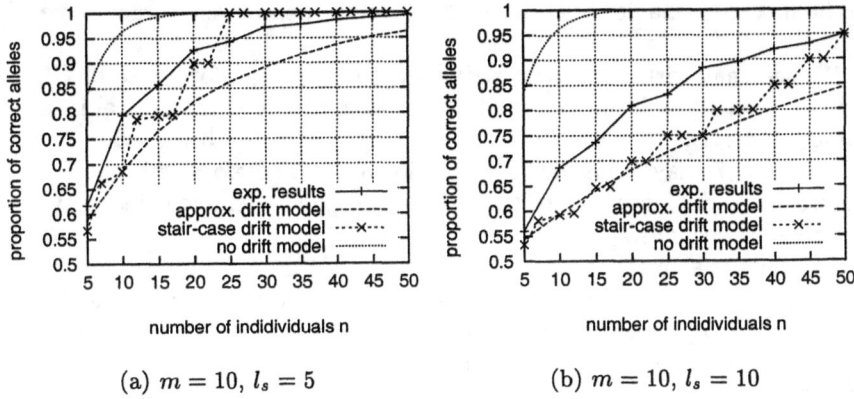

(a) $m = 10$, $l_s = 5$ (b) $m = 10$, $l_s = 10$

Fig. 3.9. Experimental and theoretical results of the proportion of correct BBs for 10 concatenated BinInt problems of length $l_s = 5$ (left) and $l_s = 10$ (right). The overall string length is $l = 50$ (top) and $l = 100$ (bottom). The solid lines with line points show the empirical results. We show predictions for considering no drift (dotted line), for the stair-case drift model (dashed line with line points), and for the approximated drift model (dashed line). Although, the GA is affected by noise from the competing $m - 1$ BBs, which superimposes the effects of drift, genetic drift has a major impact on the GA performance with increasing BB length l_s.

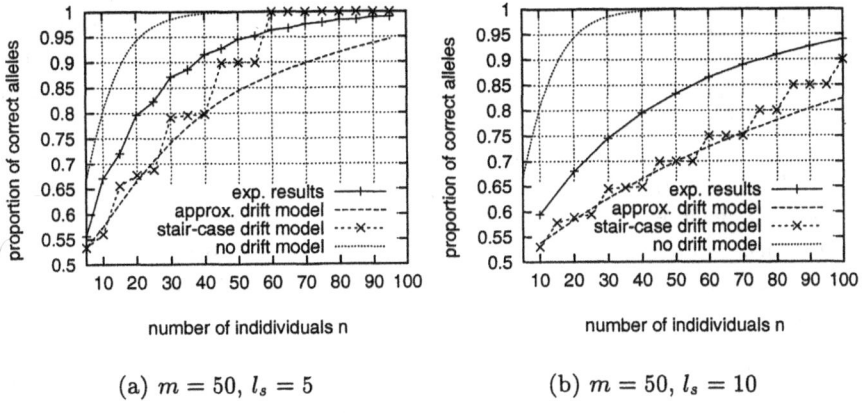

(a) $m = 50$, $l_s = 5$ (b) $m = 50$, $l_s = 10$

Fig. 3.10. Experimental and theoretical results of the proportion of correct BBs for 50 concatenated BinInt problems of length $l_s = 5$ (left) and $l_s = 10$ (right). The overall string length is $l = 250$ (left) and $l = 500$ (right). The solid lines with line points show the empirical results. We show predictions for considering no drift (dotted line), for the stair-case drift model (dashed line with line points), and for the approximated drift model (dashed line). With increasing string length l_s genetic drift has a larger impact on the GA, and the GA performance declines.

using exponentially scaled representations. Both models consider genetic drift and predict the behavior of GEAs better than the domino-convergence model alone.

We see that with increasing string length l_s, the performance of GEAs declines. This behavior is expected as we know from our theoretical investigations that t_{conv} increases linearly with l_s (see equation 3.16), whereas the drift time t_{drift} stays constant (see equation 3.17). Therefore, with increasing l_s more and more lower salient alleles are fixed due to genetic drift and GEA performance declines.

Our results show that the influence of genetic drift is reduced for an increasing number of BBs m. We know from equation 3.12 that with increasing m the probability of making the right choice between a single sample of each bit is reduced. Therefore, larger populations n are necessary which reduce the influence of genetic drift with increasing m. This relationship can be seen nicely in the presented plots. For $m = 1$ (see Figure 3.8(a)) we have no competing BBs, the necessary populations are very small, and the GA is strongly affected by genetic drift. Therefore, the no-drift model fails completely. For $m = 50$, however, there is a lot of noise from the competing BBs, and therefore larger populations are necessary. The influence of genetic drift is smaller, and the no-drift model when only considering domino convergence gives an acceptable prediction of the solution quality (see. Figure 3.10(a)).

As predicted, the stair-case model underestimates the proportion of correct alleles for small n ($n < \frac{5}{7}t_{drift}$), and overestimates it for large populations ($n > \frac{5}{7}t_{drift}$). The approximated drift model predicts the slope of the empirical results well, but due to the used domino convergence with strict sequential solving of the bits, it always underestimates the proportion of correct alleles. We believe that by introducing a convergence window, and assuming some parallel solving of the alleles, that the approximated drift model should more accurately predict the behavior of GEAs.

We have seen that the lower the number of competing BBs m is, and the more bits l_s each exponentially scaled BB has, the stronger is the impact of genetic drift on the solution quality. Although we only use a strictly serial solution process and no convergence window, the developed models considering genetic drift give us a good prediction of the expected proportion of correct alleles when using exponentially scaled representations. Population sizing models that neglect the effect of genetic drift are not able to accurately predict the expected proportion of correct BBs for a given population size.

3.2.5 Conclusions

We have illustrated the effect of non-uniformly scaled representations on the performance of genetic and evolutionary algorithms. When using small populations and easy problems, GEAs are affected by genetic drift. To be able to model the effects of genetic drift more accurately, we used the population sizing model from Harik, Cantú-Paz, Goldberg, and Miller (1999), the domino convergence model from Rudnick (1992), and the time complexity model from Thierens (1995) and Thierens, Goldberg, and Pereira (1998) and developed two population sizing models for exponentially scaled encodings considering genetic drift. The approximated genetic drift model uses an approximation (Kimura, 1964) for the probability that an allele is completely converged due to genetic drift after t generations, and gives us a lower bound for the solution quality. The stair-case drift model assumes that genetic drift occurs as long as the convergence time is larger than the expected drift time, and that the lower salient genes are fixed at the correct solution due to genetic drift with probability x_0/n.

The theoretical results reveal that genetic drift has a large impact on the probability of error and convergence time when using representations with exponentially scaled BBs. Because the alleles are solved strictly in serial, exponentially scaled representations change the dynamics of genetic search. As a result the solution quality is reduced by genetic drift, and the convergence time is increased by domino convergence. The empirical investigations show that despite the assumption that the size of the convergence window $\lambda_c = 1$, the proposed models considering genetic drift give us, in contrast to the no-drift model, accurate predictions for the solution quality. Except for very large numbers of competing BBs, or very short exponentially scaled BBs, the no-drift populations sizing model is not able to predict the expected solution quality.

When using exponentially scaled encodings, researchers should be aware of the effects of genetic drift. Exponentially scaled encodings are linear and non-redundant, but the required time to convergence increases.

3.3 Distance Distortion

During the last decades, starting with work by Bagley (1967), Rosenberg (1967) and Cavicchio (1970), researchers recognized that the concept of building blocks is important for understanding the principles of GEAs, and is a key factor in determining successful use. A well designed GEA should be able to preserve high quality building blocks, and increase their number over the generations (Goldberg, 1989c).

When using the notion of building blocks in the context of genetic representations, we must be aware that building blocks not only exist in the genotype, but also in the phenotype. A representation transforms the structure and complexity of the building blocks from the phenotype to the genotype, and therefore the structure and complexity of the building blocks can be different in the genotype and phenotype.

This section investigates how representations modify the complexity of building blocks. The investigation provides the third, and final element, towards a theory of representations for genetic and evolutionary algorithms. The distance distortion d_c of a representation measures how much the distances between individuals are changed when mapping the phenotypes to the genotypes. The results show that representations where $d_c = 0$, that means the distances between the individuals are preserved, do not modify the complexity of the problems they are used for. For all other representations where $d_c \neq 0$, the genotypic BBs can be different from the phenotypic BBs, and the complexity of the problem can be changed. Therefore, we recognize that only representations with $d_c = 0$ guarantee to solve problems of bounded complexity reliably and predictably.

We focus our investigation on how the distance distortion of a representation influences problem difficulty. We do not present models describing problem difficulty (see section 2.3) but focus on the effects of representations on problem difficulty. In the context of evolution strategies (Bäck & Schwefel, 1995) previous work developed the concept of causality (Igel, 1998; Sendhoff, Kreutz, & von Seelen, 1997b; Sendhoff, Kreutz, & von Seelen, 1997a) as a measurement of problem difficulty. Basically, both concepts, causality and distance distortion, measure the same property of representations. Both describe how well the distances between individuals are preserved when mapping the phenotypes on the genotypes. However, in this study we do not try to classify the difficulty of problems but only describe how the difficulty is changed by the encoding. Consequently, we show that by using representations, which change the distances between individuals, fully easy problems become more difficult, whereas fully difficult problems become more easy. Determining necessary conditions for representations where the distance distortion is low reveals that the high locality of an encoding is one necessary condition. High locality means that similar genotypes correspond to similar phenotypes.

In subsection 3.3.1, the influence of representations on problem difficulty is discussed. We illustrate why it is helpful to use encodings that do not modify problem difficulty for solving problems of bounded difficulty. Because otherwise, if we use representations that modify problem difficulty, GEA performance is no longer predictable which forces us to use more powerful GEAs. In subsection 3.3.2 we introduce the distance distortion of a representation and show that high locality is necessary for an encoding to have low distance distortion. In subsection 3.3.3 we show for the fully easy one-max problem how a modification of the representation where $d_c = 0$, can make the problem more difficult. The results of the theoretical investigation are verified and illustrated in subsection 3.3.4 by an empirical investigation for the one-max and deceptive trap problem. The results show that when using representations where $d_c = 0$, that the one-max problem remains fully easy, and the fully deceptive trap remains fully difficult. For most of the other representations the fully easy one-max problem becomes more difficult to solve for GEAs. The section ends with concluding remarks.

3.3.1 Influence of Representations on Problem Difficulty

This subsection discusses the influence of representations on the difficulty of a problem and why representations should preserve the complexity of building blocks when assigning the genotypes to the phenotypes.

It was already recognized by Liepins and Vose (1990) that by using different encodings the complexity of a problem can be completely changed. Changing the complexity of a problem means that the difficulty of the problem and the structure of building blocks is changed (see section 2.3). However, representations that modify BB-complexity are problematic as they do not allow us to reliably solve problems of bounded difficulty. If the complexity of BBs is changed by the encoding, some problems become easier to solve, whereas others become more difficult. To predict which problems become easier and which do not is only possible if detailed knowledge about the optimization problem exists (compare also subsection 4.4.3). To ensure that problems of bounded complexity, that means simple problems, can be reliably solved by GEAs, representations that preserve the complexity of the building blocks should be used.

Liepins and Vose (1990) proved that for a fully deceptive problem $f(x) = f_p(f_g(x))$ there is a transformation T such that the function $g(x) = f[T(x)]$ becomes fully easy. This means that every fully deceptive problem could be transformed into a fully easy problem by introducing a linear transformation T. In general, the difficulty of a problem could easily be modified by using different linear transformations T. Liepins and Vose concluded that their results underscore the importance of selecting good representations, and that good representations are problem-specific. Therefore, in order to find the best representation for a problem, it is necessary to know how to optimize the problem and what the optimal solution for the problem is.

To find the best representation for a specific problem, Liepins and Vose proposed using adaptive representations that 'simultanously search for representations at a metalevel' (Liepins & Vose, 1990, p. 110). These representations should autonomously adapt to the problem and encode it in a proper way (compare Rothlauf, Goldberg, and Heinzl (2000)). Initial steps in this direction were made by Goldberg, Korb, and Deb (1989) with the development of the messy GA. These kinds of GAs use an adaptive encoding that adapts the structure of the representation to the complexity of the problem. This approach, however, burdens the GA not only with the search for promising solutions, but also the search for a good representation.

Therefore, we go back one step and ask the question, 'what kind of encoding should be used if there is no a priori knowledge about the problem, and no adaptive encoding should be used'. Users who simply want to solve their problems by using a GEA are confronted with this kind of problem. They have no knowledge about the problem, and they do not want to do experiments to find out what structure the problem has, what the promising areas in the search space are, and what kind of representations make the problem most easy. They just want to be sure that the GEA can solve their problem as long as the complexity of their problem is bounded and the used GEA is able to solve it. One solution for their problems are representations that preserve BB-complexity. Using these types of representations means that the problem has the same difficulty in the genotypic as in the phenotypic space. Then they can be sure that the representation does not increase the complexity of the problem, and that the GEA reliably solves their problem.

Wanting representations to preserve BB-complexity raises the question of, 'why are we especially interested in encodings that preserve complexity? Is it not more desirable to construct encodings that reduce problem complexity, as Liepins and Vose propose?'. The answer is yes and no. Of course, we are interested in encodings that reduce problem difficulty. However, in general it is not possible to construct a representation that reduces the complexity for all possible problem instances. One result of the no-free-lunch theorem (Wolpert & Macready, 1995) is that if some problem instances become easier by the use of a representation that does not preserve complexity, there are other types of problem instances that necessarily become more difficult.

Therefore, we want to at least ensure that problems of bounded difficulty – these are the class of problems we are interested in – do not become more difficult to solve when using a representation. GEAs should be able to solve the problem with the same effort independently of the used representation. However, as shown in the following subsections, encodings that do not preserve problem complexity always make fully easy problems more difficult. A phenotypically easy problem could even become so difficult by using a "bad" encoding that it can not be solved efficiently. Of course, representations that do not preserve problem complexity make fully difficult problems more easy, but as we are interested in solving only problems of bounded difficulty, and not all types of problems, this is not important to us. For difficult problems, like the needle in the haystack problem, or the fully deceptive trap problem, the complexity of the problem is not bounded, and therefore, we are in general not interested in solving these kinds of problems with genetic algorithms.

The solution for all our problems would be a "perfect" encoding that preserves the complexity for problems of bounded difficulty *and* reduces the complexity for all other problems. But reaching this aim is far beyond the scope of this work. Furthermore, we believe that the no free lunch-theorem does not allow us to get such a free lunch for every problem.

Finally, we want to come back to the results of Liepins and Vose (1990) and illustrate the problems with representations that do not preserve the complexity of BBs. The transformation T, which is nothing more than a genotype-phenotype mapping, can modify the complexity of a problem in such a way that a fully difficult deceptive trap problem becomes a fully easy one-max problem. But, using the same transformation T for a fully easy one-max problem can result in a fully deceptive trap (compare also Figure 4.1(a)). Therefore, by using this representation, we are able to solve a deceptive trap, but not the one-max problem any more. If we want to solve both types of problems we must know a priori what the problem is and adjust the representation according to the problem type. However, if we do not know the problem type a priori, and if we want to make sure that we can solve at least problems of bounded difficulty reliably, we must use representations that do not modify problem difficulty.

In the following subsection, we define the distance distortion of a representation formally and show that low distance distortion is necessary for a representation to not modify problem difficulty.

3.3.2 Locality and Distance Distortion

This subsection defines the distance distortion of a representation. Furthermore, it illustrates the concept of locality and shows that high locality is necessary for a representation to have low distance distortion.

Before we are able to discuss distance distortion and locality it is necessary to define a metric on the phenotypic as well as on the genotypic search space. We need a measurement of how similar the genotypes and phenotypes of two individuals are. If the phenotypic space Φ_p has a different structure than the genotypic space Φ_g, we need two different metrics in Φ_g and Φ_p. If we limit ourself to binary representations $\Phi_g = \{0,1\}^l$ we could use, for example, the Hamming metric (Hamming, 1980; Thierens, 1992). Then, the genotypic distance between two individuals $x_g \in \Phi_g$ and $y_g \in \Phi_g$ is defined as the Hamming distance between x_g and y_g. If not only Φ_g, but also Φ_p is binary, we could also use the Hamming metric for Φ_p. The Hamming distance between x and y describes how many bits are different in the two individuals. For example, the Hamming distance between $x = 00100$ and $y = 01101$ is $d_{Hamming} = 2$. In general, however, the metric defined on Φ_p is based on the specific structure of the problem.

The locality of a representation describes how similar the phenotypic and genotypic distances are between two individuals, before, and after applying mutation. To measure locality, a mutation operator which results in a small modification of the genotype is applied to the genotype, and the distance in the phenotypic space between the parent and the offspring is observed. Therefore, the locality d_m of a representation concerning small changes can be defined as

$$d_m = \sum_{d^p_{x_i, x_j} = d^p_{min}} |d^g_{x_i, x_j} - d^g_{min}|,$$

where $d^p_{x_i, x_j}$ is the phenotypic distance between the phenotypes x_i and x_j, $d^g_{x_i, x_j}$ is the genotypic distance between the corresponding genotypes, and d^p_{min}, resp. d^g_{min} is the minimum distance between two (neighboring) phenotypes, resp. genotypes. Without loss of generality we want to assume that $d^g_{min} = d^p_{min}$. The metrics in the genotypic and phenotypic space are normalized in such a way that the smallest genotypic distance between two individuals is the same as the smallest phenotypic distance. For $d_m = 0$ the genotypic neighbors correspond to the phenotypic neighbors and the encoding has perfect locality. The locality of a representations is low (d_m is high) if mutating a genotype strongly changes the corresponding phenotype.

When using recombination-based search, the locality concerning small changes d_m must be extended towards locality concerning small and large changes. The distance distortion d_c describes how well the phenotypic distance structure is preserved when mapping Φ_p on Φ_g:

$$d_c = \frac{2}{n_p(n_p - 1)} \sum_{i=1}^{n_p} \sum_{j=i+1}^{n_p} |d^p_{x_i, x_j} - d^g_{x_i, x_j}|,$$

where n_p is the number of different individuals, $n_p = |\Phi_g| = |\Phi_p|$, and $d^g_{min} = d^p_{min}$. For $d_c = 0$ all phenotypic distances are preserved by the representation. Wee see that for $\Phi_g = \Phi_p$ high locality ($d_m = 0$) results in low distance distortion ($d_c = 0$). If for example our genotypic and phenotypic search space is binary and the locality of the genotype-phenotype mapping is perfect, then all distances between the individuals are preserved. However, if we assume that $\Phi_g \neq \Phi_p$, then high locality only is a necessary but not sufficient condition for the genotype-phenotype mapping to have low distance distortion.

Figure 3.11 illustrates the difference between representations with high versus low distance distortion. The distance distortion d_c of a representation is low if the genotypic distances correspond to the phenotypic distances. If the distances between the ghenotypes and the corresponding phenotypes are different, the distance distortion d_c of the representation is high.

It is of interest that the locality d_m and distance distortion d_c do not require the definition of genetic operators a priori. It is sufficient to define both based on the distance metrics used for Φ_g and Φ_p. The application of mutation to an individual should result in an offspring that is similar to its parent. Therefore, in many implementations, mutation creates offspring who have the lowest possible distance to the parent (for example the bit-flipping operator for binary representations). Therefore, high locality of a representation is a necessary condition for successful use of mutation-based search algorithms. Otherwise, low locality encodings do not allow a guided search and GEAs using low locality representations behave like random search.

The situation is similar when using crossover operators. The application of crossover operators should result in offspring where the distances between the offspring and its two parents are smaller than the distance between both parents. Common standard crossover operators, like n-point or uniform crossover show this behaviour. The distances between genotypic offspring and parents are always lower, or equal to, the distances between both parents. However, if a representation has high distance distortion, the genotypic distances do not correspond to the phenotypic distances. Then, the phenotypic distances between offspring and parents are not necessarily smaller than the phenotypic

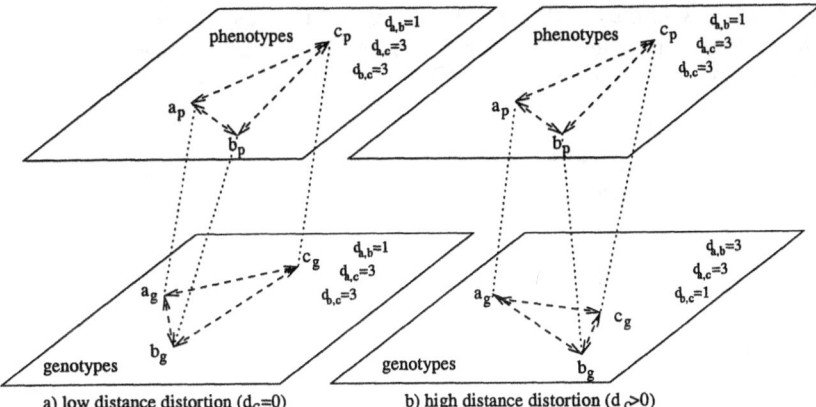

a) low distance distortion (d$_c$=0)   b) high distance distortion (d$_c$>0)

Fig. 3.11. The figures illustrate the difference between representations with low versus high distance distortion. If the distance distortion $d_c = 0$ then the distances between the corresponding genotypes and phenotypes are the same. If $d_c > 0$ genotypic distances between individuals do not correspond to phenotypic distances between individuals.

distances between both parents. The application of crossover to gentoypes does not result in offspring phenotypes that mostly consist of substructures of their parents' phenotypes. Therefore, the offspring is not similar to its parents and the use of crossover results in random search. We see that low distance distortion of a representation is a necessary condition for good performance of crossover-based GEAs.

Examining the interdependences between locality and distance distortion shows that high locality is a necessary condition for an encoding to have low distance distortion. When using standard crossover operators such as uniform or n-point crossover, the offspring could not inherit the properties of the parents if similar genotypes result in completely different phenotypes. If the encoding has low locality, the crossover operators would create offspring genotypes which are similar to the genotypes of the parents, but the resulting phenotypes would not be similar to the phenotypes of the parents. Thus, low locality of a representation would also result in high distance distortion.

In the following subsection, we show for the fully easy bit-counting problem that the problem only stays fully easy if a representation is used that preserves BB-complexity, and where the distance distortion $d_c = 0$. All other types of problems increase the complexity of the problem.

3.3.3 Modifying BB-Complexity for the One-Max Problem

In this subsection we show for the fully easy one-max problem that a representation preserves the complexity of the BBs if it preserves the distances between the individuals ($d_c = 0$). It is shown for the one-max problem that representations where $d_c = 0$ make the problem fully easy and result in highest GA performance whereas other types of representations mostly reduce GA performance.

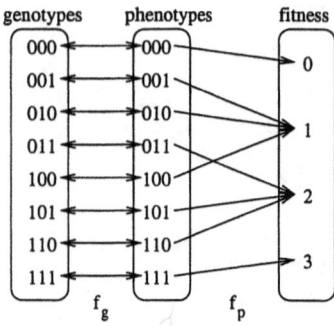

Fig. 3.12. A representation for the bit-counting function which preserves the distances ($d_c = 0$).

For the one-max, or bit-counting problem, the function $f_p : \{0,1\}^l \to \mathbb{R}$ assigns to every individual $x_p \in \Phi_p$ the fitness value $\sum_{i=0}^{l-1} x_i$. Thus, only l fitness values are assigned to 2^l phenotypes. Therefore, the fitness function f_p is affected by redundancy (see section 3.1). The genotype-phenotype mapping f_g is a one-to-one mapping, and the genotypic space Φ_g, and the phenotypic space Φ_p, have the same size $|\Phi_g| = |\Phi_p| = 2^l$ and the same properties $\Phi_g = \Phi_p$. To simplify the investigation we want to assume, without loss of generality, that the individual x_p with only ones is always represented by the individual x_g with only ones, and therefore is always the global optimum. In Figure 3.12 a 3-bit one-max problem is illustrated. The encoding used, which can be described by the genotype-phenotype mapping f_g, preserves the distances between the individuals when mapping the phenotypes to the genotypes ($d_c = 0$) as the genotype-phenotype mapping is the identity mapping $x_p = f_g(x_g) = x_g$. As a result, the phenotypic and genotypic problem complexity is the same.

In the following, we invesitgate how problem difficulty changes if we use a representation where $d_c \neq 0$. For measuring problem difficulty we use the fitness of the schemata (compare subsection 2.3.2). The fitness of a schema h of size λ is defined as $f(h) = f(u, \lambda, l)$. It has length l, u ones, $\lambda - u$ zeros in the fixed positions, and $l - \lambda$ don't care positions. For the one-max problem the schema fitness in terms of the function values $f(u) = u$ can be calculated as follows:

$$f(u, \lambda, l) = \frac{1}{2^{l-\lambda}} \sum_{i=0}^{l-\lambda} \binom{l-\lambda}{i}(i+u).$$

For all schemata of size λ the difference between the fitness $f(\lambda, \lambda, l)$ of the best schemata with λ ones, and the fitness $f(\lambda - 1, \lambda, l)$ of its strongest competitor with $\lambda - 1$ ones is

$$\Delta f = \frac{1}{2^{l-\lambda}} \sum_{i=0}^{l-\lambda} \binom{l-\lambda}{i} > 0.$$

Thus, all schemata h that contain the global optimum x_{opt} are superior to their competitors, and the one-max problem is phenotypically fully easy.

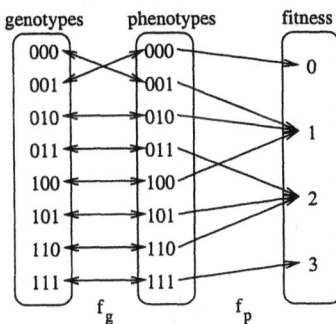

Fig. 3.13. A representation for the bit-counting function which does not preserve the distances between the individuals $(d_c \neq 0)$, and therefore modifies BB-complexity. The distance distortion $d_c = \frac{2}{7*8} * 12 = 3/7 \neq 0$.

In the following, we investigate how the problem complexity changes if the distances between the individuals are changed. The genotype-phenotype mapping of two genotypes x_g and y_g which correspond to the phenotypes x_p and y_p, is changed such that afterwards x_g represents y_p, and y_g represents x_p (see Figure 3.13). Beginning with the low distance distortion encoding illustrated in Figure 3.12, there are three different possibilities to modify the distances by changing the mapping of two individuals:

- Both individuals x_g and y_g have the same number of ones $(u_{x_g} = u_{y_g})$
- Both individuals have a different number of ones, and the number of different positions d_{x_g,y_g} in the two individuals x_g and y_g is the same as the number of different ones $(d_{x_g,y_g} = u_{x_g} - u_{y_g})$
- Both individuals have a different number of ones, and the number of different positions is higher than the number of different ones $(d_{x_g,y_g} > u_{x_g} - u_{y_g})$

In the following, we investigate how the complexity of the problem is changed for these three situations:

If f_g is modified for two genotypes x_g and y_g that have the same number of ones in the string, then the corresponding fitness values remain unchanged $(f(x_g) = f(y_g))$. Therefore, the fitness of the schemata, and the complexity of the problem both remain constant. For example, we can change the mapping of the genotypes 1001 and 0101 for a 4 bit one-max problem. Both individuals have 2 ones in the string and their fitness is 2. The complexity of the problem remains unchanged.

f_g could be modified for two individuals x_g and y_g that have a different number of ones, and therefore different fitness values. We assume that the number of different positions in the two individuals is the same as the number of different ones $(d_{x_g,y_g} = u_{x_g} - u_{y_g})$. Before the change, the individual x_g has l ones and therefore fitness l; y_g has h ones and fitness h. We want to assume $h > l$. After the change, x_g has fitness h although it has only l ones, whereas y_g has only a fitness of l but h ones. Before the modification, all schemata that lead to the global solution are superior to their competitors. Subsequently, after the modification of the mapping the fitness of all schemata h that contain y_g but not x_g, is reduced by $(h-l)/(2^{l-\lambda})$, whereas the fitness of all misleading schemata containing only x_g is increased by this amount. Schemata that contain x_g as well as y_g are not changed. As a result, the average fitness of high quality schemata is reduced, whereas the fitness of misleading schemata is increased. Let us illustrate this with a small 3-bit example. f_g from Figure 3.12 should be modified for the genotypes 001 and 101. Therefore, $x_g = 001$ corresponds to $x_p = 101$ and $y_g = 101$ corresponds to $y_p = 001$ Then, individual $x_g = 001$ has fitness 2, and individual $y_g = 101$ has fitness 1. The fitness of the schema 1** is reduced, whereas the fitness of schema 0** increases. For size two schemata, the fitness of 10* and 1*1 decreases, whereas the fitness of 00* and 0*1 increases. As a result, the problem becomes more difficult to solve for a GA.

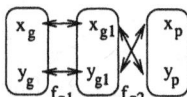
Fig. 3.14. A decomposition of f_g.

Finally, we could decompose f_g into two mappings f_{g1} and f_{g2} if the number of different positions in the two genotypes x_g and y_g is higher than the number of different ones $d_{x_g,y_g} > u_{x_g} - u_{y_g}$ (see Figure 3.14). f_{g1} maps x_g to x_{g1}, and y_g to y_{g1}. x_{g1} (resp. y_{g1}) should have the same number of ones as x_g (resp. y_g) $(u_{x_{g1}} = u_{x_g}, u_{y_{g1}} = u_{y_g})$, but some positions are different in the two individuals x_{g1} and y_{g1} $(d_{x_{g1},y_{g1}} = u_{x_p} - u_{y_p})$. Therefore, as the number of ones stays constant, f_{g1} does not change the fitness of the schemata (compare item 1). For x_{g1} and x_p (resp. y_{g1} and y_p), the number of different ones is the same as the number of different positions. Thus, f_{g2} has the same properties as discussed in the previous item and increases the fitness of misleading schemata, as well as reduces the fitness of the high-quality schemata.

We see that most modifications of a genotype-phenotype mapping f_g where $d_c = 0$, make the one-max problem more difficult to solve. Only when the mapping between genotypes and phenotypes is changed that have the same number of ones in the string, is the structure of the BBs preserved, and we get the same performance as for the one representation where $d_c = 0$. The above proof can be applied in the same way to a fully deceptive trap problem. Then, most of the encodings where $d_c \neq 0$, reduce the fitness of the misleading schemata, and increase the fitness of the high-quality schemata, which makes the problem easier.

In the following subsection, we present an empirical verification of the results.

3.3.4 Empirical Results

In this subsection we present an empirical investigation into how the problem complexity is changed for the one-max and the deceptive trap problem if representations are used that do not preserve the distances between the individuals when mapping the phenotypes to the genotypes. We show empirically that for representations where $d_c = 0$ the fully easy one-max problem remains fully easy. Most of the representations where $d_c \neq 0$ make the one-max problem more difficult to solve for GAs. The situation is vice versa for the fully difficult deceptive trap where the use of representations where $d_c \neq 0$ always makes the problem easier to solve for GEAs.

For a non-redundant genotype-phenotype mapping f_g there are $2^l!$ different possibilities to assign the 2^l genotypes to the 2^l phenotypes (Assigning the genotypes to the phenotypes can be interpreted as a permutation of 2^l numbers). Any of these possibilities represents a specific mapping like for example the binary encoding.

Table 3.3. 24 possibilities to assign four genotypes $\{x_0^g,\ x_1^g,\ x_2^g,\ x_3^g,\}$ to four phenotypes $\{x_0^p,\ x_1^p,\ x_2^p,\ x_3^p,\}$.

x_p	0	1	2	3	4	5	6	7	8	9	10	11	12	13	14	15	16	17	18	19	20	21	22	23
x_0^p	x_0^g	x_0^g	x_0^g	x_0^g	x_0^g	x_0^g	x_1^g	x_1^g	x_1^g	x_1^g	x_1^g	x_1^g	x_2^g	x_2^g	x_2^g	x_2^g	x_2^g	x_2^g	x_3^g	x_3^g	x_3^g	x_3^g	x_3^g	x_3^g
x_1^p	x_1^g	x_1^g	x_2^g	x_2^g	x_3^g	x_3^g	x_0^g	x_0^g	x_2^g	x_2^g	x_3^g	x_3^g	x_0^g	x_0^g	x_1^g	x_1^g	x_3^g	x_3^g	x_0^g	x_0^g	x_1^g	x_1^g	x_2^g	x_2^g
x_2^p	x_2^g	x_3^g	x_1^g	x_3^g	x_1^g	x_2^g	x_2^g	x_3^g	x_0^g	x_3^g	x_0^g	x_2^g	x_1^g	x_3^g	x_0^g	x_3^g	x_0^g	x_1^g	x_1^g	x_2^g	x_0^g	x_2^g	x_0^g	x_1^g
x_3^p	x_3^g	x_2^g	x_3^g	x_1^g	x_2^g	x_1^g	x_3^g	x_2^g	x_3^g	x_0^g	x_2^g	x_0^g	x_3^g	x_1^g	x_3^g	x_0^g	x_1^g	x_0^g	x_2^g	x_1^g	x_2^g	x_0^g	x_1^g	x_0^g

In Table 3.3 we illustrate for $l = 2$ that there are $2^2! = 24$ possibilities to assign the four genotypes $\{x_0^g = 00, x_1^g = 01, x_2^g = 10, x_3^g = 11\}$ to the four phenotypes $\{x_0^p = 00, x_1^p = 01, x_2^p = 10, x_3^p = 11\}$. Each of the 24 genotype-phenotype mappings represents a specific representation, it assigns the fitness values to the genotypes in a different way and it results in a different difficulty of the problem.

In the following we want to investigate how problem difficulty changes for the one-max and deceptive trap problem if we use different types of representations which do not preserve the distances between the individuals.

We start by illustrating the phenotype-fitness mapping f_p for the one-max and deceptive trap problem. This phenotype-fitness mapping f_p describes which fitness value is assigned to which phenotype. In the following we want to assume that this mapping does not change and that it is independent of the used genotype-phenotype mapping. The fitness function f_p for the fully easy l-bit one-max problem is defined as

$$f_p(x_p) = \sum_{i=0}^{l-1} x_i,$$

and the l-bit deceptive trap function is defined as:

$$f_p(x_p) = \begin{cases} l - 1 - \sum_{i=0}^{l-1} x_i & \text{for } \sum_{i=0}^{l-1} x_i < l \\ l & \text{for } \sum_{i=0}^{l-1} x_i = l \end{cases}$$

Now we have to calculate how many different genotype-phenotype mappings exist. If we use genotypes and phenotypes of length l the number of possible genotype-phenotype mappings is $2^l!$. To reduce this number, we assume without loss of generality that the phenotype x_p with only ones, which has a fitness value $f_p = l$, is always assigned to the individual x_g with only ones. Then, the number of different genotype-phenotype mappings is reduced to $(2^l - 1)!$. For example if we look at Figure 3.12 or 3.13 we have $2^l = 8$ genotypes and $2^l = 8$ phenotypes. Therefore, we have $8! = 40340$ different possibilities to assign the genotypes to the phenotypes. If we assign $x_g = 111$ always to $x_p = 111$ then there are only $7! = 5040$ different possibilities to assign the genotypes to the phenotypes. Every genotype-phenotype mapping represents a different representation.

Furthermore, we have seen in the previous subsection that for the used phenotype-fitness mapping f_p (one-max and deceptive trap problem) there are some genotype-phenotype mappings that do not modify the BBs. These mappings have the same properties and f_g differs only for individuals that phenotypically have the same number of ones. There are $\prod_{i=1}^{l} \binom{l}{i}!$ encodings of that kind. For example in Figure 3.12 we can change the genotype-phenotype mapping and assign $x_g = 001$ to $x_p = 010$ and $x_g = 010$ to $x_p = 001$. Although we use a different representation the assignment of the fitness values to the genotypes has not changed. This effect is a result of the redundancy of the used one-max and deceptive trap problem which both only considers the number of ones in the phenotype.

If we use these results we can calculate how many groups of different genotype-phenotype mappings exist that result in a different structure of the BBs. If we have $(2^l - 1)!$ different genotype-phenotype mappings and use a l-bit one-max or deceptive problem, then there are $\prod_{i=1}^{l} \binom{l}{i}!$ mappings that do not change the structure of the BBs. Therefore, we have

$$\frac{(2^l - 1)!}{\prod_{i=1}^{l} \binom{l}{i}!}$$

groups of different genotype-fitness mappings. Each group consists of $\prod_{i=1}^{l} \binom{l}{i}!$ different genotype-phenotype mappings which do not affect the structure of the BBs and where only the mapping is changed between genotypes and phenotypes that have the same number of ones.

When using a 3-bit one-max or deceptive trap problem then there are $\frac{(2^3-1)!}{3!3!} = 5040/36 = 140$ groups of different genotype-fitness mappings with different properties. Each of these 140 different groups result in a different genotype-fitness mapping and therefore in a different structure of the BBs. We use for our investigation 10 concatenated building blocks of size 3. Therefore, the overall string length $l = 30$ and the fitness of an individual is calculated as the sum over the fitness of the ten BBs.

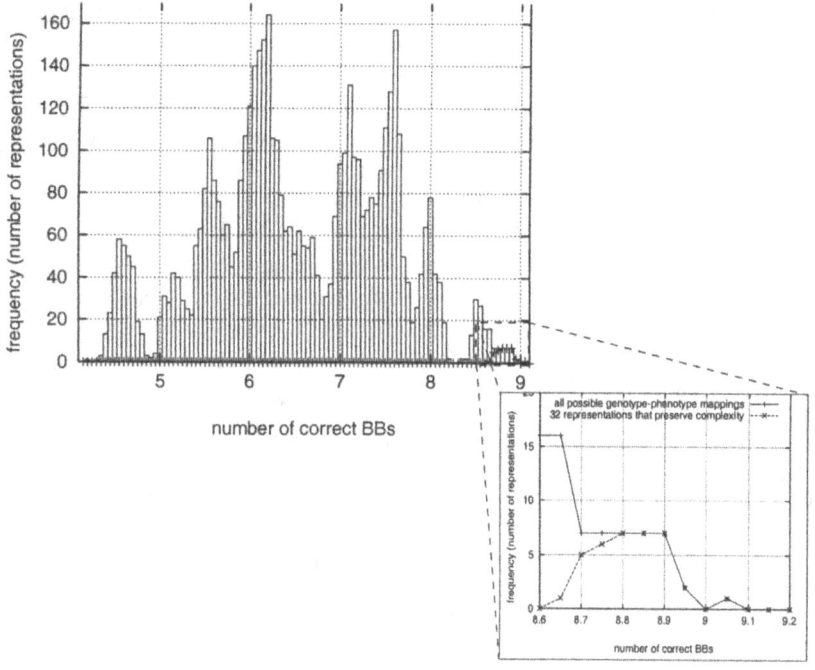

Fig. 3.15. Experimental results of the frequency of the number of correct BBs at the end of a run for all possible encodings of a 3-bit one-max problem. We present results for 10 concatenated 3-bit problems. The optimal solution is always 111 so there are $(2^3 - 1)! = 5040$ different possible genotype-phenotype mappings. We use a GA with tournament selection without replacement, uniform crossover and a population size of $n = 15$. We perform 200 runs for every possible encoding. Only for these 36 representations where the distance distortion is zero or where the structure of the BBs is preserved, the fully easy one-max problem remains fully easy. All other encodings do not preserve BB-complexity and make the problem more difficult to solve for GAs.

To illustrate how genotype-phenotype mappings change the complexity of a problem we measure how many of the ten BBs a GA finds dependent on the used representation. As we use a 3 bit problem there are 5040 different genotpye-phenotype mappings (representations). The distances between the genotypes and phenotypes are preserved by only one out of the 5040 different genotype-phenotype mappings. The encoding is shown in Figure 3.12. However, due to the strucutre of the one-max and deceptive trap problem, there are 35 other genotype-phenotype mappings which, although they change the distances between the genotypes and phenotypes, do not change the structure of the BBs. These types of encodings assign genotypes and phenotypes that have the same number of ones in the string in a different way.

Figure 3.15 presents the results of our experiments for the one-max problem. We show the distribution of the number of correct BBs at the end of a GA run when using different types of genotype-phenotype mappings. The plot shows results for all 5040 different genotype-phenotype mappings. The ordinate counts the number of genotype-phenotype mappings that allow a GA to correctly identify a certain number of BBs. We used a GA with tournament selection without replacement of size 2, uniform crossover, no mutation and a population size of $n = 15$. We performed 200 runs for each specific genotype-phenotype mapping, and each run was stopped after the population was fully converged. The average number of BBs found at the end of the run gave us a measurement of the problem difficulty for the GA using one specific representation. The more BBs which could be correctly identified at the end of the run, the easier the problem was for GAs.

How can we interpret the data in Figure 3.15? Every bar indicates the number of different genotype-phenotype mappings that allow a GA to correctly identify a specific number of BBs. For example, the bar of height 95 at position 7.0 means that a GA correctly identifies on average between 6.975 and 7.025 BBs for 95 different genotype-phenotype mappings. The bar at position 4.85 means that there are 4 different genotype-phenotype mappings that allow a GA to correctly identify on average between 4.825 and 4.875 BBs. The plot shows that by using a GA with only 15 individuals we find independently of the used representation at least 4.2 BBs, and we are not able to correctly identify more than 9 out of ten BBs. Furthermore, it is surprising that we have no normal distribution over the number of correct BBs but that there are clusters. For example there are many representations that allow a GA to find on average between 5.8 and 6.3 correct BBs but there are only a few representations that allow a GA to correctly identify on average between 6.5 and 6.8 BBs. The answer can be found in the relatively low number of different genotype-fitness mappings. We have found that there are 140 different groups of genotype-fitness mappings. Therefore, although we have 5040 different genotype-phenotype mappings, there are only 140 different levels of problem complexity possible. The observed clusters are probably a result of these small number of different levels of problem complexity.

It is more interesting to ask how the distance preserving genotype-phenotype mapping from Figure 3.12 performs? And how the performance is of the other 35 genotype-phenotype mappings that, although they do not preserve the distances, also preserve the complexity of the BBs and result in the same genotype-fitness mapping? The small plot in Figure 3.15 answers these questions. The bold line shows the performance of a GA using these 36 different representations which all preserve the structure of the BBs. The use of representations that do not change the complexity of the one-max problem results in the highest GA performance. For example there are 7 different representations that allow a GA to correctly identify between 8.825 and 8.875 BBs. All 7 encodings belong to the group of 36 encodings that all preserve the structure of the BBs. Furthermore, we see that all representations that allow a GA to correctly identify on average more than 8.75 BBs belong to this group. These 36 encodings that preserve the complexity of the BBs (the encoding with $d_c = 0$ is one of them) result in the highest proportion of correct BBs. When using these encodings the one-max problem remains fully easy and the size of the BBs stays $k = 1$.

Changing the encoding, that means assigning the elements of Φ_g in a different way to the elements in Φ_p, always results in a representation where $d_c \neq 0$. If the genotype-fitness mapping is different from the encoding where $d_c = 0$ then the BB-complexity increases and the problem becomes more difficult to solve. A GA has more difficulties in solving the problem, and the proportion of correct BBs is lower. The plot illustrates nicely that the used representation can change the complexity of the one-max problem dramatically. However, only those encodings that preserve the complexity of BBs allow a GA to unleash its full power for the fully easy one-max problem.

In Figure 3.16 we present results for 10 instances of a 3-bit deceptive trap. The GA parameters chosen are the same as for the one-max problem. The plots show that, as expected, the GA performs worst for representations that preserve the distances and the structure of the BBs. All other encodings that do not preserve the fully misleading character of the problem make the problem easier to solve for GAs.

We have empirically shown that only representations that preserve the distances between the individuals and do not change the structure of the BBs guarantee that fully easy problems remain fully easy. Therefore, representations where the distance distortion $d_c = 0$ preserve BB-complexity and are a good choice if we want GAs to reliably solve problems of bounded complexity. As soon as a representation does not preserve BB-complexity, some of the easy problems become more difficult and therefore can no longer be solved by the GA. Indeed, some of the difficult problems could become solvable by using representations that do not preserve BB-complexity, but in general we are not interested in solving these types of problems.

3.3.5 Conclusions

This section focused on the influence of representations on problem difficulty. We started in subsection 3.3.1 by illustrating the influence of representations on problem complexity. We reviewed that by the use of representations fully

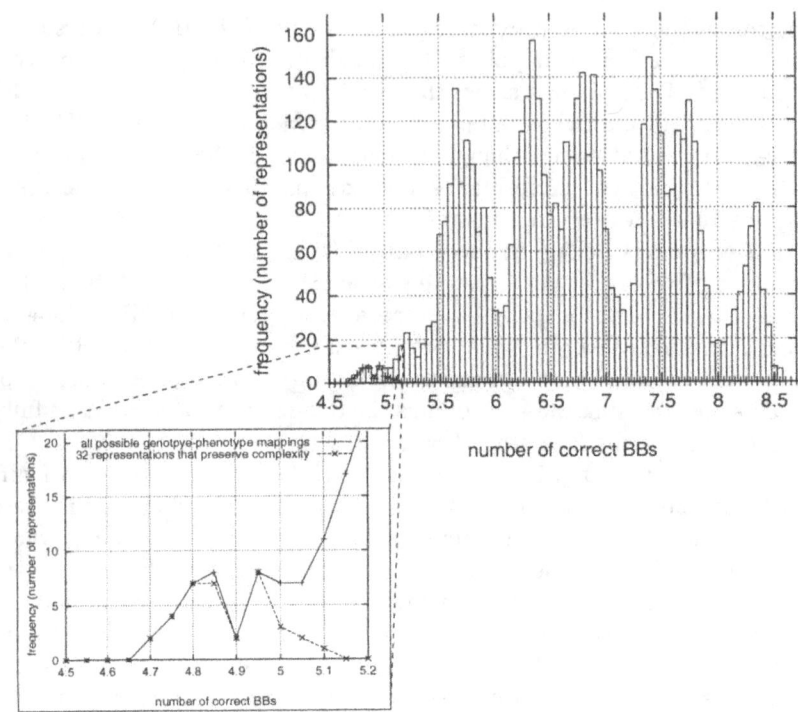

Fig. 3.16. Experimental results of the frequency of the number of correct BBs at the end of the run for all possible encodings of a 3-bit deceptive-max problem. We present results for 10 concatenated 3-bit problems. The optimal solution is always 111 so there are $(2^3 - 1)! = 5040$ different possible encodings. The GA uses tournament selection without replacement, uniform crossover and a population size of $n = 15$. We perform 200 runs for every possible encoding. Only for those 36 representations which preserve the distances, or do not change the complexity of the BBs, does the fully difficult deceptive trap remain fully difficult. For all other representations, the BB-complexity is reduced and the problem becomes easier to solve for GAs. The figures illustrate that the problem difficulty is strongly determined by the selected type of encoding.

easy problems can become fully difficult, and vice versa. One possibility to guarantee that GEAs are able to reliably solve problems of bounded difficulty is to use representations that preserve the distances between the individuals when mapping the phenotypes to the genotypes. In subsection 3.3.2 we introduced the concept of distance distortion d_c, and illustrated that high locality is necessary for a representation to have low distance distortion, as well as to be able to preserve the complexity of the BBs. Furthermore, we showed theoretically in subsection 3.3.3 that when using representations where $d_c = 0$, the phenotypically fully easy one-max problem remains genotypically fully easy. Most modifications of the distances between the individuals makes the fully easy one-max problem more difficult to solve for GEAs. Finally, subsection 3.3.4 verified the theoretical results and presented an empirical study for the one-max and deceptive trap problem on how the problem complexity of the problem depends on the used encoding. The results showed that only for representations where $d_c = 0$, or for representations that preserve the structure of the BBs, that the complexity of the two problems remain unchanged. The one-max problem remains fully easy and the deceptive trap remains fully difficult. For all representations that do not preserve BB-complexity, the one-max problem becomes more difficult and the deceptive trap easier.

We presented in this section the third and final element towards a theory of representations. We investigated how the complexity of a problem is influenced by the distance distortion of a representation. If the distance distortion $d_c = 0$, then the genotypic and phenotypic complexity of the problem is the same. For $d_c \neq 0$ the distances between the individuals are modified and the problem difficulty can be different for the genotypes and phenotypes.

In general, we want GEAs to be able to solve a class of problems of bounded complexity fast and reliably. However, the results have shown that the use of representations where $d_c \neq 0$, in general modifies the complexity of BBs and can only increase the problem complexity for fully easy problems. Therefore, easy problems that are solvable using a complexity-preserving representation could become unsolvable when using a representation that modifies BB-complexity. To make sure a GA can reliably solve the problems of bounded complexity it is designed for, we recommend the use of representations that do not change the distances between the individuals and therefore, do not modify BB-complexity.

This section illustrated nicely that representations can dramatically change the complexity of a problem. The presented work has shown that even fully difficult problems can be solved easily if a proper representation is used. However, using the same representation for a fully easy one-max problem can make the problem fully difficult and unsolvable. Therefore, the use of representations that modify BB-complexity could be advantageous if we know that the problem is very difficult. But in general users do not have this information and therefore, they should be careful when using these types of representations.

3.4 Summary and Conclusions

In section 3.1 we described, analyzed, and modeled the effect of linear redundant encodings on the performance of GEAs. We showed that a representation incorporating uniform redundancy does not change the behavior of GEAs. However, for non-uniform redundant encodings, the GEA performance could be changed dramatically. Then, an overrepresentation of the good solutions results in a higher GEA performance, whereas, if the good solutions are strongly underrepresented, a failure of GEAs is inescapable. By modeling redundancy as a problem of BB-supply, we were able to present a model of convergence based on the Gambler's ruin model from Harik, Cantú-Paz, Goldberg, and Miller (1999).

This was followed in section 3.2 by an investigation into how the scaling of the BBs of an encoding influences the performance of GEAs. We extended previous work (Rudnick, 1992; Thierens, 1995; Thierens, Goldberg, & Pereira, 1998; Harik, Cantú-Paz, Goldberg, & Miller, 1997) and formulated a more exact convergence model considering genetic drift for exponentially scaled representations. Using this model, we were able to more accurately predict the behavior of GEAs using exponentially scaled representations.

Finally, we presented the third and final element of a theory of representations, namely the influence of representations on problem complexity. Section 3.3 shows that representations where the distance distortion d_c is equal to zero preserve the distances between the individuals and do not modify the difficulty of a problem. If representations where $d_c \neq 0$ are used, the complexity of BBs can be changed. Therefore, fully easy problems become more difficult, and difficult problems easier. We have discussed why representations that keep easy problems easy and make difficult problems easier are nice to have, but not possible without having an exact knowledge about the optimization problem a priori. Furthermore, we identified the high locality of a representation to be a necessary condition for an encoding to have low distance distortion.

In this chapter we identified three important elements towards a general theory of representations. We identified redundancy, the scaling of BBs, and the distance distortion as having a major influence on the performance of genetic and evolutionary representations. We were able to show that redundant encodings do not modify the performance of a GEA as long as the representation is uniformly redundant. Our investigation into non-uniformly scaled representations has shown that these types of encoding prolong the search process and increase the problems of GEAs with genetic drift. Finally, we have seen that representations where the distance distortion is unequal to zero do not preserve BB-complexity in general and make phenotypically easy problems more difficult. Therefore, to make sure that GAs are able to reliably solve easy problems and problems of bounded complexity, the use of distance-preserving representations is recommended.

Even by only presenting some basic elements of a general theory of representations we are able to analyze and predict the behavior and performance of existing representations significantly better. The presented theory gives us a deeper understanding on how existing representations influence the performance of GEAs, as well as allows us to design new representations more efficiently. By using the presented theory, on the one hand we can develop very general and robust representations that can be applied to problems of unknown complexity, and on the other hand very problem-specific representations which could fail for some problems, but perform very well for a specific problem.

Although the provided elements of representation theory already allow a much more guided design and analysis of representations, further research is still necessary to develop a general representation theory. Especially, the relationship between the presented elements of theory of representations should be investigated more deeply. We believe that as we are able to easily separate the effects of redundant and exponentially scaled representations that there is not much interconnection and overlapping between these two elements of theory. However, for the distance distortion and its influence on BB-complexity, the situation is different. We have seen that the modification of problem complexity is strongly influenced by redundancy or scaling. Therefore, further research is necessary to identify the exact relations between the presented elements of theory.

Finally, we want to encourage researchers to do more basic research towards the development of a general theory of representations. We believe that we provided some important parts, but there is still a long way to go. However, the path is worth following, as a general theory of representations would allow us to unleash the full power of genetic and evolutionary search and help us to solve problems fast, accurately and reliably.

4. Time-Quality Framework for a Theory-Based Analysis and Design of Representations

Over the last decades, researchers gained more and more knowledge about the principles of genetic and evolutionary algorithms (GEAs) and were able to formulate a theory describing the behavior of GEAs more precisely (Goldberg, 1989c; Bäck, Fogel, & Michalewicz, 1997). The existing elements of GEA theory explain quite accurately the influence of many important GEA parameters, as well as selection, recombination, or mutation methods on the performance of GEAs. By using the existing GEA theory, straight forward design and the development of new, competent GAs (Goldberg, Deb, Kargupta, & Harik, 1993; Harik & Goldberg, 1996; Pelikan, Goldberg, & Cantú-Paz, 1999; Pelikan, Goldberg, & Lobo, 1999) became possible. However, concerning representations for GEAs, a framework which describes the influence of representations on the performance of GEAs is still missing, although it is well known that the used representation has a strong influence on GEA performance. Such a framework could help us to develop new representations in a more theory-guided manner and would be an important step towards a general theory of representations for GEAs.

The purpose of this chapter is to develop a framework for a theory-based analysis and design of representations for GEAs based on the elements of theory we presented in the previous chapter. The framework should allow us to model and predict the influence of different types of representations on the performance of genetic and evolutionary search. It should describe how redundancy, scaling, and distance distortion of a representation influence the time to convergence and the expected solution quality. By using the framework, we would be able to theoretically compare the efficiency of different representations, as well as to design new representations in a theory-guided way.

The chapter starts with a brief overview of the determinants – time and quality – of GEA performance. In section 4.2, the elements of the framework, namely redundancy, scaling, and distance distortion/locality, are presented. We review their influence on representations, formulate how the three properties of representations can be measured, and describe how genetic and evolutionary search is affected. In section 4.3 the framework itself is described. We formulate how the probability of error α and the convergence time t_{conv} depend on the different elements of the framework. Because we are not yet able to consider the effect of scaled representations in general, the section is split up into two parts concerning uniformly and non-uniformly scaled representations. For both types of scaled representations we describe how the solution quality and the time to convergence depends on the redundancy and distance distortion of an encoding. This is followed in section 4.4 by some implications of the framework on the design of representations. We show how the use of representations with different properties affects the supply of BBs, the dynamics of genetic search, or the size of BBs. The chapter ends in section 4.5 with a summary and concluding remarks.

4.1 Solution Quality and Time to Convergence

The following section briefly reviews determinants for GEA performance. It focuses on solution quality and time to convergence.

For comparing the efficiency of different GEAs using different types of representations, a measurement of GEA performance is necessary. Widely used determinants for GEA performance are the solution quality and the time to convergence. In general, the solution quality and convergence time depend on the used genetic operators, the GEA parameters, the used representation, and the optimization problem itself.

The solution quality of GEAs can be measured by the probability P_n of GEA success. GEA success means that the optimal solution is found by the GEA. When using the more common probability of GEA failure α, GEA success is defined as $P_n = 1 - \alpha$. Earlier work by Harik et al. (1997) has shown that when using selectorecombinative GAs, the probability of error $\alpha = 1 - P_n$ goes with $O(e^{-n})$. With decreasing α the population size n increases exponentially. Therefore, instead of using α, the population size n that is necessary for solving a problem can also be used for comparing GEA performance. Measuring GEA performance becomes more complicated if the best solution is not known a priori. Then, P_n cannot be calculated and the best fitness at the end of a GEA run must be used. It corresponds to the probability of GEA success P_n, and is determined by the used population size n.

The time to convergence t_{conv} describes how many generations selectorecombinative GEAs need to converge completely. A population is converged if there is no genetic diversity in the population after t_{conv} generations and all individuals in the population represent the same phenotype. It was shown (Thierens & Goldberg, 1993; Miller & Goldberg, 1996b; Miller & Goldberg, 1996a) that the convergence time mainly depends on the length of the string l and the used selection scheme. As soon as the population size n is large enough to solve the problem reliably, the convergence time t_{conv} does not depend on n any more.

To compare the overall performance of different GEAs the number of fitness evaluations n_f can be used. For a given solution quality $P_n \gg 0$ the total number of fitness calls can be calculated as

$$n_f = n * t_{conv}.$$

After we have discussed solution quality and convergence time, we focus in the following section on elements of the time-quality framework.

4.2 Elements of the Framework

After the illustration in the previous section on how the performance of GEAs can be measured, we focus in this section on the elements of representations that influence GEA performance. The purpose of this section is to describe the elements of representation theory that are used in the framework. We review the elements, describe how we can measure them, and illustrate their effects on GEAs. In chapter 3 we presented redundancy, scaling, and distance distortion as elements of representation theory. Although we believe that these three elements are some important elements of the time-quality framework, there could still be others. Finding them is left to further research.

The section consists of three subsections which discuss the single elements, namely redundancy, scaling, and distance distortion. In each subsection we briefly describe what we mean, illustrate how redundancy, scaling, or distance distortion can be measured, and finally describe how GEAs are affected.

4.2.1 Redundancy

Section 3.1 has shown that the use of redundant encodings affects the performance of GEAs. In the context of representations, redundancy means that on average one phenotype is represented by more than one genotype. Therefore, $|\Phi_g| > |\Phi_p|$ when using redundant representations. Consequently, a representation is not redundant if $|\Phi_g| = |\Phi_p|$. Then, the number of genotypes is the same as the number of phenotypes. Because we assume that every phenotype must be represented by at least one genotype, the size of the genotypic space can not be smaller than the size of the phenotypic space.

To model the effects of redundancy we have introduced k_r as the order of redundancy (see subsection 3.1.1). It measures the amount of redundant information in the encoding (in bit). There are k_r bits and 2^{k_r} different possibilities (individuals) to encode 1 Bit of information (2 possibilities). Using no redundancy in an encoding results in $k_r = 1$. Furthermore, r is defined as the number of genotypic BBs of size kk_r that represent the optimal phenotypic BB of size k. Therefore, for non-redundant encodings $k_r = 1$ and $r = 1$. We know from equation 3.2 that for redundant encodings

$$r \in [1, 2, \ldots, 2^{kk_r} - 2^k + 1]. \tag{4.1}$$

In general, there are 2^k different phenotypes and they are represented by 2^{kk_r} different genotypes. Using uniformly redundant representations results in

$$r_{uniform} = 2^{k(k_r-1)}$$

and $x_0/n = r/2^{kk_r} = 1/2^k$. x_0 denotes the the initial supply of BBs. Therefore, a representation is non-uniformly redundant if $r/2^{kk_r} \neq 1/2^k$. For $r/2^{kk_r} > 1/2^k$ the optimal solution is overrepresented, and for $r/2^{kk_r} < 1/2^k$ the optimum is underrepresented.

Our investigation into the effects of redundancy on GEAs in subsection 3.1.2 has shown that the supply of BBs in the initial population is influenced by the use of non-uniformly redundant encodings. If the optimal solution is overrepresented by the used representation the performance of GEAs is increased, that means lower run duration t_{conv} and lower probability of error α. The situation is reversed if the optimal solution is underrepresented t_{conv} and α increases.

4.2.2 Scaling

In section 3.2, we discussed the effects of exponentially scaled representations on the performance of GEAs. Representations are uniformly scaled, if all alleles have the same contribution to the fitness function. Therefore, GEAs using uniformly scaled representations solve all alleles implicitly in parallel. In contrast, a representation is non-uniformly scaled if some alleles have a higher contribution to the fitness than others. As a result domino convergence occurs and the alleles are solved sequentially according to their salience. The most salient alleles are solved first, whereas the lowest salient alleles are solved last.

To more formally describe the scaling of a representation, a measurement of how strong a representation is scaled is necessary. Therefore, we describe by the order of scaling s the difference in salience for the different alleles. When using a bitstring representation of length l and ordering the alleles according to their contribution to the fitness in ascending order, we define the order of scaling $s \in [1, \infty[$ as

$$s = \frac{1}{l-1} \sum_{i=1}^{l-1} \frac{x_{i+1}^c}{x_i^c},$$

where x_i^c denotes the contribution of the ith most salient allele to the fitness, and $x_{i+1}^c \geq x_i^c$, for $i \in [1, l-1]$. Therefore, x_1^c denotes the contribution of the lowest salient allele and x_l^c denotes the contribution of the most salient allele. When using uniformly redundant encodings the contribution of all alleles to the fitness function is the same which results in $x_i^c = \text{const}$, for $i \in [1, l]$. Therefore, $s = 1$ for uniformly scaled representations. When using exponentially scaled representations the order of scaling $s > 1$ is constant. (When using binary encoded strings, $s = 2$.)

The order of scaling s influences the dynamics of genetic search. With increasing s, the alleles are solved more and more sequentially. Rudnick (1992) proposed the use of a convergence window for modeling the dynamic solving process when using non-uniformly scaled representations. The convergence window is a set of contiguous alleles that are not yet fully converged but have started to converge. The size $\lambda_c \in [1, l]$ of the convergence window is equal to l for uniformly scaled encodings and equal to one for $s \to \infty$. $\lambda_c = 1$ results in strictly sequential solving of the alleles, whereas for $\lambda_c = l$ all alleles are solved in parallel.

With increasing order of scaling s the size λ_c of the convergence window is reduced. Earlier work (Thierens et al., 1998; Lobo et al., 2000) shows in correspondence to the results of subsection 3.2.4, that the assumption of a convergence window of size $\lambda_c = 1$ results for $s = 2$ (exponentially scaled representations) in a good approximation of the dynamics of GEA search. However, for a more general theory of scaled representations a more detailed analysis of the interdependencies between λ_c and s is necessary.

4.2.3 Distance Distortion

We have seen in section 3.3 that when using a representation, the distances between the phenotypes can be different from the distances between the genotypes. Therefore, the complexity of the building blocks is different in the genotypic and phenotypic space.

We have illustrated that high locality concerning small, as well as large changes, guarantees that an encoding preserves the complexity of a problem[1]. If the distance distortion d_c is not zero, that means the distances between the individuals are changed by the representation, the complexity of the optimization problem can be modified.

Our investigation into the modification of BB-complexity has shown that high locality of a representation guarantees that the complexity of the problem remains unchanged when using mutation-based search. We defined (see subsection 3.3.2) the locality d_m of a representation concerning small changes as

$$d_m = \sum_{d^p_{x_i, x_j} = d^p_{min}} |d^g_{x_i, x_j} - d^g_{min}|,$$

where $d^p_{x_i, x_j}$ is the phenotypic distance between the phenotypes x_i and x_j, $d^g_{x_i, x_j}$ is the genotypic distance between the corresponding genotypes, and d^p_{min}, respective d^g_{min} is the minimum distance between two (neighboring) phenotypes, respectively genotypes. Without loss of generality we want to assume that $d^g_{min} = d^p_{min}$. For $d_m = 0$ the genotypic neighbors correspond

---

[1] In particular, high locality is necessary for mutation-based search, preservation of distances helps crossover-based search, and perfect locality is necessary for preservation of distances.

to the phenotypic neighbors, the encoding has perfect locality, and the complexity of the phenotypic problem is not modified when using mutation-based search approaches.

When using recombination-based search, the locality concerning small changes d_m must be extended towards locality concerning small and large changes. The distance distortion d_c describes for non-redundant representations how well the phenotypic distance structure is preserved when mapping Φ_p on Φ_g:

$$d_c = \frac{2}{n_p(n_p - 1)} \sum_{i=1}^{n_p} \sum_{j=i+1}^{n_p} |d_{x_i,x_j}^p - d_{x_i,x_j}^g|,$$

where $n_p = |\Phi_g| = |\Phi_p|$, and $d_{g,min} = d_{p,min}$. For $d_c = 0$ all phenotypic distances are preserved by the representation and the complexity of the phenotypic BBs remains unchanged in the genotypic space.

We have seen in section 3.3 that by using representations that do not preserve the distances between the individuals when mapping the phenotypes onto the genotypes ($d_c \neq 0$), the complexity of the problem which can be measured by the size k of the BBs, can be changed. Distinguishing between the size of BBs in the phenotypic space k_p, and the size of BBs in the genotypic space k_g allows us to model the influence of the distance distortion d_c on the performance of GEAs more exactly. Section 3.3 has shown that there are encodings where the distance distortion $d_c = 0$ and the phenotypic distances between individuals correspond to the genotypic distances. If we use non-redundant representations and non-redundant phenotype-fitness mappings then the complexity of building blocks is preserved by those types of representations. The problem has the same genotypic as phenotypic complexity.

$$k_g = k_p \text{ , if } d_c = 0.$$

The situation becomes more difficult if $d_c \neq 0$. As soon as the distances between the individuals are not preserved by the representation, the complexity of the BBs is changed and $k_g \neq k_p$. Every phenotypic problem with complexity k_p can be transformed by the use of a representation into a genotypic problem with complexity $k_g \in [1, l]$. For every problem there is always a representation that results in a fully easy problem $k_g = 1$ as well as a representation that results in a fully difficult, misleading trap with $k_g = l$ (Liepins & Vose, 1990). Therefore, when using representations where $d_c \neq 0$, the genotypic size of BBs k_g depends not only on the genotype-phenotype mapping f_g, but also on the specific structure of the phenotypic problem f_p. Therefore, we get

$$k_g = \begin{cases} k_p & \text{, for } d_c = 0, \\ k_g(f_g, f_p) \text{ , with } 1 \leq k_g \leq l & \text{, for } d_c \neq 0, \end{cases} \tag{4.2}$$

where l denotes the length of the binary string.

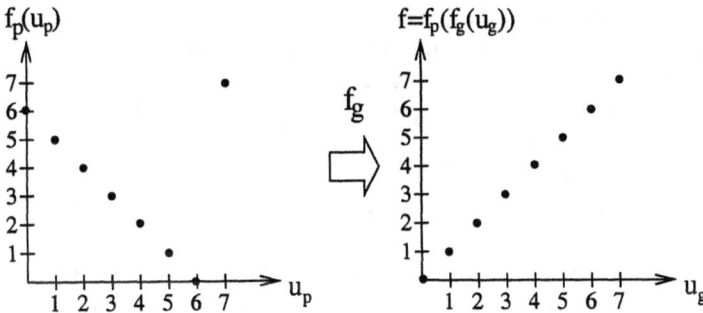

(a) The genotype-phenotype mapping f_g defined in equation 4.3 makes the fully deceptive phenotypic trap (left) fully easy (right).

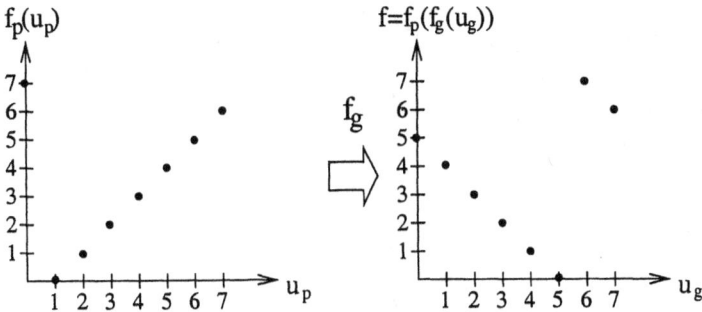

(b) Because the optimal solution is located differently ($u_p = 0$) the genotype-phenotype mapping f_g from above does not reduce the complexity of the BBs significantly.

Fig. 4.1. We show how a specific representation f_g (see equation 4.3) modifies the complexity of different phenotypic fully deceptive problems. If the optimal solution is located at $u = 7$ the problem becomes fully easy ($k_g = 1$). However, if the optimal solution is located at $u = 0$ the complexity of the problem remains approximately unchanged. We see that for designing a genotype-phenotype mapping f_g that makes difficult problems easier, the structure of the phenotypic optimization problem f_p must be known. Therefore, theory-guided design of representations that reduce BB-complexity by modifying the distance structure between the individuals is tricky.

We want to illustrate with a small example why k_g does not only depend on f_g and k_p, but also on f_p. Subsection 3.3.1 (see also Liepins and Vose (1990)) has shown that every fully deceptive trap ($k_p = l$) can be transformed into a fully easy problem ($k_g = 1$) by a linear transformation. For this purpose we want to define the genotype-phenotype mapping f_g as

$$u_g = \begin{cases} l - u_p - 1 & \text{if } u_p \neq l, \\ l & \text{if } u_p = l, \end{cases} \qquad (4.3)$$

where l is the length of the string, u_g is the number of ones in the genotype and u_p is the number of ones in the phenotype. This encoding does not preserve the distances between the individuals ($d_c \neq 0$). By the genotype-phenotype mapping f_g a phenotypically fully deceptive trap ($k_p = 7$) becomes fully easy (see Figure 1(a)). However, using the same mapping f_g for a different phenotypic problem of the same complexity (Figure 1(b)) does not significantly reduce k_g. Although the size of BB $k_p = 7$ is the same, and only the position of the optimal solution has changed, the problem is still almost fully deceptive ($k_g \approx l$) after applying f_g. We see that for predicting k_g the knowledge of k_p is not enough when using representations with $d_c \neq 0$. It is necessary to know f_p as well as f_g to predict the influence of representations with low locality on the difficulty of the problem.

4.3 The Framework

This section provides the time-quality framework modeling the influence of representations on the performance of genetic and evolutionary algorithms.

The framework allows us to theoretically predict and compare the performance of GEAs using different types of representations. Therefore, thorough analysis and theory-guided design of representations becomes possible. Although the presented framework is not yet complete and there are still some gaps, rough approximations, unclear interdependencies, and also more as yet unknown elements, we believe that the framework is an important step towards a more general theory of representations.

The framework itself is based on the characteristics of the used encoding we introduced in section 4.2. There, we have seen that the redundancy of a representation can be described by the order of redundancy k_r and the number of copies r which are given to the optimal solution. Furthermore, the modification of BB-complexity is determined by the distance distortion d_c, which measures how well the phenotypic distances between the individuals correspond to the genotypic distances. Finally, the scaling of a representation can be described by using the order of scaling s. Currently there is no general model available for the influence of s on the performance of GEAs. Therefore, we want to focus in this framework on uniformly scaled representations ($s = 1$) and exponentially scaled representations with $s \geq 2$.

The structure of the section follows the still missing general model of the influence of scaling. Therefore, the section is split into two parts. In subsection 4.3.1 we present the part of the framework for uniformly scaled representations and in subsection 4.3.2 we focus on exponentially scaled representations.

4.3.1 Uniformly Scaled Representations

In the following subsection, we present the part of the framework that describes the influence of representations on GEA performance if the representations are uniformly scaled. We describe how the probability of error α and the time to convergence t_{conv} depend on redundancy and distance distortion.

Based on the work from Harik, Cantú-Paz, Goldberg, and Miller (1997) we get for the probability of error

$$\alpha = 1 - \frac{1 - (q/p)^{x_0}}{1 - (q/p)^n},$$

where x_0 is the expected number of copies of the best BB in the randomly initialized population, $q = 1 - p$ is the probability of making the wrong decision between two competing BBs, and n is the population size. From equation 3.5 we know that

$$x_0 = n \frac{r}{2^{kk_r}},$$

where k is the phenotypic size of BBs, r is the number of genotypic BBs of length kk_r that represent the best phenotypic BB, and k_r is the order of redundancy. After some approximations (see subsection 3.1.3) we finally model in equation 3.8 the influence of redundant encodings on the population size n as

$$n = -\frac{2^{k_r k - 1}}{r} \ln(\alpha) \frac{\sigma_{BB} \sqrt{\pi m'}}{d}, \tag{4.4}$$

where $m' = m - 1$ with m is the number of BBs, d is the signal difference, and σ_{BB}^2 is the variance of the BBs. The probability α of GA failure can be calculated as:

$$\alpha = \exp\left(-\frac{ndr}{2^{k_r k - 1} \sigma_{BB} \sqrt{\pi m'}}\right) \tag{4.5}$$

We have described in subsection 4.2.3 that the problem difficulty measured by the size of BBs k is modified by the distance distortion d_c of the representation. In equation 4.2 the genotypic size k_g of the BBs is calculated as

$$k_g = \begin{cases} k_p & \text{, if } d_c = 0, \\ k_g(f_g, f_p) \text{ ,where } 1 \le k_g \le l & \text{, if } d_c \ne 0, \end{cases}$$

where f_g is the genotype-phenotype mapping (the used representation), and f_p is the optimization problem with the size k_p of the phenotypic BBs. Substituting k_g into 4.5 we get for uniformly scaled representations:

$$\alpha = \exp\left(-\frac{ndr}{2^{k_r k_g - 1}\sigma_{BB}\sqrt{\pi m'}}\right). \tag{4.6}$$

The probability of error α goes with $O\left(\exp\left(\frac{-r}{2^{k_r k_g}}\right)\right)$. We see that using redundant representations ($k_r > 1$) without increasing r has the same influence on GEA performance as increasing the size of BBs k_g.

From subsection 3.1.4 we get for the time to convergence for a uniformly scaled representation (see equation 3.9)

$$t_{conv} = \frac{\sqrt{l}}{I}\left(\frac{\pi}{2} - \arcsin\left(2\frac{x_0}{n} - 1\right)\right), \tag{4.7}$$

where l is the length of the phenotypes, and I is the selection intensity. Substituting x_0 from equation 3.5 into 4.7 yields

$$t_{conv} = \frac{\sqrt{l}}{I}\left(\frac{\pi}{2} - \arcsin(\frac{r}{2^{k_r k - 1}} - 1)\right).$$

When considering the effect of the distance distortion d_c (see equation 4.2) we finally get for the time to convergence

$$t_{conv} = \frac{\sqrt{l}}{I}\left(\frac{\pi}{2} - \arcsin(\frac{r}{2^{k_r k_g - 1}} - 1)\right). \tag{4.8}$$

t_{conv} increases with larger k_g and decreasing $r/2^{k_r}$. With $0 < \frac{r}{2^{k_r k_g}} < 1$ we can calculate upper and lower bounds for the expected time to convergence as

$$0 < t_{conv} < \frac{\sqrt{l}}{I}\pi$$

If $r/2^{k_r k_g} \approx 1$ most of the randomly created genotypic individuals represent the phenotypic optimum. Therefore, GEAs converge very fast and $t_{conv} \to 0$. If either k_g is a large number or $r/2^{k_r}$ is small then there is only a small fraction of optimal BB in the initial population and GEAs need many generations to converge.

4.3.2 Exponentially Scaled Representations

In the following subsection, we describe the influence of redundancy and distance distortion on the performance of genetic and evolutionary algorithms if the representations are exponentially scaled. In contrast to the previous subsection where the size of the convergence window λ_c is equal to the string length and all alleles are solved in parallel, we assume that the alleles are solved strictly in serial and $\lambda_c = 1$.

As illustrated in section 3.2 and subsection 4.2.2 we can use the domino convergence model for estimating the performance of GEAs using exponentially scaled representations. We assume that the alleles are solved strictly in serial and there are no interdependencies between the l_s alleles in an exponentially scaled BB. However, it is possible to concatenate m exponentially scaled BBs of length l_s. When using exponentially scaled representations the maximum size of BBs is $k = 1$. All schemata of order $k = 1$ that contain the best solution have higher fitness than their competitors. Therefore, it makes no sense to consider the effect of distance distortion on GEA performance when using exponentially scaled representations. Subsection 4.2.3 has shown that increasing the distance distortion d_c modifies the size of BBs k and results in interdependencies between the alleles. However, if $k_g > 1$ the domino convergence model can not be used any more, because we can then not assume that the alleles are still solved sequentially. Therefore, we assume in the following that $k = 1$ and the representation does not modify the size of BBs when mapping the phenotypes onto the genotypes.

When using redundant representations we know from equation 3.5 that

$$\frac{x_0}{n} = \frac{r}{2^{kk_r}},$$

where x_0 is the expected number of copies of the best BBs in the initial population, n is the population size, k is the size of BBs, $m' = m - 1$ with m is the number of BBs, k_r is the order of redundancy, and r is the number of genotypic BBs of size kk_r that represent the best phenotypic BB.

As we have seen in subsection 3.2.2 the probability p of making the right choice between a single sample of each BB remains constant for the l_s bits in the exponentially scaled BB if we assume that all alleles which are not yet touched by the solving process remain in their initial state. Substituting x_0/n into equation 3.11, we get

$$p = N \left(\frac{1}{\sqrt{2\frac{r}{2^{k_r}} \left(1 - \frac{r}{2^{k_r}}\right) \left(\frac{4}{3}m - 1\right)}} \right). \tag{4.9}$$

As illustrated above $k = 1$ and there are m competing BBs with l_s exponentially scaled alleles. Furthermore, with $x_0 = \frac{nr}{2^{k_r}}$ we get from equation 3.13 for the probability of error

$$\alpha = \frac{(1/p - 1)^{x_0} - (1/p - 1)^n}{1 - (1/p - 1)^n} = \frac{(1/p - 1)^{\frac{nr}{2^{k_r}}} - (1/p - 1)^n}{1 - (1/p - 1)^n}. \tag{4.10}$$

In the following, we want to approximate equation 4.10 in analogy to subsection 3.1.3. If we assume that x_0 is small we get from equation 4.10

$$\alpha \approx \left(\frac{1-p}{p}\right)^{x_0}.$$

When using the first two terms of the power series expansion of the normal distribution for approximating equation 4.9 we get

$$\alpha \approx \exp\left(x_0 \ln\left(\frac{1-x}{1+x}\right)\right),$$

where $x = 1/\sqrt{\pi \frac{x_0}{n}(1 - \frac{x_0}{n})(\frac{4}{3}m - 1)}$. Because x is a small number we can assume that $\ln(1-x) \approx -x$ and $\ln(1+x) \approx x$. Using these approximations we get

$$\alpha \approx \exp\left(-x_0 \frac{2}{\sqrt{\pi \frac{x_0}{n}(1 - \frac{x_0}{n})(\frac{4}{3}m - 1)}}\right).$$

If we approximate $\frac{x_0}{n}(1 - x_0/n)$ by x_0/n we get for the probability of error

$$\alpha \approx \exp\left(-\frac{nr}{2^{k_r - 1}\sqrt{\pi \frac{r}{2^{k_r}}(\frac{4}{3}m - 1)}}\right).$$

Simplifying this equation yields finally

$$\alpha \approx \exp\left(-\frac{2n\sqrt{r}}{\sqrt{2^{k_r}\pi(\frac{4}{3}m - 1)}}\right). \tag{4.11}$$

Using this rough approximation we appreciate that α is reduced with increasing $r/2^{k_r}$ and n. α is also reduced with a smaller number m of competing BBs. The reader should notice that α does not depend on the length l_s of an exponentially scaled BB, as we assumed that the alleles remain in their initial state as long as they are not reached by the search window.

We have seen in section 3.2 that genetic drift reduces the performance of GEAs when using exponentially scaled representations. Genetic drift can be considered by either the approximated drift model (see equation 3.19) or the stair-case drift model (see equation 3.20). By substituting the probability of error α either from equation 4.10 or from 4.11 into either equation 3.19 or 3.20 we get the average percentage of incorrect alleles $\bar{\alpha}$. For example, we can calculate the overall percentage $\bar{\alpha}$ of incorrect alleles using the approximated drift model as:

$$\bar{\alpha} = \frac{1}{l_s}\sum_{\lambda=0}^{l_s-1}\left(\left(1 - s'(\lambda\frac{\pi}{2}\sqrt{\pi m})\right)\alpha + \frac{1}{2}s'(\lambda\frac{\pi}{2}\sqrt{\pi m})\right), \tag{4.12}$$

with

$$s'(t) = \begin{cases} 0 & \text{for } t < -n\ln(2/3), \\ 1 - \frac{3}{2}\exp(-t/n) & \text{for } t > -n\ln(2/3). \end{cases}$$

With increasing l_s, more and more of the lower salient alleles are fixed randomly and $\bar{\alpha}$ is reduced.

The time to convergence for the m alleles of the same salience can be calculated by using equation 3.9 as

$$t_{conv} = \frac{\sqrt{m}}{I}\left(\frac{\pi}{2} - \arcsin\left(\frac{r}{2^{k_r-1}} - 1\right)\right). \tag{4.13}$$

As before we assume that $k = 1$. After m alleles of the same salience are converged the GEAs tries to solve the next m alleles with the next lower salience. Because each of the m BBs consists of l_s alleles with different salience and the solving process is strictly serial, we get for the overall time to convergence

$$t_{conv} = l_s\frac{\sqrt{m}}{I}\left(\frac{\pi}{2} - \arcsin\left(\frac{r}{2^{k_r-1}} - 1\right)\right), \tag{4.14}$$

The time to convergence increases linearly with the length of an exponentially scaled BB l_s. With larger $r/2^{k_r}$ the time to convergence is reduced.

4.4 Implications for the Design of Representations

In the previous section we have outlined a time-quality framework modeling the influence of representations on the performance of genetic and evolutionary algorithms.

The purpose of this section is to describe some of the important implications of the framework on the behavior of genetic and evolutionary algorithms. We show how the influence of different types of representations on the performance of GEAs can be described by using the presented framework. Based on the framework, we see that representations that overrepresent a specific solution can result in high GEA performance, but are not robust concerning the location of the optimal solution. When using exponentially scaled representations, the framework tells us that there is a trade-off between the accuracy of the solution quality and convergence time. Because representations that do not preserve the distances between the individuals affect the size of BBs, the behavior of GEAs using representations with $d_c \neq 0$ is difficult to predict.

The section starts by illustrating the effects of non-uniformly redundant representations. We have seen in section 3.1 and subsection 4.2.1 that redundancy affects the supply of BBs in the initial population. Therefore, representations that overrepresent a specific solution result in high GEA performance but are not robust. Subsection 4.4.2 illustrates that the scaling of a representation influences the dynamics of genetic search. GEAs using exponentially scaled representations deliver rough approximations of the optimal solution after a few generations, but the overall time to convergence is increased in comparison to uniformly scaled representations. Finally, we show in subsection 4.4.3 the effects of representations that change the distances between the individuals when mapping the genotypes on the phenotypes. If the distance distortion $d_c \neq 0$, the genotypic problem complexity depends on the representation and the optimization problem. Therefore, the performance of GEAs for a specific problem is difficult to predict when using representations that modify the distances.

4.4.1 Uniformly Redundant Representations Are Robust

Section 3.1 has illustrated the effects of redundancy on the performance of GEAs. The results have shown that the quality of the solutions and the time to find them can be increased if we focus the genetic search on some specific areas of the search space.

We described in subsection 4.2.2 the influence of redundancy by r denoting the number of genotypic BBs that represent the optimal phenotypic BB and k_r denoting the order of redundancy. Therefore, $r/2^{k_r}$ can be used for characterizing redundancy in an encoding.

Our framework in the previous section tells us how the solution quality (α and t_{conv}) depends on $r/2^{k_r}$. For uniform redundant representations (see equation 4.6 and 4.8)

$$\alpha = \exp\left(-\frac{ndr}{2^{k_r k_g - 1}\sigma_{BB}\sqrt{\pi m'}}\right),$$

and

$$t_{conv} = \frac{\sqrt{l}}{I}\left(\frac{\pi}{2} - \arcsin(\frac{2r}{2^{k_r k_g}} - 1)\right).$$

When neglecting the effect of genetic drift we get for exponentially scaled representations (see equation 4.11 and 4.14)

$$\alpha \approx \exp\left(-\frac{2n\sqrt{r}}{\sqrt{2^{k_r}\pi(\frac{4}{3}m - 1)}}\right),$$

and

$$t_{conv} = l_s \frac{\sqrt{m}}{I}\left(\frac{\pi}{2} - \arcsin\left(\frac{r}{2^{k_r - 1}} - 1\right)\right)$$

We see that α goes for uniformly scaled representations with $O(\exp(-r/2^{k_r}))$ and for exponentially scaled representations with $O(\exp(-\sqrt{r/2^{k_r}}))$. The time to convergence t_{conv} is reduced for both types of representations with increasing $r/2^{k_r}$. Therefore, GEA performance increases with larger $r/2^{k_r}$. As a result designing efficient representations seems to be quite an easy task.

Initially it appears that we simply have to increase $r/2^{k_r}$ and are rewarded with high performing GEAs. Therefore, we have to investigate if there are any problems associated with increasing $r/2^{k_r}$.

When using redundant representations, the order of redundancy k_r does not depend on the structure of the optimal solution. However, r depends by definition on the structure of the optimal solution. r measures how many genotypic BBs of size kk_r represent the optimal phenotypic BB of size k. On average $r_{avg} = 2^{k(k_r-1)}$ genotypic BBs represent one of the 2^k phenotypic BBs. Therefore, if $r > r_{avg}$ for some phenotypic individuals, there must also be some individuals with $r < r_{avg}$. That means if some individuals are overrepresented by a specific representation there must be others which are underrepresented.

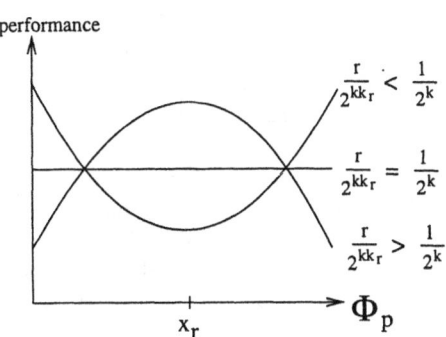

performance

$$\frac{r}{2^{kk_r}} < \frac{1}{2^k}$$

$$\frac{r}{2^{kk_r}} = \frac{1}{2^k}$$

$$\frac{r}{2^{kk_r}} > \frac{1}{2^k}$$

x_r

Φ_p

Fig. 4.2. We show how the performance of GEAs using redundant representations depends on the location of a specific individual x_r in the search space Φ_p. r determines the number of genotypes that represent a specific phenotype x_r. The performance of GEAs is independent of r if all phenotypes are uniformly represented ($r = 2^{k(k_r-1)}$ for all phenotypes). If x_r is overrepresented ($r > 2^{k(k_r-1)}$) GEAs perform better when searching for individuals similar to x_r, and worse for individuals with a larger distance to x_r. If x_r is underrepresented the situation is reversed.

We have learned from the framework that solution quality increases with increasing $r/2^{k_r}$. If we have uniform redundancy $r = 2^{k(k_r-1)}$, GEAs perform the same as without redundancy. For uniformly redundant representations, the performance of GEAs is independent of the location of the optimal solution. If a specific phenotype x_r is overrepresented $r > 2^{k(k_r-1)}$, GEAs searching for optimal solutions that are similar to x_r perform better. However, when using this representation and searching for solutions that have a large distance to x_r, GEAs perform worse. The situation is vice versa if x_r is underrepresented. We see that by increasing $r/2k_r$, we reduce the robustness of the representation. A representation is denoted to be robust if the performance of a GA is independent of the location of the optimal solution in the search space. We illustrate this behavior of redundant encodings in Figure 4.2. The Figure shows how the performance of GEAs depends on the over- or underrepresentation of the phenotype x_r.

We see that when designing representations, redundancy is helpful if the number of copies r that are given to the optimal solution x_r is above average. However, to systematically increase the number of copies of the optimal solution, it is necessary to know where the optimal solution is located in the search space Φ_p. Otherwise, if we do not know where the optimal solution can be found, it is not possible to increase the number of copies of the best solution or solutions that are similar to the best solution in a systematic way by using redundant representations. Problem-specific knowledge is necessary to increase the value of r. If we do not have any problem-specific knowledge about the structure of the problem, either non-redundant, or uniformly redundant representations, should be used. Both types of encodings guarantee that GEAs perform robustly, that means independently of the structure of the optimal solution.

4.4.2 Exponentially Scaled Representations Are Fast, but Inaccurate

In section 4.3 we examined the effect of uniformly and non-uniformly scaled representations on the performance of GEAs. We saw (see also subsection 4.2.2) that a different scaling of representations modifies the dynamics of genetic search. When using uniformly redundant representations all alleles are solved in parallel, whereas for exponentially scaled BBs the alleles are solved strictly serially. In the following we want to illustrate that GEAs using exponentially scaled representations deliver fast solutions which are inaccurate.

In our framework we have presented two different models for scaled representations. For uniformly scaled representations, we assumed that all alleles are solved in parallel, and that the size of the convergence window is the same as the string length. We get from equation 4.8 for $l = l_s m$ and $k_g = 1$

$$t_{conv}^{uniform} = \frac{\sqrt{l_s m}}{I} \left(\frac{\pi}{2} - \arcsin\left(\frac{r}{2^{k_r} - 1} - 1\right) \right).$$

When using exponentially scaled representations we use the domino convergence model and the size of the convergence window $\lambda_c = 1$. Therefore, we get from equation 4.14 for the overall time to convergence

$$t_{conv}^{exp} = l_s \frac{\sqrt{m}}{I} \left(\frac{\pi}{2} - \arcsin\left(\frac{r}{2^{k_r - 1}} - 1\right) \right).$$

In each of the m exponentially scaled BBs of size l_s the alleles are solved strictly sequentially.

We see that when using exponentially scaled representations, the first alleles are converged to the correct solution after a short time and we get a first rough approximation of the correct solution. Furthermore, the alleles are solved from the most salient to the least salient. When using exponentially scaled representations the most salient allele has the same contribution to the fitness function as all lower salient alleles together. Because the low salient alleles do not significantly change the fitness of the optimal solution, we get an acceptable approximation after a few generations.

The situation is different when examining the number of generations that are necessary until the whole string is converged. GEAs using uniformly scaled representations converge faster and find the optimal solution after $t_{conv}^{uniform} = t_{conv}^{exp}/\sqrt{l_s}$ generations. We can compare the different times to convergence

$$t^{exp} < t_{conv}^{uniform} < t_{conv}^{exp},$$

where $t^{exp} = t_{conv}^{exp}/l_s$ denotes the time after m alleles of the same salience are converged (moving the convergence window to the next lower salient allele). GEAs always need more time to completely converge when using exponentially scaled representations than when using uniformly scaled representations.

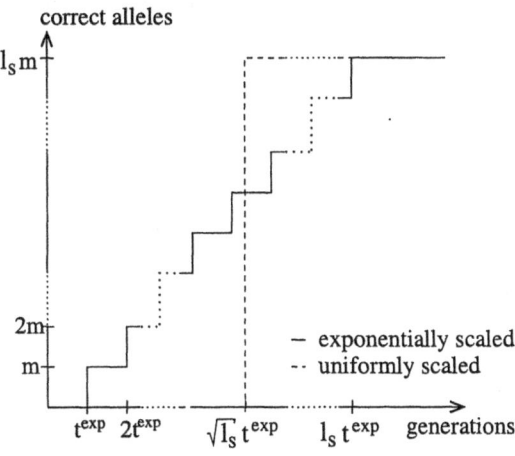

Fig. 4.3. Number of correctly identified alleles over the number of generations for GEAs using uniformly versus non-uniformly scaled representations. The number of correctly identified alleles corresponds to the accuracy of the solution. GEAs using non-uniformly scaled representations provide an inaccurate solution to the problem more rapidly, but need longer to find the exact solution.

We want to illustrate the influence of scaling on the dynamics of genetic search in Figure 4.3. The Figure shows the number of correctly identified alleles over the number of generations using uniformly scaled versus non-uniformly scaled representations. The number of correctly identified alleles is a measurement for the accuracy of the solution we get. The plots show that GEAs using non-uniformly scaled representations steadily improve the solution quality. After a few generations GEAs already provide us with a correct, but yet inaccurate solution to the problem. GEAs using uniformly scaled representations do not give us approximations after a few generations, but allow us to find the exact optimum faster.

We see that non-uniformly scaled representations rapidly deliver correct, but inaccurate solutions, whereas uniformly scaled representations do not produce early results, but give us the optimal solution faster. If we do not want to spend much time and we are not interested in high accuracy, non-uniformly scaled representations are a considerable choice. On the other hand, if we need exact solutions, we should use uniformly scaled representations. GEAs using uniformly scaled representations can find the exact optimum in a shorter length of time.

4.4.3 BB-Modifying Representations Are Difficult to Predict

Section 3.3 illustrated how the complexity of an optimization problem can be modified by using representations that do not preserve the distances between the individuals when mapping the phenotypes on the genotypes. The genotypic size of BBs k_g is determined by the distance distortion d_c (see subsection 4.2.3). If the distances are preserved by the representation ($d_c = 0$) the genotypic problem complexity is the same as the phenotypic problem complexity. However, for $d_c \neq 0$ the genotypic size of BBs k_g depends not only on the used representation but also on the specific optimization problem.

We have already illustrated in subsection 4.3.2 that when looking at exponentially scaled representations it makes no sense to consider the effect of d_c on GEA performance. The performance of GEAs is modeled using the domino convergence model, which assumes strictly serial solving of the alleles. However, if the representation modifies the size of BBs and $k_g \neq 1$, there are interdependencies between the alleles, and the domino convergence model can not be used any more. Therefore, we want to focus in the following on uniformly scaled representations.

For uniformly scaled representations, the probability of error (see equation 4.6) can be approximated as

$$\alpha = \exp\left(-\frac{ndr}{2^{k_r k_g - 1}\sigma_{BB}\sqrt{\pi m'}}\right),$$

and the time to convergence (equation 4.8)

$$t_{conv} = \frac{\sqrt{l}}{I}\left(\frac{\pi}{2} - \arcsin(\frac{2r}{2^{k_r k_g}} - 1)\right),$$

where

$$k_g = \begin{cases} k_p & \text{, if } d_c = 0, \\ k_g(f_g, f_p) \text{ ,where } 1 \leq k_g \leq l & \text{, if } d_c \neq 0. \end{cases}$$

The probability of error goes with $O(\exp(1/2^{k_g}))$, and the time to convergence t_{conv} increases with increasing k_g. However, the genotypic size of BBs k_g is influenced by the distance distortion d_c. If $d_c = 0$ the genotypic size of BBs k_g is the same as the phenotypic size k_p and the complexity of the problem is preserved when mapping the phenotypes onto the genotypes. The situation becomes different if $d_c \neq 0$. Then, the genotypic size of BBs k_g depends on the used representation (the genotype-phenotype mapping f_g), and on the optimization problem f_p (the phenotype-fitness mapping). We have illustrated in the small example shown in subsection 4.2.3 (see Figure 4.1) that predicting the performance of GEAs using representations that modify the distances between the individuals is difficult. It depends on where the optimal solution is located, on the genotype-phenotype mapping f_g, and the structure of the problem f_p.

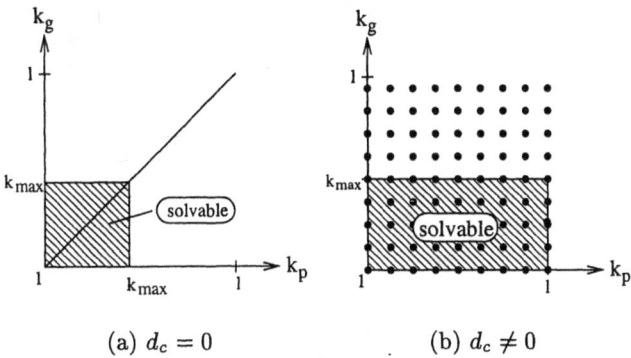

(a) $d_c = 0$ (b) $d_c \neq 0$

Fig. 4.4. We illustrate how the genotypic size of BBs k_g depends on the phenotypic size of BBs k_p when using different types of representations. If representations that preserve the distances between the individuals ($d_c = 0$) are used, easy problems up to complexity k_p can be reliably solved (Figure 4.4(a)). If $d_c \neq 0$ (Figure 4.4(b)), the genotypic size of BBs k_g does not correspond to k_p, but depends on the used representation and the optimization problem. Then, we cannot predict the genotypic size of BBs k_g without knowing f_p.

We illustrate the problem with distance modifying representations in Figure 4.4. In Figure 4.4(a) the fitness distortion is zero and the size of BBs is the same in the genotypic and the phenotypic space. If we assume that the used GEAs can solve a problem up to $k_g = k_{max}$, we can be sure that all problems with $k_p < k_{max}$ can be solved reliably. The situation becomes different if $d_c \neq 0$ (compare Figure 4.4(b)). Then, the genotypic size of BBs k_g depends not only on k_p, but also on the representation f_g itself, and the structure of the optimization problem f_p. Using a specific f_g with $d_c \neq 0$ can result for the same k_p in different k_g. Therefore, we can not predict which types of problems remain solvable and which are no longer solvable. Originally easy problems could become fully difficult and fully difficult problems could become fully easy. To predict the performance of GEAs using representations that modify the distances between the individuals when mapping the phenotypes on the genotypes is not possible if we have no knowledge about f_p.

We summarize that we can not predict the performance of GEAs using representations that modify the distances between the individuals if we have no exact knowledge about the optimization problem f_p and the used representation f_g. However, if we know that a problem is too difficult to be solved by the used GEAs, a representation with $d_c \neq 0$ can sometimes advantageously use this problem-specific information. Then, the representation modifies the genotypic size of the BBs k_g and there is a chance that the problem becomes so easy that it can be solved. When a problem is fully difficult, all representations that do not preserve BB-complexity make the problem easier and more likely to be solved by GEAs. If we have no information a priori about a problem, or if we know that a problem is easy to solve for GEAs, we strongly favor the use of representations that preserve the distances between the individuals ($d_c = 0$). These types or representations allow GEAs to reliably solve easy problems up to some complexity bound.

4.5 Summary and Conclusions

We presented in this chapter a time-quality framework for a theory based analysis and design of representations for genetic and evolutionary algorithms. The chapter started with the determinants of GEA performance. The performance of GEAs is determined by the expected quality of the solutions and the number of generations that are necessary to find them. This was followed in section 4.2 by a description of the three elements the framework consists of. We presented how redundancy, scaling, and distance distortion of a representation are measured and how they affect GEA performance. In section 4.3 we presented the main part of the chapter: the framework. Based on the work outlined in chapter 3, we showed how the probability of error α and the time to convergence t_{conv} is influenced by different types of representations. Finally, we presented in section 4.4 some implications of the framework on the design of representations.

This chapter presented a framework for a theory-guided analysis and design of representations for genetic and evolutionary algorithms. Based on the three elements of representation theory outlined in chapter 3, the framework theoretically describes how different types of representations influence GEA performance. The presented framework provides us with some important benefits. It gives us a theoretical model for a better understanding of the influence of representations on GEA performance. Furthermore, it allows us to model and predict the performance of GEAs using a specific representation for different types of optimization problems. Therefore, a theory-based use, analysis, and design of representations becomes possible by using the outlined framework.

Based on the results from chapter 3, the framework shows that redundant representations increase GEA performance, if the optimal solution is overrepresented by the representation. However, if some specific individuals are overrepresented, others remain underrepresented, and the performance depends on the structure of the optimal solution. Only uniformly redundant representations are robust concerning the structure of the optimal solution.

By modifying the scaling of a representation the dynamics of genetic search are changed. If GEAs use exponentially scaled representations the domino convergence model can be used because the alleles are solved serially according to their salience. Therefore, the most salient alleles are solved after a few generations and a rough approximation of the optimal solution is available. However, to solve all alleles, GEAs using exponentially scaled representations need a larger number of generations compared to using uniformly scaled representations.

The presented framework reveals that the distance distortion of a representation is crucial for the performance of GEAs. The distance distortion describes how well the distances between the individuals are preserved when mapping the phenotypes on the genotypes. The analysis shows that it influences the genotypic size of BBs. If the distance distortion is zero, meaning that the genotypic distances correspond to the phenotypic distances, the complexity of the problem is preserved and easy problems remain easy. If the distances are modified, the size of BBs is modified, and for predicting the resulting problem complexity, exact knowledge about the optimization problem and the used representation is necessary.

We believe that the representation framework developed in this chapter is an important step toward a more general theory of representations for GEAs. Although it is not yet completed and there are still many open questions and shortcomings, it provides us with a much better understanding of the principles of representations and allows a more theory-guided design of representations. We want to encourage researchers to use the presented framework as a basis for a more detailed and extensive investigation into representations. At the end of this long journey a theory of representations might be found which allows us to use genetic and evolutionary algorithms more powerfully and efficiently.

5. Analysis of Binary Representations of Integers

In the previous chapter we presented a framework which describes the effects of representations on the performance of GEAs. We verified the elements of the framework for problems where the phenotypes and genotypes are both bitstrings. The question is still open as to whether the framework also holds true for problems where the genotypes and phenotypes use different types of representations. This question can be answered by examining problems where the genotypes are still binary but the phenotypes are integers. Integer optimization problems are common in many real-world applications. Although the most natural way for representing integer problems is to use an integer representation, previous work has shown that by using representations with lower cardinality of the alphabet (for example binary representations) the possible number of schemata can be increased in comparison to integer strings (Goldberg, 1990b).

Consequently, researchers developed different types of binary representations for integers. The most common are binary, gray, or unary representations. Previous work has shown that these three representations have different properties and influence GEA performance differently (Caruana & Schaffer, 1988; Whitley, 1999; Whitley, Rana, & Heckendorn, 1997; Whitley, 2000à).

The purpose of this chapter is to use the framework of representations we presented in the previous chapter to explain the performance differences of selectorecombinative GEAs when using different binary representations for integers. For our investigation into the influence of representations on GEA performance, we use two integer problems, an easy one-max problem, and a difficult deceptive trap. For encoding integer problems, we use binary, gray, and unary representations.

The analysis of the unary encoding using the previously presented elements of representation theory reveals that the encoding is redundant, but does not represent the phenotypes uniformly. Therefore, the performance of GEAs depends on the structure of the optimal solution. If the good solutions are overrepresented by the encoding, GEAs perform well, whereas, if the good solutions are underrepresented, GEAs fail.

The binary encoding uses exponentially scaled alleles to represent integer values. Therefore, the convergence behavior is affected by domino convergence and genetic drift. However, the analysis shows that genetic drift only results in a reduction of GEA performance for easy problems and small populations. An investigation into modification of problem difficulty reveals that the redundant unary encoding preserves the problem complexity perfectly when mapping the phenotypes to the genotypes, in contrast to the binary and gray encoding. When encoding a fully easy phenotypic problem using the unary encoding the resulting genotypic problem is still fully easy (Deb & Goldberg, 1994). In contrast, binary and gray encoding increase the difficulty of fully easy problems and reduce the difficulty of fully difficult problems.

Although, gray encoding was designed to overcome the problems with the Hamming cliff (Schaffer et al., 1989) and to have perfect locality, an analysis of its distance distortion reveals that $d_c \neq 0$. Therefore, the difficulty of problems concerning selectorecombinative GEAs is changed. A schema analysis for the integer one-max problem reveals that the structure of BBs is preserved even worse than when using binary encoding. As a result, in comparison to the binary encoding, the difficulty of the integer one-max problem increases for selectorecombinative GAs. This result is not contradictory to the Free-Lunch theorem from Whitley (1999) and Whitley (2000a) regarding the gray encoding but confirms the results therein. The difference can be found in the used search method. We investigate the influence of gray encoding on recombination-based search approaches, whereas Whitley (1999) looks at mutation-based search methods. He basically counts the number of local optima. Because gray encoding preserves the distances between all neighboring phenotypes, problem difficulty for mutation-based search approaches remains unchanged and the number of local optima introduced by gray encoding is smaller. Therefore, the performance of mutation-based search approaches on easy problems must be higher when using gray than when using binary encodings.

After a brief presentation of the two integer problems in section 5.1, section 5.2 presents the used gray, binary and unary encodings and analyzes their properties. This is followed in section 5.3 by a theoretical comparison of the three encodings using the elements of theory presented in chapter 3. We illustrate how the unary encoding is affected by redundancy, how the exponential scaling of BBs influences the performance of the binary encoding, and how gray encoding does not preserve problem difficulty well. Based on the elements of representation theory, we are able to make theoretical predictions about GEA performance. These predictions are finally proven in section 5.4 by empirical results. The chapter ends with concluding remarks.

5.1 Two Integer Optimization Problems

In this section, we present the two integer problems we want to use for a comparison of different representations defined on binary strings.

To be able to make a fair comparison between different representations, we are not allowed to change the complexity of the problem when using different representations. Therefore, the problem must be defined on the integer phenotypes independently of the used binary representation. Using the genotypic space $\Phi_g = \{0,1\}^l$ and the phenotypic space $\Phi_p = \mathbb{N}$, the genotype-phenotype mapping is defined as

$$f_g(x_g) : \{0,1\}^l \rightarrow \mathbb{N}.$$

When assuming that the fitness function f_p assigns a real number to every individual in the phenotypic space, we get for the phenotype-fitness mapping:

$$f_p(x_p) : \mathbb{N} \rightarrow \mathbb{R}.$$

For the optimization problem itself we want to use integer-specific variations of the one-max and the fully-deceptive trap problem. Traditionally, these problems are defined on binary strings, but we want to define them in a similar way for integers. The integer one-max problem is defined as

$$f_p(x_p) = x_p,$$

and the integer deceptive trap is

$$f(x_p) = \begin{cases} x_p & \text{if } x_p = x_{p,max}, \\ x_{p,max} - x_p - 1 & \text{else,} \end{cases}$$

where $x_p \in \mathbb{R}_0^+$. The one-max problem for integer problems is a fully easy problem, whereas the integer deceptive trap should be fully difficult to solve for GEAs.

For measuring the similarity of individuals, we need to define a metric on Φ_g and Φ_p (see subsection 3.3.2). In the following we want to use the Hamming distance (Hamming, 1980) on Φ_g. Thus, the distance between two genotypes x and y of length l is defined as $d_{x,y} = \sum_{i=0}^{l-1} |x_i - y_i|$. The distance measures the number of alleles that are different in both individuals. The more bits two individuals have in common, the more similar they are. The Hamming metric is chosen with respect to the bit-flipping operator. Using this mutation operator results in an individual that has the smallest possible genotypic distance from its parent. Following the Hamming metric for the genotypes, we measure the distance between two phenotypes x and y as $d_{x,y} = |x - y|$. The distance between two individuals (they are integers) is simply the difference between both integers.

5.2 Binary String Representations

After we have defined our optimization problem in the previous section, we present possible binary representations for integers.

Before presenting the used binary representations we briefly discuss why we only use binary representations. Instead of using binary strings with cardinality $\chi = 2$, higher χ-ary alphabets could also be used for the genotypes.

In these cases, a χ-ary alphabet is used for the string of length l instead of a binary alphabet. Therefore, instead of encoding 2^l different individuals with a binary alphabet, we are able to encode χ^l different possibilities. However, Goldberg (1990b) has shown that schema processing is maximum with binary alphabets. Thus, we want to focus on representations for binary strings.

Focusing on binary representations, we still have an almost infinite number of different encodings we can use. If we use a redundancy-free encoding and want to encode 2^l phenotypes with 2^l possible genotypes, then there are $(2^l)!$ different possibilities for the genotype-phenotype mapping f_g (see subsection 3.3.4). Nevertheless, for our comparison we want to focus on the three most widely used representations defined on binary strings:

- binary representation,
- gray representation, and
- unary representation.

In contrast to the unary encoding, the binary and gray encoding allows us to encode information redundancy-free. For the encoding of s possibilities, both encodings use $\log(s)$ bits. The redundant unary encoding, however, uses $s-1$ bits for encoding s different possibilities. In the following we want to briefly review the important properties of the three different encodings:

When using the *binary representation*, each integer value $x_p \in \Phi_p = \{1, 2, \ldots, x_{p,max}\}$ is represented by a binary string x_g of length $l = \log_2(x_{p,max})$. The genotype-phenotype mapping f_g is defined as

$$x_p = f_g(x_g) = \sum_{i=0}^{l-1} 2^i x_{g,i},$$

with $x_{g,i}$ denoting the ith bit of x_g.

Since the bits in the string are exponentially scaled, we must use the domino convergence model and GAs are affected by genetic drift (see section 3.2). The bits are solved sequentially, and the low salient bits can be fixed randomly before they are reached by the solving process. Furthermore, the encoding has problems associated with the Hamming cliff (Schaffer, Caruana, Eshelman, & Das, 1989). The Hamming cliff describes the effect that some neighboring phenotypes (the phenotypes have a distance of one) are represented by completely different genotypes (the distance between the genotypes is much larger than one). Therefore, the locality of the encoding is reduced. As a result, especially mutation-based search approaches have problems when using this encoding because they rely on a high locality of the encoding. We have seen in subsection 3.3.2 that high locality is a necessary condition for a representation to preserve BB-complexity. Therefore, the ability of the binary encoding to preserve problem complexity is reduced in comparison to BB-preserving representations. However, the encoding also has some very interesting properties: It is linear, very compact and redundancy-free. For a brief example of the binary encoding, the reader is referred to Table 5.1

To overcome problems with the Hamming cliff and the different scaling of the alleles in binary strings, the *gray encoding* was developed (Caruana & Schaffer, 1988; Schaffer, Caruana, Eshelman, & Das, 1989). When using gray encoding the average contribution to the represented integer is the same for each allele in the binary string.

The gray encoded string itself can be constructed in two steps. At first, the phenotype is encoded using the binary representation, and subsequently the binary encoded string can be converted into the corresponding gray encoded string. The binary string $x \in \{0,1\}^l = \{x_1, x_2, \ldots, x_l\}$ is converted to the corresponding gray code $y \in \{0,1\}^l = \{y_1, y_2, \ldots, y_l\}$ by the mapping $\gamma : \mathbb{B}^l \to \mathbb{B}^l$:

$$y_i = \begin{cases} x_i & \text{if } i = 1, \\ x_{i-1} \oplus x_i & \text{otherwise,} \end{cases}$$

where $\oplus$ denotes addition modulo 2. The decoding of a gray encoded string is as follows:

$$x_i = \bigoplus_{j=1}^{i} y_j,$$

for $i = \{1, \ldots l\}$. As mentioned before, a gray encoded string has the same length as a binary encoded string and the encoding is redundancy-free. Furthermore, the representation overcomes the problems with the Hamming cliff. Every two phenotypes that have phenotypic distance 1 are encoded by genotypes that have the Hamming distance 1. Therefore, the locality is perfect. This property gives gray encoding an advantage over the binary encoding when using mutation-based operators like the bit-flipping operator (Whitley, 1999). As before, Table 5.1 illustrates for a small example the use of the gray encoding.

Finally, the *unary encoding* can be used for encoding integers. A phenotype x_p is encoded by the number of ones u in the corresponding genotype x_g. With the length of the string l and $x_{g,i}$ as the ith bit of x_g we get

$$x_p = f_g(x_g) = \sum_{i=0}^{l-1} x_{g,i}.$$

In contrast to the binary and gray encoding, a string of length $l = s - 1$ is necessary for representing s different phenotypes x_p. Therefore, the genotype-phenotype mapping is no longer a one-to-one mapping and we incorporate redundancy. When encoding the phenotypic space $\Phi_p = \{0, 1, \ldots, l\}$ using a unary string of length l, each of the $l + 1$ phenotypes $x_p \in \Phi_p$ is represented by $\binom{l}{x_p}$ different genotypes x_g. The number of genotypes that represent x_p is illustrated for $l = 7$ in Figure 5.1. Some phenotypes are represented by only one genotype ($x_p = 0$ and $x_p = 7$), whereas $x_p = 3$ and $x_p = 4$ are represented by 35 genotypes. The encoding is uniformly scaled and it preserves the complexity of a problem. Using the concepts of locality and distance distortion

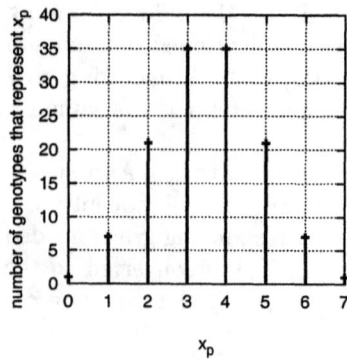

Fig. 5.1. Redundancy of the unary encoding for $l = 7$

for the unary encoding is difficult and meaningless, as the encoding is redundant. For every two neighboring phenotypes, two corresponding neighboring genotypes exist. However, there are also some corresponding genotypes that are not neighbors. We can illustrate this with a small example. $x_p = 2$ and $y_p = 3$ are phenotypically neighbors. The corresponding genotypes $x_g = 0011$ and $y_g = 0111$ are neighbors, too. However, other corresponding genotypes like $x_g = 0011$ and $y_g = 1110$ are not neighbors. The situation is similar for the distance distortion. To every phenotypic distance there is a corresponding genotypic distance, but due to redundancy there are also distance that do not correspond to each other.

Finally, we want to give a brief example for the three different types of encodings. Table 5.1 illustrates how the integers $[0, 7]$ can be represented by the binary, gray, and unary encoding. Because each integer number x_p is represented by $\binom{7}{x_p}$ different unary strings we leave for illustrative purposes some unary strings out of the example.

Table 5.1. An example for using binary, gray, and unary encodings

x_p	binary	gray	unary
0	000	000	0000000
1	001	001	0000001, 0000010, ...,0100000, 1000000
2	010	011	0000011, 0000101, ...,1010000, 1100000
3	011	010	0000111, 0001011, ...,1101000, 1110000
4	100	110	0001111, 0010111, ...,1110100, 1111000
5	101	111	0011111, 0101111, ...,1111010, 1111100
6	110	101	0111111, 1011111, ...,1111101, 1111110
7	111	100	1111111

5.3 A Theoretical Comparison

In the following section, we use the framework of representations we presented in chapter 4 to theoretically compare the performance of the binary, gray and unary representation. The framework allows us to make predictions about the performance of GEAs which will be empirically proven in section 5.4. In particular we illustrate the effects of non-uniform redundancy on the unary encoding, how the genetic search process is prolonged by the effect of exponentially scaled BBs for the binary encoding, and how the complexity of the problem is not well preserved by the gray encoding.

5.3.1 Redundancy and the Unary Encoding

We know from section 3.1 that redundancy reduces the performance of GEAs if the encoding underrepresents the good solutions. Furthermore, we know from the previous section (see Figure 5.1) that the unary representation is a redundant encoding. Therefore, to predict the performance of GEAs using the unary encoding for our two integer problems illustrated in section 5.1, we have to investigate if the good solutions are underrepresented.

Before discussing the performance of GEAs using the unary encoding, we discuss some of the advantageous properties of the encoding. The encoding is linear and it does not change problem complexity when mapping the phenotypes to the genotypes. Therefore, easy problems remain easy and difficult problems remain difficult (Goldberg, 1989b; Deb & Goldberg, 1993; Deb & Goldberg, 1994). The only real handicap of the encoding seems to be the non-uniform redundancy.

In the following, we want to predict the performance of GEAs using the unary encoding for the one-max and deceptive trap problem from section 5.1. We assume that $|\Phi_p| = s$. Therefore, for both problems, the integer one-max and the integer deceptive trap problem, the length of the unary encoded string is $l = s-1$. Thus, 2^{s-1} different genotypes only encode s different phenotypes. $\log_2(s)$ Bits of information content (see subsection 3.1.1) are encoded by $s-1$ bits. The reader should notice that when using a redundancy-free encoding like the gray or binary encoding we would need only $\log_2(s)$ bits. Therefore, we get for the order of redundancy (see subsection 3.1.1)

$$k_r = \frac{s-1}{\log_2(s)} \text{ , for } s > 1.$$

On average, k_r bits of a unary encoded bitstring are necessary for encoding one Bit of information content. This means, on average 2^{k_r} different genotypes represent only two different phenotypes. When using the unary encoding for the integer one-max or deceptive trap defined in subsection 5.1 the optimal phenotype $(x_p = l)$ is represented by only one genotype (a string of only ones). Therefore, the number of genotypic BBs that represent the best phenotypic BB is $r = 1$.

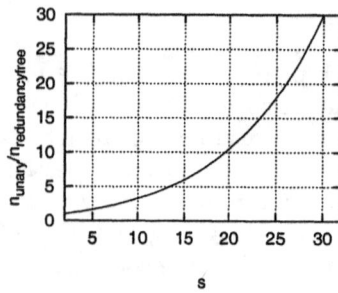

Fig. 5.2. Necessary population size n_{unary} when using unary encoding.

From equation 3.8, we get for the population size $n = O\left(\frac{2^k r}{r}\right)$ when using redundant encodings. Therefore, the necessary population size n when using unary encoding for the two integer problems presented in section 5.1 is increased in comparison to an encoding with uniform or no redundancy as

$$n_{unary} = n_{redundancyfree} * 2^{\frac{s-1}{\log_2(s)}-1}.$$

The equation shows that with increasing string length $l = s - 1$, the necessary population size when using unary encoding increases exponentially. This effect is illustrated in Figure 5.2. For even small problems the necessary population size n_{unary} is unreasonably high. Obviously the use of the unary encoding results for the proposed integer one-max and deceptive trap problem in a low GA performance.

However, we know from section 3.1 that the performance of non-uniformly redundant encodings depend on the specific problems they are used for. GEAs using redundant encodings only perform badly if the good solutions are underrepresented. Therefore, we want to investigate in the remaining subsection for which problems the unary encoding performs well.

For both problems, the integer one-max and deceptive trap problem, the optimal solution $x_p = l$ is strongly underrepresented by only one genotype. The encoding performs badly for our two test problems, but in general the optimal solution is not necessarily a string with only ones (or only zeros), and therefore, GEAs could perform well using unary encoding for other problems. If we use a different integer fitness function f_p (that means we solve a different problem) and the optimal solution would be, for example, $l/2$, then the optimal solution would be strongly overrepresented and GEAs using this representation would be very effective.

We want to illustrate the problem of the unary encoding with underrepresented solutions in Figure 5.3 more clearly. On average

$$\gamma_{avg} = \frac{1}{s} \sum_{x_p=0}^{s-1} \binom{s-1}{x_p}$$

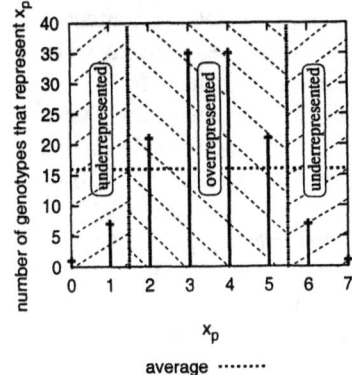

average ········

Fig. 5.3. Areas of over- and underrepresented phenotypes when using unary encoding

genotypes represent one phenotype. Therefore, if $\binom{s-1}{x_{p,opt}} < \gamma_{avg}$ the optimal solution $x_{p,opt}$ is underrepresented and the performance of the GA is reduced. However, if $\binom{s-1}{x_{p,opt}} > \gamma_{avg}$ the optimal solution is overrepresented and redundancy helps the GA in finding the optimum.

We see that the performance of a GA using unary encoding depends on the structure of the optimization problem we want to solve. The unary encoding can be a very good choice due to its many advantageous properties if the optimal solution is not strongly underrepresented. However, GEAs can never unfold their maximum power if the use of the unary encoding results in an underrepresentation of the good solutions.

5.3.2 Scaling, Modification of Problem Difficulty, and the Binary Encoding

In section 3.2, we illustrated how the search process of GEAs is prolonged by using non-uniformly scaled encodings. If integers are encoded using the binary representation, domino convergence occurs and GEAs are affected by genetic drift. As a result, the probability of GEA failure α is increased for small population sizes n. However, scaling only affects GEAs for the easy one-max problem because the optimal solution can be found even with small populations. For more difficult problems, like the deceptive trap, the necessary population size is large enough that no drift occurs.

Therefore, considering the aspect of scaling, the performance of GEAs using the binary encoding is expected to be slightly reduced for the easy one-max problems, but not for the more difficult deceptive trap problem. Here the necessary population size is large enough that no drift occurs. For further theoretical models describing the effects of exponentially scaled encodings, we would refer the interested reader to the results from Thierens (1995), Thierens, Goldberg, and Pereira (1998), or Lobo, Goldberg, and Pelikan (2000).

The performance of GEAs using binary encoding is not only affected by the exponential scaling of BBs, but also by problems associated with the Hamming cliff (Caruana & Schaffer, 1988; Caruana, Schaffer, & Eshelman, 1989; Schaffer, Caruana, Eshelman, & Das, 1989). Binary encodings have the effect that genotypes of some phenotypical neighbors are completely different. As an example we can chose the phenotypes $x_p = 15$ and $y_p = 16$. Both individuals have distance of one, but the resulting genotypes $x_g = 01111$ and $y_g = 10000$ have the largest possible genotypic distance $d_{x,y} = 5$. As a result the locality of the binary representation is partially low. We have found in subsection 3.3.2 that an encoding preserves the difficulty of a problem if it has perfect locality and if it does not modify the distances between the individuals. The non-redundant binary encoding does not preserve the distances and therefore we expect that the binary encoding changes the structure and complexity of the BBs. How exactly the structure of the BBs is changed will be measured in the following subsection (compare Table 5.3).

5.3.3 Modification of Problem Difficulty and the Gray Encoding

In section 3.3, we illustrated the effects of representations that do not preserve the distances between the individuals when mapping the genotypes to the phenotypes. The non-redundant gray encoding has high locality ($d_m = 0$) but changes the distances between the individuals ($d_c \neq 0$). Therefore, the complexity of BBs when mapping phenotypic integers on genotypic bitstrings is modified. As a result, fully easy integer problems remain not fully easy for selectorecombinative GEAs. Only when using mutation-based search approaches does the complexity of the problem remain unchanged when mapping the phenotypes on the genotypes.

We have noticed that in contrast to the binary encoding, the gray encoding is not affected by scaling, nor has it problems with the Hamming cliff. In addition, the encoding has perfect locality concerning small changes which makes it very attractive for mutation-based search approaches. Every neighbor of a phenotype is also a neighbor of the corresponding genotype. This performance advantage of gray encoding in comparison to binary encoding has already been described in other work (Whitley, Rana, & Heckendorn, 1997; Rana & Whitley, 1997; Whitley & Rana, 1997; Rana & Whitley, 1998; Whitley, 1999; Whitley, 2000a; Whitley, 2000b). This work formulated a Free-Lunch theorem for the use of gray encoding and mutation-based search approaches. GEAs using mutation as the main search operator perform better on easy problems (these are the problems which we are interested in) when using gray encoding than when using binary encoding. The proof actually shows that the number of local optima introduced by using gray encoding is smaller than when using binary encoding. This theoretical proof can be confirmed by using the proposed framework about representations, and combining the results about locality and distance distortion from section 3.3 with the statements about problem difficulty from section 2.3. Gray encoding is an encoding with perfect locality concerning small changes $(d_m = 0)$. All phenotypes with distance $d^p = 1$ are also neighboring genotypes $(d^g = 1)$. Therefore, the difficulty of a problem remains unchanged if search operators are used that only perform a small step in the search space. A mutation step results in an individual where the phenotypic offspring has the same distance to its parent as the genotypic offspring. For mutation operators, the distance structure remains the same for the genotypes and phenotypes. Therefore, the difficulty of problems remains unchanged when using mutation-based search. As a result, easy problems and problems of bounded difficulty are easier to solve when using mutation-based search with gray encoding than with binary encoding.

However, in this work we focus on crossover-based and not mutation-based search methods. Therefore, the correct method to measure problem difficulty is to use schema analysis (compare subsection 2.3.2). If the complexity of schemata, and not the neighborhood structure, is modified, then the performance of selectorecombinative GAs changes. The following analysis of the schemata fitness reveals for the integer one-max and deceptive trap problem that in comparison to the binary encoding, gray encoding does not preserve the complexity of BBs as well. This leads to a lower GA performance.

To investigate how well BB-complexity is preserved, we analyze the fitness of the schemata for a 3-bit problem $(s = 2^3 = 8)$ using gray versus binary encoding. In Table 5.2 we present the binary and gray encoded genotypes, and the resulting fitness values for the integer one-max and deceptive trap problem.

Table 5.2. Using binary and unary encoding for an integer one-max and deceptive trap problem $(s = 8)$. The resulting length of the genotypes $l = 3$.

genotype x_g	binary	000	001	010	011	100	101	110	111
	gray	000	001	011	010	110	111	101	100
phenotype x_p	integer	0	1	2	3	4	5	6	7
fitness	$f_{onemax}(x_p)$	0	1	2	3	4	5	6	7
	$f_{deceptive}(x_p)$	6	5	4	3	2	1	0	7

In Table 5.3, we present the average fitness of the schemata for the integer one-max and the deceptive trap problem using binary or gray encoding for $l = 3$. Reviewing problem complexity reveals that the problem is fully deceptive if all schemata of lower order containing the global optimum are inferior to their competitors (Deb & Goldberg, 1994). Analogously, the problem is fully easy if all schemata containing the global optimum are superior to their competitors.

The analysis shows that for the fully easy integer one-max problem with binary encoding all schemata containing the global optimum $x_g = 111$ are superior to their competitors. Therefore, the fully easy integer one-max problem remains fully easy, and the binary encoding preserves the difficulty of this problem well. However, the schema analysis for the gray encoding reveals that the schemata containing the global optimum $x_g = 100$ are not always superior to their competitors. Therefore, the problem is not fully easy anymore, and the gray encoding does not preserve the easiness of the one-max problem.

Table 5.3. Schemata fitness for the integer one-max and deceptive trap problem using binary versus gray encoding. The one-max problem remains fully easy when using the binary representation. Using gray encoding makes the problem more difficult as some of the high quality schemata have the same fitness as the misleading schemata. The situation for the deceptive trap is the opposite one. The fully difficult deceptive trap becomes easier to solve when using gray encoding.

		order	3	2			1			0
integer one-max problem (s = 8) — binary	schema		111	11*	1*1	*11	**1	*1*	1**	***
	fitness		7	6.5	6	5	11	4.5	5.5	3.5
	schema			01*	0*1	*01	**0	*0*	0**	
	fitness			2.5	2	3	3	2.5	1.5	
	schema			10*	1*0	*10				
	fitness			4.5	5	4				
	schema			00*	0*0	*00				
	fitness			0.5	1	2				
integer one-max problem (s = 8) — gray	schema		100	10*	1*0	*00	1**	*0*	**0	***
	fitness		7	6.5	5.5	3.5	5.5	3.5	3.5	3.5
	schema			11*	1*1	*11	0**	*1*	**1	
	fitness			4.5	5.5	3.5	1.5	3.5	3.5	
	schema			01*	0*1	*01				
	fitness			2.5	1.5	3.5				
	schema			00*	0*0	*00				
	fitness			0.5	1.5	3.5				
integer deceptive trap problem (s = 8) — binary	schema		111	11*	1*1	*11	**1	*1*	1**	***
	fitness		7	3.5	4	5	4	2.5	2.5	3.5
	schema			01*	0*1	*01	**0	*0*	0**	
	fitness			3.5	4	3	3	3.5	4.5	
	schema			10*	1*0	*10				
	fitness			1.5	1	2				
	schema			00*	0*0	*00				
	fitness			**5.5**	5	4				
integer deceptive trap problem (s = 8) — gray	schema		100	10*	1*0	*00	1**	*0*	**0	***
	fitness		7	3.5	4.5	6.5	2.5	4.5	4.5	3.5
	schema			11*	1*1	*11	0**	*1*	**1	
	fitness			1.5	0.5	2.5	4.5	2.5	2.5	
	schema			01*	0*1	*01				
	fitness			3.5	4.5	2.5				
	schema			00*	0*0	*00				
	fitness			5.5	4.5	2.5				

The schemata analysis of the integer trap problem reveals that due to problems with the Hamming cliff the problem remains not fully deceptive when using the binary encoding. Some of the schemata containing the global optimum $x_g = 111$ are superior to their competitors (*11 and **1). However, when using gray encoding even more schemata containing the global optimum are not inferior to their competitors (1*0, *00, *0*, **0). The phenotypically fully difficult problem is not fully difficult anymore.

In contrast to the binary and gray encoding, the unary encoding preserves the complexity of BBs. The integer one-max problem remains fully easy, and the integer deceptive trap problem remains fully deceptive when using unary encoding (Deb & Goldberg, 1993; Deb & Goldberg, 1994). The binary and gray encoding do not preserve the distances between the individuals and therefore change the complexity of integer problems.

5.4 Empirical Results

In this section we present an empirical verification of the performance differences between the three different representations we illustrated in the previous section.

We compare the performance of a GA using binary, gray, and unary encodings for the integer one-max and deceptive trap problems as defined in section 5.1. We performed 250 runs for each problem instance and each run was stopped after the population was fully converged. That means that all individuals in the population are the same. For the integer one-max problem we used uniform crossover, and for the integer deceptive trap we used two-point crossover. As selection method we used tournament selection without replacement of size two. We used no mutation as we wanted to focus on the influence of representations on selectorecombinative GEAs.

The Figures 5.4, 5.5, 5.6, and 5.7 present results for the integer one-max problem, and the Figures 5.8 and 5.9 for integer deceptive trap problems. The plots show for different representations the proportion of correct BBs at the end of the run (left) and the run duration t_{conv} (right) with respect to the population size n. For the one-max problem, we concatenated 20 problems (BBs) of order 2 ($s = 2^2 = 4$, see Figure 5.4), 3 ($s = 8$, Figure 5.5), 4 ($s = 16$, Figure 5.6), and 5 ($s = 32$, Figure 5.7)[1]. The fitness of an individual is calculated as the sum of the fitness of the 20 concatenated BBs. Because large integer deceptive traps are not solvable by GAs in a reasonable time, we only present results for the deceptive trap problem of order 2 ($s = 4$, Figure 5.8), and 3 ($s = 8$, Figure 5.9). Using binary or gray encoding results for the order 2 problems in a string length $l = 40$, for order 3 in $l = 60$, for order 4 in $l = 80$, and for order 5 in $l = 100$. When using unary encoding we need $20 * 3 = 60$ bits for order 2, $20 * 7 = 140$ bits for order 3, $20 * 15 = 300$ bits for order 4, and $20 * 31 = 620$ bits for order 5 problems.

Due to the problems of the unary encoding with redundancy, which result in an underrepresentation of the optimal solution, GAs using unary encoding perform increasingly badly with increasing problem size. Therefore,

---

[1] The order r of a problem is defined as $r = \log_2 s$ and describes the length of the corresponding binary or gray encoded string.

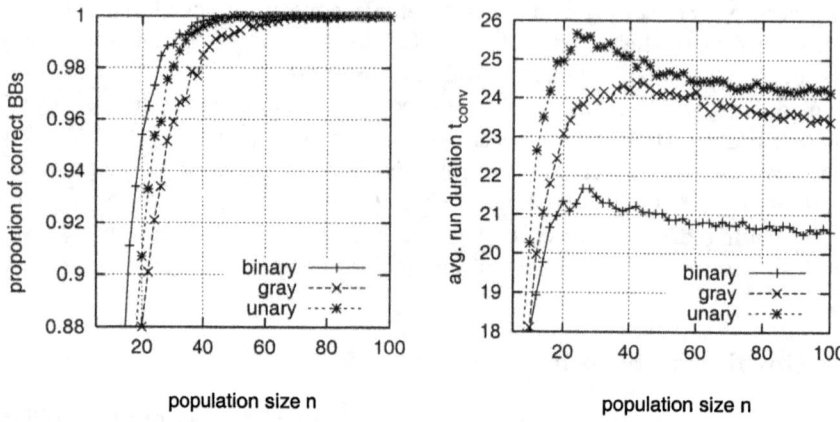

Fig. 5.4. Integer one-max problem of order 2. We concatenated $m = 20$ BBs and the size of the search space $|\Phi_p| = 2^2 = 4$. We show the average proportion of correct BBs at the end of run (left) and the average length of the runs (right). Due to the low complexity of the problems all three representations perform about the same. However, binary encoding is much faster in finding the good solutions.

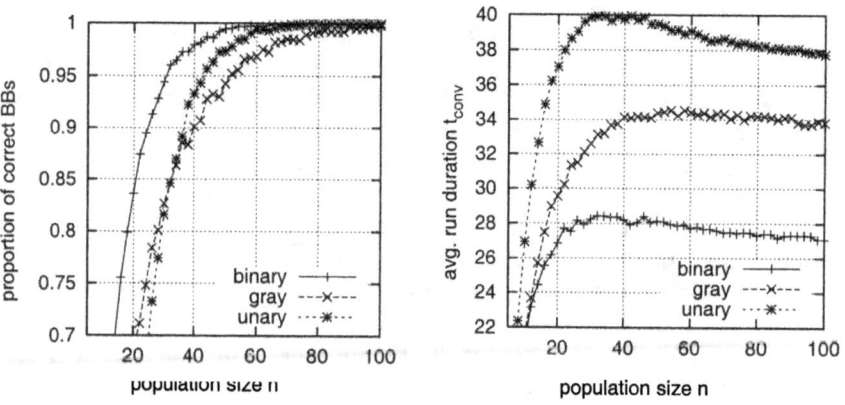

Fig. 5.5. Integer one-max problem of order 3. We concatenated $m = 20$ BBs and the size of the search space $|\Phi_p| = 2^3 = 8$. We show the average proportion of correct BBs at the end of run (left) and the average length of the runs (right). Binary encoding performs the best. Because the optimal solutions are underrepresented GAs using the unary encodings perform worse than when using gray encoding for small population sizes.

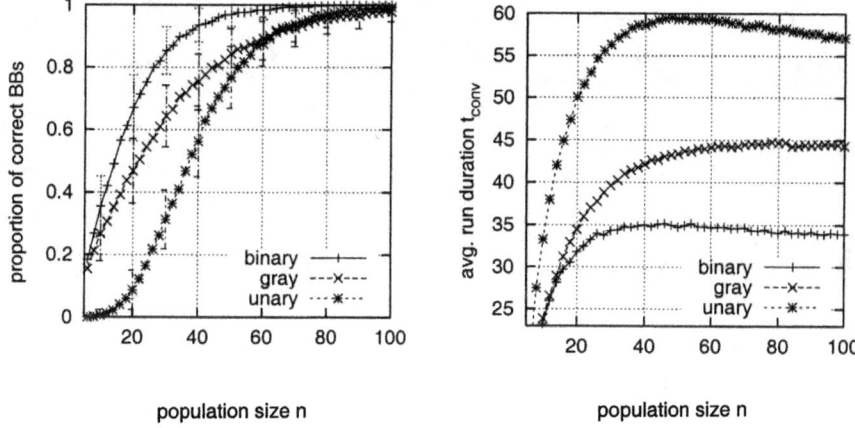

Fig. 5.6. Integer one-max problem of order 4. We concatenated $m = 20$ BBs and the size of the search space $|\Phi_p| = 2^4 = 16$. We show the average proportion of correct BBs at the end of run (left) and the average length of the runs (right). Because the binary encoding preserves BB-complexity better than gray encoding a GA using binary representations performs best. Because of problems with redundancy the unary encoding performs worst and needs the most fitness evaluations. The error bars indicate the standard deviation of some results.

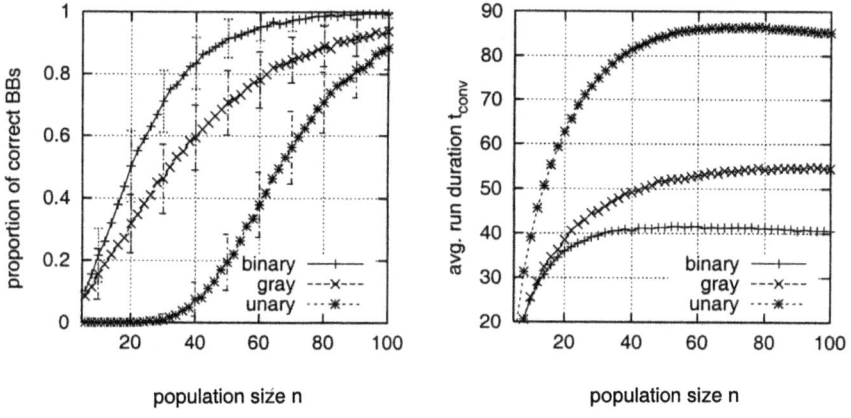

Fig. 5.7. Integer one-max problem of order 5. We concatenated $m = 20$ BBs and the size of the search space $|\Phi_p| = 2^5 = 32$. We show the average proportion of correct BBs at the end of run (left) and the average length of the runs (right). As before, binary encoding perform best. It becomes obvious that with increasing problem size GEAs using unary encoding have increasing difficulty in finding the good solutions. Furthermore, the performance differences between the binary and gray encoding become larger with increasing problem size. The error bars indicate the standard deviation of some results.

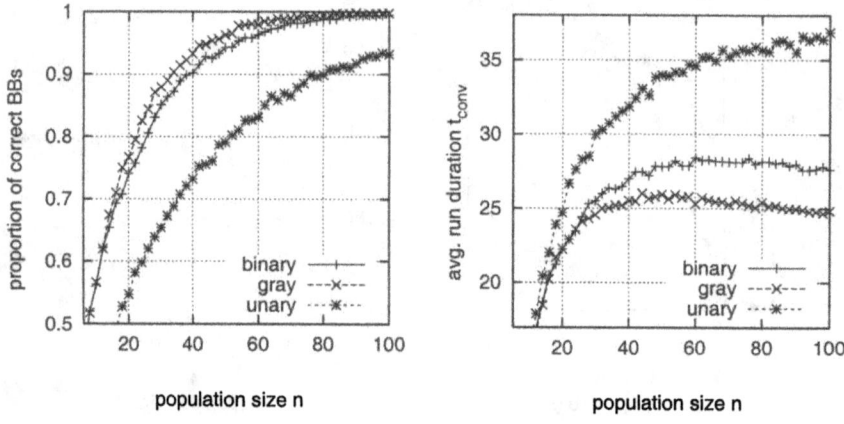

Fig. 5.8. Integer deceptive trap problem of order 2. We concatenated $m = 20$ BBs and the size of the search space $|\Phi_p| = 2^2 = 4$. We show the average proportion of correct BBs at the end of run (left) and the average length of the runs (right). Gray encoding performs slightly better than binary encoding as it preserves the structure of the BBs worse. Because unary encoding strongly underrepresents the optimal solution it performs the worst.

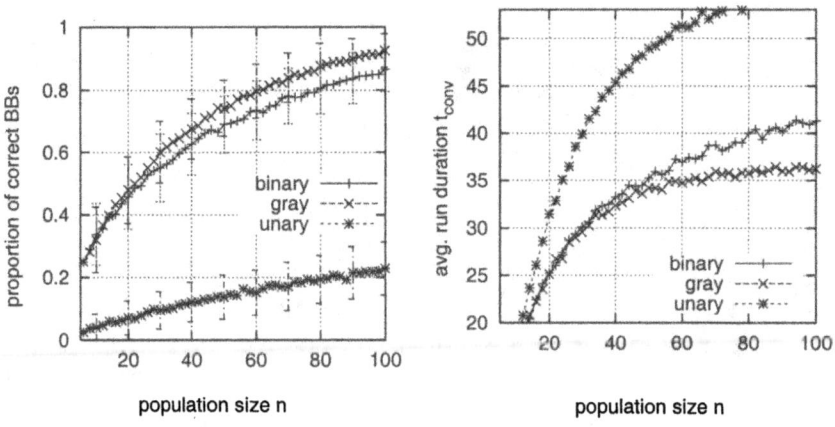

Fig. 5.9. Integer deceptive trap problem of order 3. We concatenated $m = 20$ BBs and the size of the search space $|\Phi_p| = 2^3 = 8$. We show the average proportion of correct BBs at the end of run (left) and the average length of the runs (right). Gray encoding performs significantly better than binary encoding as it makes fully difficult problems easier to solve. Unary encoding fails as it has problems with redundancy. The error bars indicate the standard deviation of some results.

for one-max problems of order more than three the GA performance is significantly worse than when using gray or binary encoding. Although, the one-max problem remains fully easy, GEA performance is reduced because the optimal solution is strongly underrepresented. Only for the almost trivial one-max problem of order 2 or 3 has the unary encoding a comparable performance. The plots nicely illustrate that only for small one-max problems the benefits from the preservation of BB-complexity can compensate the performance reduction caused by the underrepresentation of the optimal solution. For deceptive traps of order more than 2, unary encoding fails completely because the problem remains fully difficult and the optimal solution is underrepresented. Furthermore, the plots show that due to the preservation of BB-complexity, a GA using unary encoding performs in comparison to gray or binary encoding relatively better for the easy one-max than for the deceptive trap. The failure of the encoding for the deceptive trap can be better understood if we recognize that an order 3 problem results in a fully deceptive BB of length $l = 7$. This problem is only solvable with much larger population sizes.

As expected, the gray encoding performs worse than the binary encoding for the one-max problem, and better for the deceptive trap problem. Because for the integer one-max and deceptive trap problem the gray encoding preserves BB-complexity less than the binary encoding, the fully easy integer one-max problem becomes more difficult to solve, whereas the fully difficult deceptive trap is easier to solve for a GA using the gray encoding.

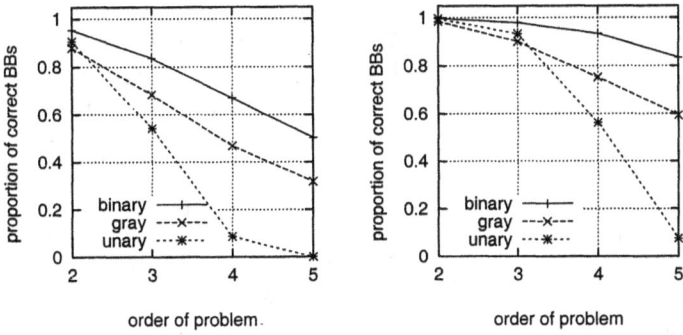

Fig. 5.10. Proportion of correct BBs at the end of the run versus the order of the problem for a population size of 20 (left) and 40 (right) for binary, gray, and unary encoding. The figures are plotted for the integer one-max problem. It can be seen that with increasing order of the problem the performance of the unary representation decreases exponentially. When using gray or binary encoding the performance of the GA declines much less and almost linearly with increasing problem size.

Finally, the influence of exponentially-scaled BBs on the performance of GEAs using the binary encoding becomes clear for the one-max problem. For small population sizes, genetic drift has a larger impact on the solution quality. Therefore, for the easy one-max and small population sizes, GAs using binary encoding perform only slightly better than gray encoding. For larger population sizes, however, the effect of genetic drift is reduced and GAs using binary representation perform relatively better.

We see that the empirical results verify nicely the theoretical predictions from the previous section. Figure 5.10 summarizes some of the results for the one-max problem and shows the proportion of correct BBs at the end of the run over the order of the problem. Due to the underrepresentation of the optimal solutions the performance of a GA using unary encoding drops approximately exponentially with increasing problem size. A GA using binary encoding performs best in comparison to gray and unary encoding, as the exponential scaling of BBs affects a GA only for small populations, and the encoding preserves BB-complexity better than the gray encoding. As a result, the easy one-max problem remains easier with the binary encoding than with the gray encoding. Although gray encoding preserves BB-complexity worst, it still significantly outperforms unary encoding which fails for the one-max and deceptive trap due to the underrepresentation of the optimal solution.

5.5 Conclusions

Section 5.1 started this chapter by presenting two integer problems. The integer one-max problem represents a fully easy problem, whereas the integer fully deceptive trap is an example of a fully difficult problem. We use both problems as test problems for comparing the performance of different integer representations. In section 5.2, we presented the binary, gray and unary encoding. These encodings are common representations for integer problems. We reviewed how they encode integers, and illustrated their important properties. This is followed in section 5.3 by a theoretical comparison of the expected performance of GEAs using the three different representations. We showed that using the unary encoding reduces GEA performance as the optimal solution is underrepresented. Therefore, the necessary population size for solving the two integer problems is increased. Using binary encoding results in a more compact, redundancy-free representation, but the alleles are exponentially scaled. Therefore, genetic drift occurs for small and easy problems and larger population sizes are necessary. Furthermore, due to problems with the Hamming cliff, the binary encoding does not preserve BB-complexity well. As a result the performance of GEAs is reduced in comparison to BB-preserving encodings for fully easy problems, and increased for fully difficult problems. For the easy integer one-max problem the situation becomes even worse with GEAs using gray encoding. Although gray encoding was developed to overcome the problems with the Hamming cliff, and the exponential scaling of BBs, an analysis of the average fitness of the schemata for the integer one-max problem shows that the encoding preserves BB-complexity even less than binary encoding. Thus, a decrease in performance is unavoidable. However, for the integer deceptive trap we expect better performance using gray instead of binary encoding. To verify the theoretical predictions made in section 5.3, we performed an empirical investigation into the performance of GAs using the different encodings in section 5.4. The results confirmed the predictions.

Throughout the entire chapter, we used the representation framework from chapter 4 for the analysis of binary representations for integers. The analysis has shown that the pieces of representation theory can effectively be used for predicting the performance of GEAs. We were able to explain the differences in performance of the binary, gray, and unary representations by using the outlined theory about redundant encodings, non-uniformly scaled BBs, and representations that modify problem difficulty. In particular, we gained the following insights:

We have seen that the binary encoding has non-uniformly scaled BBs. However, the influence of exponentially scaled BBs on GEA performance can be neglected as it only affects the performance of GEAs for easy problems and small population sizes. Although, the time to convergence is increased from $O(\sqrt{l})$ to $O(l)$, the negative effects of the exponential scaling of BBs on GEA performance is an effect we can easily overcome by using larger population sizes. This is different with the effect of redundancy on the performance of GEAs. Although, the redundant unary encoding has many desirable properties, GEAs using unary encoding fail for the integer one-max and deceptive trap. Because the optimal solution is strongly underrepresented for these two types of problems, GEAs can barely recover even with larger population sizes. Therefore, GEAs using the unary encoding perform significantly worse in comparison to GEAs using the non-redundant binary or gray encoding. Finally, the investigation of the preservation of problem difficulty has revealed for the integer one-max and deceptive trap problem that the unary encoding preserves BB-complexity perfectly, the binary encoding preserves it slightly worse, and the gray encoding the worst. Thus, for selectorecombinative GAs using gray encoding, the integer one-max problem become more difficult in comparison to binary encoding, whereas the difficult deceptive trap problem become easier.

To give a final recommendation for selectorecombinative GEAs is difficult. Both encodings, the binary and the gray encoding, change the distances between the individuals and therefore change the complexity of the optimization problem. Thus, the resulting problem difficulty depends not only on the used representation but also on the considered optimization problem (compare subsection 4.4.3 and 4.2.3). We have seen that some easy problems like the integer one-max problem are easier using the binary encoding than using the gray encoding. However, there will be other easy problems that will become easier using the gray encoding.

When using mutation-based GEAs instead of crossover-based GAs, the gray encoding is the best choice (Whitley, 1999). For this type of search process, the schemata analysis is less meaningful (see subsection 2.3.1) and the high locality of the gray encoding allows an effective local search. Problem difficulty for mutation-based search approaches remains unchanged because gray encoding preserves the distances between all neighboring individuals. As a result, the performance of mutation-based search approaches on easy problems, and problems of bounded complexity, is higher when using gray rather than binary encodings.

6. Analysis of Tree Representations

In the previous chapter, we illustrated that our framework modeling the influence of representations on the performance of GEAs not only works for binary phenotypes, but also for problems where the phenotypes are integers. However, it is possible to go one step further and to look at problems where the phenotypes and the genotypes use completely different types of representations. One example for these type of problems are tree optimization problems. Trees are a special type of graph. Representations for trees must incorporate the additional restriction of a graph to be a tree. Therefore, if the genotypes are still binary strings, there is a large semantic gap between tree structures and bitstrings. In contrast to general network problems, where a representation simply has to indicate which links are used for the graph, no natural or intuitive "good" tree representations exist which are accessible for GEAs. As a result, researchers have proposed a variety of different tree representations with different properties. However, up till now no theory-based analysis exists about how GEA performance is influenced by the different types of tree representations.

The purpose of this chapter is to fill this gap and to analyze, based on the time-quality framework from chapter 4, the influence of some of the most widely used tree representations on GEA performance. We use the existing theory about redundant representations, exponentially scaled BBs, and encodings that modify problem difficulty to predict GEA behavior. Because analyzing all known tree representations is beyond the scope of this work, we focus on three most widely used tree representations: Prüfer number encoding (Prüfer, 1918), link and node biased encoding (Palmer, 1994), and characteristic vector encoding (Celli, Costamagna, & Fanni, 1995; Berry, Murtagh, McMahon, & Sugden, 1997; Ko, Tang, Chan, Man, & Kwong, 1997; Dengiz, Altiparmak, & Smith, 1997c; Dengiz, Altiparmak, & Smith, 1997b; Dengiz, Altiparmak, & Smith, 1997a; Berry, Murtagh, McMahon, Sugden, & Welling, 1999; Premkumar, Chu, & Chou, 2001). Analyzing these three representations shows that Prüfer numbers have low locality, that the link and node biased encoding is not uniformly redundant, and that the redundant characteristic vector encoding is affected by stealth mutation.

The chapter is structured as follows. In the following, we introduce the tree design problem and develop some basic requisites for graph problems.

This is followed in section 6.2 by an investigation into the Prüfer number encoding. It concentrates on the Prüfer numbers' missing high locality which is necessary for an encoding to preserve the structure of the BBs. In section 6.3 we show that the redundant link and node biased encoding is biased towards stars as long as a node-specific bias is used and biased towards the minimum spanning tree if the biases are small. Finally, section 6.4 presents another redundant encoding, namely the characteristic vector encoding. This encoding is uniformly redundant and the performance of GEAs is independent on the structure of the optimal solution. However, GEA behavior is modified by stealth mutation. The chapter ends with concluding remarks.

6.1 The Tree Design Problem

This section provides the background for analyzing how tree representations affect GEA performance. After a brief definition of the network design problem in subsection 6.1.2 we focus on metrics and distances for graphs. This is followed by an illustration of different tree structures like stars or lists. To be able to measure the phenotypic difficulty of a tree problem, we introduce in subsection 6.1.4 a schema analysis for graphs. Based on the schema analysis, we present in subsection 6.1.5 scalable test problems for graphs. The one-max tree problem, which is similar to the well known one-max problem, is a fully easy problem, whereas the deceptive trap for trees is fully difficult. Finally, the section ends with a review of former design criteria for tree encodings as provided by Palmer (1994).

6.1.1 Definition

This subsection provides the necessary definitions for analyzing tree problems.

We define a network as a graph with n nodes and a maximum of $n(n-1)$ links connecting the nodes. If the network is fully connected it has at least $n-1$ links. In the following, we assume that all links are undirected (they can be used in both directions) and that the network is always fully connected. Therefore, the maximum number of possible links is $n(n-1)/2$. The position of the nodes in the graph is given a priori and the distances between two different nodes a and b are defined by using the Euclidian distance metric as

$$d_{a,b} = \sqrt{(x_a - x_b)^2 + (y_a - y_b)^2}, \tag{6.1}$$

where x denotes the abscissa and y the ordinate of a node in a Cartesian coordinate system.

The basic purpose of the network is to transport objects, for example goods or information, from some nodes in the network to other nodes. Therefore, a rule is necessary for how to transport the objects through the network.

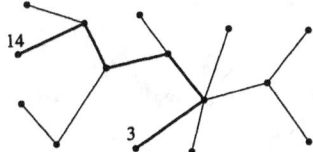

Fig. 6.1. A 15 nodes tree with the path connecting nodes 3 and 14 emphasized.

The rule for how to route the traffic through the network is based on the used routing algorithm. If the number of links in a fully connected network is larger than $n - 1$, the routing of the traffic through the network can be dynamically changed dependent on the current traffic load, the delay, the failure of nodes or links, or other criteria. In contrast, if the number of links in a fully connected network is equal to $n - 1$ there is only one unique path from every node to every other node and no dynamic routing is necessary.

Trees are defined as an undirected and connected graph with no cycles. For a tree with n nodes there are exactly $n - 1$ links. It was discovered by Cayley (1889) that for a graph with n nodes, there are exactly $n^{(n-2)}$ possible trees. A tree structure has some remarkable benefits: It represents the network structure with the lowest number of possible links to still obtain a connected graph. Furthermore, no dynamic routing is necessary as there is only one possible path for the traffic between any two nodes (compare Figure 6.1). Finally, the size of the search space $|\Phi_{tree}| = n^{n-2}$ is much smaller than for general networks $|\Phi| \lessapprox 2^{n(n-1)/2}$.[1] However, the use of trees also has some drawbacks: Trees are very vulnerable to link or node failures. If one link or one node fails, the tree divides up into two unconnected subtrees which can not communicate with each other. However, despite this fact, trees are widely used for communication networks (Minoux, 1987; Abuali, Wainwright, & Schoenefeld, 1995; Elbaum & Sidi, 1996; Güls, 1996; Tang, Man, & Ko, 1997; Brittain, Williams, & McMahon, 1997; Streng, 1997; Gargano, Edelson, & Koval, 1998; Gerstacker, 1999; Brittain, 1999; Chu, Premkumar, Chou, & Sun, 1999; Chu & Premkumar, 1999; Knowles, Corne, & Oates, 1999; Grasser, 2000; Gaube, 2000; Edelson & Gargano, 2000; Edelson & Gargano, 2001). The network design problem itself is defined as follows: Based on the

- number of network nodes n,
- locations of the n nodes,
- traffic demands between all n nodes,
- available capacities for the links,
- cost of the links dependent on the capacity and length,

we determine the

- topology (structure) of the network,

---

[1] We assume that there is only one possible capacity for a link. For different types of lines with k different capacities the number of possible network structure increases to $|\Phi| \lessapprox k^{n(n-1)/2}$.

- capacity of the links,
- routing of the traffic through the network.

The general aim of the design process is to minimize the overall cost of the network with the constraint that all traffic demands between the nodes must be satisfied.

If we focus on tree structures, the capacity of the links, and the routing of the traffic, is determined by the topology. This means for trees that the optimization problem simplifies down to finding the optimal structure of the tree.

6.1.2 Metrics and Distances

As illustrated in subsection 3.3.2, a metric is necessary for the genotypic and phenotypic space Φ_g and Φ_p to define genetic operators like mutation or recombination. The application of the mutation operator to a genotype should result in the smallest possible change in the individual, and should generate an offspring with distance 1 for the genotypes and the phenotypes. The recombination operator should ensure that the offspring inherit substructures from the parents. In terms of metric, the distance between an offspring and its parents should be lower than the distance between the two parents.

In accordance with chapter 3, the Hamming metric (Hamming, 1980) is used for the genotypes. Thus, the Hamming distance between two genotypes $x \in \{0,1\}^l$ and $y \in \{0,1\}^l$ of length l is defined as

$$d_{x,y} = \sum_{i=0}^{l-1} |x_i - y_i|.$$

The distance d measures the number of alleles that are different in both individuals. Similarly, the distance between two different phenotypes (trees) is measured by using the Hamming distance d^h for trees. The Hamming distance between two trees measures the number of different links in the two trees. Therefore, the minimum Hamming distance between two different trees is $d^h = 2$.

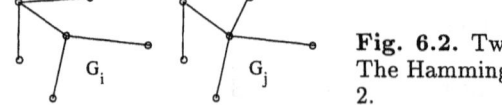

Fig. 6.2. Two graphs G_i and G_j with $d_{i,j} = 1$. The Hamming distance between the two graphs is 2.

As illustrated in Figure 6.2 the minimal Hamming distance between two trees is two, although they have $n - 2$ links of all $n - 1$ links in common. To simplify the metric, we define the distance $d_{i,j} \in \{0, 1, 2, \ldots n - 2\}$ between two trees G_i and G_j by half of the number of different links ($d_{i,j} = \frac{1}{2} d_{i,j}^h$). It can be calculated as

$$d_{i,j} = \frac{1}{2} \sum_{a=1}^{n-1} \sum_{b=0}^{a-1} |l_{ab}^i - l_{ab}^j|,$$

where l_{ab}^i is 1 if the link from node a to node b exists in tree G_i and 0 if it does not exist in G_i. Then, the number of links that the two trees G_i and G_j have in common can easily be calculated as $n - 1 - d_{i,j}$. A mutation of a tree should result in the exchange of one link, and the distance between parent and child is $d_{parent,child} = 1$.

6.1.3 Tree Structures

When focusing on trees, different basic topological structures can be identified. In general we can distinguish between

- star networks,
- list networks, and
- tree networks.

In the following, we briefly characterize the properties of the different tree structures. Figure 6.3 illustrates the different tree types. The degree of a node is defined as the number of links which are connected to the node.

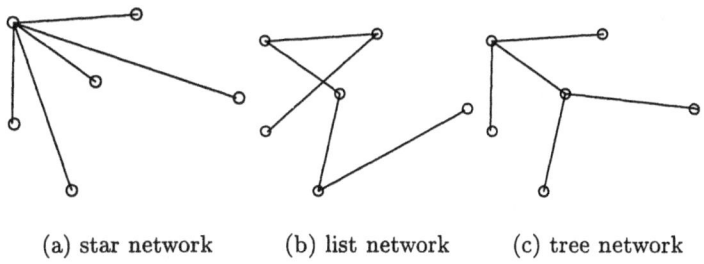

(a) star network   (b) list network   (c) tree network

Fig. 6.3. Different tree structures

A star (Figure 6.3(a)) has one center and all other nodes are connected to the center. Therefore, the center of the network has degree $n - 1$ and all other nodes have degree 1. For a network with n nodes there are n different stars. A failure of a link or a node (except the center) disconnects only the affected node. However, if the center node fails, no further communication over the network is possible.

For a list network (Figure 6.3(b)) two nodes have degree one (leaf nodes), and all other nodes have degree 2. There are many more possible list networks than star networks as the number of possible lists is $\frac{1}{2}n!$. A link or a node failure results in two separate sublists.

Finally, there are arbitrary trees (Figure 6.3(c)) which have no special structure except that they are trees. The degree of a node can vary from 1 to $n-1$. As for star and list networks, the sum over the degrees d_i of all n nodes can be calculated as $\sum_{i=1}^{n} d_i = 2(n-1)$. In the following, we denote an arbitrary network as a tree if we have no particular assumptions about the structure of the graph.

6.1.4 Schema Analysis for Graphs

In this subsection we define schemata for graphs in analogy to schemata defined on bitstrings (compare subsection 2.2.3). Schema analysis is helpful in determining whether graph problems are easy or difficult to solve for selectorecombinative GAs.

When assuming that GEAs process schemata, the analysis of schema fitness is the appropriate method to measure problem difficulty (see subsection 2.3.2). The BB hypothesis (see subsection 2.2.3) defines building blocks to be highly fit schemata of short defining length and low order. Consequently, problems are fully easy if all schemata of order one that contain the optimum have higher fitness than their competitors. Problems are difficult if all lower order schemata containing the global optimum are inferior to some of their competitors. The one-max problem is an example of a fully easy problem, whereas the fully deceptive trap of order k is an example for a fully difficult problem.

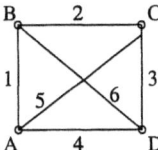

Fig. 6.4. Labeling of links for $n = 4$

We can measure problem complexity of network problems by introducing schema analysis for graphs. To formally define schemata, we have to label the possible links in a graph with numbers $\{1, 2, \ldots n(n-1)/2\}$. Figure 6.4 illustrates an example of labeling the links in a network with $n = 4$ nodes. Then, a schema is a string of length $l = n(n-1)/2$ and the symbol at the ith position describes the existence of a link. 1 indicates that the link is established, 0 indicates no link, and * indicates don't care (dashed line). Don't care means that the link is either established or not. Figure 6.5 illustrates some possible schemata for a 4 node network using the labeling from Figure 6.4.

When using schemata for trees, there is the additional restriction that each tree has exactly $n-1$ links and it must be connected. Therefore, there must be $n-1$ ones in each solution string, and the string must encode a connected tree. This means that the average fitness of a schema must be

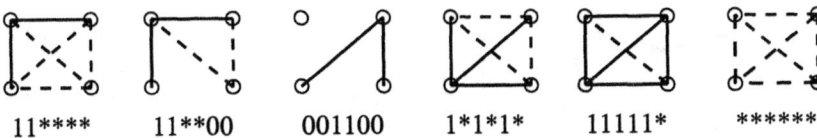

Fig. 6.5. Some schemata for graphs

calculated only from the trees that are represented by the schema. Other non-trees that are represented by the schema do not affect the schema fitness. This implies, for example, that schemata with more than $n-1$ ones, or more than $\frac{1}{2}n(n-1)-(n-1)$ zeros, do not exist because they do not encode a valid tree.

Using schema analysis, the complexity of a network problem can easily be measured by the maximum order of the building blocks k. If a problem is fully easy then all lower order schemata that contain the global optimum are superior to their competitors. All building blocks have order one ($k=1$). A problem is fully deceptive if all lower order schemata that contain the optimum are inferior to their competitors. To find the optimum, GEAs must be able to find BBs of order $k = n(n-1)/2$. In general, the order of the largest BB determines the complexity of a problem. In contrast to binary strings, the length of the schemata has no meaning in the context of graphs for the complexity of a problem as the labeling of the nodes does not affect problem complexity.

We have seen in section 3.3 that representations which do not preserve the distances between the individuals modify problem difficultly. With the analysis of graph schemata, we can compare the complexity of graph problems defined on the phenotypes to the complexity of the corresponding binary problems defined on bitstrings. This allows us to more easily recognize whether a tree encoding modifies BB-complexity. In the following, we consequently define a fully easy and a fully difficult scalable test problem based on the schema analysis for graphs.

6.1.5 Scalable Test Problems for Graphs

To test the performance of optimization algorithms for the topological design of trees, standard test problems should be used. In the following, motivated by the previous subsection, we define a fully easy and a fully difficult scalable tree problem.

The one-max tree problem is based on the one-max problem for binary representations (Ackley, 1987). An optimal solution is chosen either randomly or by hand. The structure of this optimal solution can be determined: It can be a star, a list, or a random tree with n nodes.

For the calculation of the fitness of the individuals, the distance d_{ij} between two trees G_i and G_j is used (compare section 6.1.2). Using this metric, the fitness of an individual x_i can be defined as the distance to the optimal

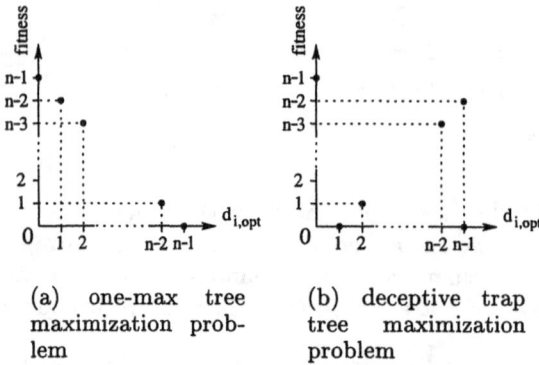

(a) one-max tree maximization problem

(b) deceptive trap tree maximization problem

Fig. 6.6. Scalable graph optimization problems

solution G_{opt}. The fitness of an individual varies between 0 and $n-1$ for a n-node network. When defining a minimization problem the fitness of an individual G_i is defined as the distance $d_{i,opt}$ to the optimal solution G_{opt}. Therefore, $f_i^{min} = d_{i,opt}$, and $f_i^{min} \in \{0, 1, \ldots, n-2\}$. An individual has fitness (cost) of $n-2$ if it only has one link in common with the best solution. If the two individuals do not differ ($G_i = G_{opt}$), the fitness (cost) of G_i is $f_i^{min} = 0$. If our example tree from Figure 6.7 is chosen as the optimal solution and we have a minimization problem, the star with center D would have fitness (cost) of 1, because the two trees differ at two edges[2] ($d_{i,opt} = 1$).

When defining a maximization problem, the fitness f_i^{max} of an individual G_i is defined as the number of edges it has in common with the best solution G_{opt} (compare Figure 6.6(a)). Therefore, $f_i^{max} = n-1 - d_{i,opt}$. If we have a maximization problem, and our example network from Figure 6.7 is chosen as the optimal solution, the star with center D would have fitness $f^{max} = 3$ because the two networks have three links in common, and the distance between the two networks is 1.

Because both test problems are similar to the standard one-max-problem it is easy to solve for mutation-based GEAs, but somewhat harder for recombination-based GAs (Goldberg, Deb, & Thierens, 1993). The knowledge about the standard one-max problem can be used for this one-max tree problem.

Analyzing the schemata as illustrated in the previous subsection shows that all building blocks of the one-max tree problem have order 1. All schemata that contain the global optimum G_{opt} are superior to their competitors. Therefore, the problem is fully easy. For a network with 4 nodes, the schemata are already of length $l = 6$, and there are $3^6 = 729$ different schemata. Due to the limited space, we want to leave the explicit calculation

---

[2] A-C respectively A-D

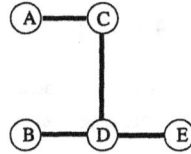

Fig. 6.7. A five node tree

Table 6.1. An example of calculating the average schema fitness for a 4 node one-max tree maximization problem where 111000 is the optimal solution. All schemata (in our example 11**0*) which contain the global optimum have higher fitness than their competitors. The problem is fully easy.

schema	schema fitness	represented trees	fitness of trees
110***	**2.33**	111000	3
		110100	2
		110001	2
11**1*	2	110010	2
01**0*	1.67	011100	2
		011001	2
		010101	1
10**0*	1.67	101100	2
		101001	2
		100101	1
01**1*	1.33	011010	2
		010110	1
		010011	1
10**1*	1.33	101010	2
		100110	1
		100011	1
00**0*	1	001101	1
00**1*	0.67	001110	1
		001011	1
		000111	0

of all schema fitnesses to the reader. We only illustrate in Table 6.1 the fitness calculation for schemata where the first, second, and fifth position are fixed for a 4 node one-max problem. The optimal tree is defined as $x_{opt} = 111000$. Obviously, the schema 11**0* which contains the global optimum is superior to all its competitors.

In analogy to this fully easy one-max tree problem, we define a fully difficult deceptive trap problem for graphs. As previously, we first choose an optimal solution G_{opt} with fitness $n - 1$ (assuming a maximization problem) either by hand or randomly. Then, the fitness of all other individuals $G_i \neq G_{opt}$ is defined as $f_i = d_{opt,i} - 1$. The fitness function is illustrated in Figure 6.6(b). This problem is fully difficult as all schemata with $k < n(n - 1)/2$ containing the global optimum are inferior to their misleading competitors. Mutation-based search approaches have great problem in finding the global optimum, whereas GEAs using crossover can find the optimum with proper population size.

Using again the example network from Figure 6.7 as the optimal solution, a star with center D has fitness 0 because the distance to the optimal solution is 1. The optimal solution itself has fitness 4.

6.1.6 Tree Encoding Issues

In this subsection we review the tree encoding issues as described by Palmer (1994) and Palmer and Kershenbaum (1994b) and relate them to the insights into the basic elements of representation theory we gained in chapter 3. According to Palmer, tree representations should possess the following properties:

- A representation should be able to represent all possible trees.
- It should be unbiased in the sense that all trees are equally represented.
- A representation should be capable of representing only trees.
- The construction of the phenotype from the genotype and vice versa should be easy.
- A representation should possess locality concerning small changes.
- The schemata should encourage short, low order schemata.

In the following, we discuss these issues and relate them to the theory of representations outlined in chapter 3 and 4.

The issue that a representation should be able to represent all possible trees is almost trivial. As long as we have no special knowledge about the problem we want to solve, it makes no sense to use a representation that might not represent some of the possible solutions. Otherwise, it could happen that GEAs search for the optimal solution, but the optimal solution can never be reached because it can not be encoded. If we have knowledge about the optimization problem, we can weaken this issue and demand representations to at least encode all the solutions we are interested in, and those which could be the optimal solution.

A representation is unbiased if all tress are represented by the same number of genotypes. Problems with biased encodings can be explained by the more general concept of redundant encodings illustrated in section 3.1. If some phenotypes are over- or underrepresented, the encoding is biased, and the performance of GEAs is changed. The influence on the performance of GEAs by biased encodings can be modeled by using the Gambler's ruin model (Harik, Cantú-Paz, Goldberg, & Miller, 1999). Section 3.1 has shown that as long as the high quality solutions are overrepresented, a bias increases performance. If the high quality solutions are underrepresented, a decline of GEA performance is unavoidable. Therefore, with respect to a robust encoding which can be used for problems of unknown complexity, it is desirable to use unbiased encodings.

Some tree representations can also represent non-trees. These kind of representations are affected by two problems: Firstly, it could be difficult to generate valid initial populations. Secondly, the application of genetic operators can result in invalid solutions. The question arises of how to handle invalid solutions, and what to do with non-trees. In general, there are two possibilities[3]: Invalid solutions can either be repaired, or they can be left unchanged in the population and hopefully they will disappear by the end of the run.

Repairing invalid solutions means that some of the trees are represented not only by valid individuals but also by some invalid solutions. Therefore, the representation is redundant. Phenotypes are not uniformly represented by the genotypes if the repair process is somehow shifted. Only a completely unbiased repair process which does not favor some tree structures guarantees an unbiased population and uniform redundancy.

To keep invalid solutions in the population could sometimes be helpful for GEAs (Orvosh & Davis, 1993). Nevertheless, it must be ensured that the optimal solution at the end of the run is valid. Otherwise, the application of GEAs to tree design problems is useless as it does not result in valid solutions. To drive GEAs towards valid solutions, researchers often use penalties for invalid solutions. However, additional penalties change the fitness function and with it the behavior of GEAs. Therefore, they should be used very carefully.

An easy construction of the phenotype from the genotype, and vice versa, is necessary for an efficient implementation of GEAs. However, it depends on the complexity of the fitness function whether the computational effort for the genotype-phenotype mapping significantly affects the run duration of the computer experiments. In contrast to costly fitness evaluations, a slightly more complicated genotype-phenotype mapping could often be neglected.

The problem of locality is part of the larger question of how well the encoding preserves the complexity of a problem. As illustrated in subsection 3.3.2, perfect locality is a necessary condition for the distance distortion $d_c = 0$. If the distance distortion $d_c \neq 0$ the problem difficulty can be changed when mapping the phenotypes to the genotypes.

Finally, Palmer noted in his doctoral thesis Goldberg's basic design principle of meaningful building blocks (compare section 2.4.1) and demanded encodings to encourage short, low order schemata. Otherwise, "long schemata cause genetic algorithms to drift" (Palmer, 1994, p. 40). However, as illustrated in section 3.2, drift is caused by non-uniformly scaled alleles and domino convergence, and not by the length and the size of the building blocks. The size and length of the building blocks determine the complexity of a problem for selectorecombinative GAs.

---

[3] Of course, there is a third possibility: To remove the individual from the population. However, we do not consider this case.

We recognize that when analyzing the design issues from Palmer that many aspects can theoretically be explained by the framework presented in chapter 4.

6.2 Prüfer Numbers

Prüfer numbers are a widely used representation for trees. The purpose of this section is to use the framework from chapter 4 for an investigation into the properties of Prüfer numbers. The analysis focuses on the low locality of the encoding and shows how GEA performance is affected.

The section starts with an historical review of the use of the Prüfer number encoding in the context of genetic and evolutionary search. The review shows a strong increase in interest into the encoding over the last 5 to 10 years. This is followed by the construction and deconstruction process of Prüfer numbers. Subsection 6.2.3 illustrates the benefits and drawbacks of the encoding. The use of the Prüfer number encoding is very charming due to its advantageous properties, although the low locality of the encoding has already been identified by Palmer (1994) to be its main drawback. Subsequently, in subsection 6.2.4 we present a deeper investigation into how exactly the low locality damages the performance of GEAs. In analogy to section 3.3, we illustrate why high locality is necessary for an encoding to preserve problem difficulty and perform random walks through the search space. After an analysis of the neighborhood structure of Prüfer numbers, we finally present empirical results for different tree structures using mutation and recombination-based evolutionary search methods. The section ends with concluding remarks.

6.2.1 Historical Review

In this subsection we give a brief historical review of the development and use of the Prüfer number encoding in the context of genetic and evolutionary algorithms.

Cayley (1889) identified the number of distinct spanning trees on a complete graph with n nodes as n^{n-2} (Even, 1973, pp. 103-104). Later, this theorem was very elegantly proven by Prüfer (1918) by the introduction of a one-to-one correspondence between spanning trees and a string of length $n - 2$ over an alphabet of n symbols. This string is denoted as Prüfer number, and the genotype-phenotype mapping is the Prüfer number encoding. It is possible to derive a unique tree with n nodes from the Prüfer number of length $n - 2$ and vice versa (Even, 1973, pp. 104-106). Of course there are other one-to-one mappings from strings of $n - 2$ labels onto spanning trees on the n labeled links. One example is the Blob Code which was developed and proposed by Picciotto (1999). Julstrom (2001) compared this encoding to Prüfer numbers and found for easy problems a higher performance of GEAs using the Blob Code than Prüfer numbers.

Later, in the context of genetic and evolutionary algorithms, several researchers used the Präfer number encoding for the representation of trees. Palmer used the encoding in his doctoral thesis at the beginning of the nineties (Palmer, 1994; Palmer & Kershenbaum, 1994a; Palmer & Kershenbaum, 1994b), and compared the performance of Präfer numbers with some other representations for the optimal communication spanning tree problem. However, he noticed that the Präfer number encoding has low locality and therefore is not a good choice for encoding trees in the context of genetic and evolutionary algorithms. The low performance of the encoding was confirmed by Julstrom (1993) who used Präfer numbers for the rectilinear steiner problem, and also observed low GEA performance using this encoding.

About the same time Abuali, Schoenefeld, and Wainwright (1994) used Präfer numbers for the optimization of probabilistic minimum spanning trees (PMST) with genetic algorithms. The investigation focused more on the influence of different operators than on the performance of Präfer numbers. However, at the end of the work, the conclusion was drawn that in contrast to Palmer and Julstrom, Präfer numbers "lead to a natural GEA encoding of the PMST problem" (Abuali, Schoenefeld, & Wainwright, 1994, p. 245). Some years later, similar results were reported by Zhou and Gen (1997) who successfully used the Präfer encoding for a degree constraint minimum spanning tree problem. The degree constraint was considered by repairing invalid solutions that violate the degree constraints. Furthermore, Präfer numbers were used for spanning tree problems (Gen, Zhou, & Takayama, 1998; Gen, Ida, & Kim, 1998), the time-dependent minimum spanning tree problem (Gargano, Edelson, & Koval, 1998), the fixed-charge transportation problem (Li, Gen, & Ida, 1998) and a bicriteria version of it (Gen & Li, 1999), and a multiobjective network design problem (Kim & Gen, 1999). Most of this work reported good results when using Präfer numbers, and labeled the encoding to be (very) suitable for encoding spanning trees. As an example of positive results we want to cite Kim and Gen (1999), who wrote:

"The Präfer number is very suitable for encoding a spanning tree, especially in some research fields, such as transportation problems, minimum spanning problems, and so on."[4]

However, other relevant work by Krishnamoorthy, Ernst, and Sharaiha (1999), who used Präfer numbers for the degree constraint spanning tree problem, from Julstrom (2000) who compared a list of edges encoding with Präfer numbers, or from Gottlieb and Eckert (2000) who used Präfer numbers for the fixed charge transportation problem showed that Präfer numbers result in a low GEA performance. A summarizing study by Gottlieb, Julstrom, Raidl, and Rothlauf (2001) compared the performance of Präfer numbers for four different network problems and concluded that Präfer numbers always

---

[4] Special thanks to Bryant A. Julstrom for his help with finding this statement.

perform worse than other encodings, and are not suitable for encoding trees when using genetic and evolutionary algorithms.

To explain the differences between the good and bad results obtained by GEAs using Prüfer numbers Rothlauf and Goldberg (1999) investigated the locality of the encoding more closely. It was shown that Prüfer numbers only have high locality if they encode stars. For all other tree types the locality is low which leads to a degradation of GEAs (see also Rothlauf and Goldberg (2000) and Rothlauf, Goldberg, and Heinzl (2001)). Therefore, the differences in performance could be well explained if one assumes that the performance of GEAs depends on the structure of the optimal solution. Obviously, researchers who report good solutions when using Prüfer numbers used problems where the optimal solution is more star-like and therefore easy to find for GEAs. However, when using Prüfer numbers for more general, non-star like problems, a strong decrease in GEA performance is inescapable. The results from Rothlauf and Goldberg (1999) were confirmed by Gottlieb and Raidl (2000) who investigated the effects of locality on the dynamics of evolutionary search.

We have seen that the performance of genetic and evolutionary algorithms using Prüfer numbers is a strongly discussed topic. Some researchers report good results and favor the use of Prüfer numbers. Other researchers, however, point to the low locality of the encoding, report worse results and advise us not to use Prüfer numbers. A closer investigation into how locality depends on the structure of the tree could solve these contradictory results. As the work from Rothlauf and Goldberg (2000) indicates that the locality of Prüfer numbers strongly depends on the structure of the tree, GEAs show good results if the good solutions are star-like, and worse results for all other types. In subsection 6.2.4 we review the main results from Rothlauf and Goldberg (1999) and Rothlauf and Goldberg (2000) and extend it with some work.

6.2.2 Construction

In this subsection, we review the construction rule for the Prüfer number encoding. We present both sides of the story: How a Prüfer number can be constructed from a tree, and how a tree can be constructed from a Prüfer number.

The Construction of the Prüfer Number from a Tree

The degree of a node denotes the number of links that are connected to the node. Thus, as a fully connected tree has exactly $n - 1$ links and the degree of a node lies between 1 and $n - 1$. A node has degree one if it is a leaf node. It has degree $n - 1$ if it is the center of a star. There are always at least two nodes which have degree 1.

The Prüfer number itself encodes an n-node tree with a string of length $n - 2$, and each element of the string is of base n. As the mapping is one-to-one, a Prüfer number is a unique encoding of a tree, and there are n^{n-2} different possible Prüfer numbers (Cayley, 1889; Prüfer, 1918).

For the construction of the Prüfer number from a tree, we label all nodes with numbers from 1 to n. Then, the Prüfer number can be constructed from a tree by the following algorithm:

1. Let i be the lowest numbered node of degree 1 in the tree.
2. Let j be the one node which is connected to i (there is exactly one). The number of the jth node is the furthest right digit of the Prüfer number.
3. Remove node i and the link (i, j) from the tree and from further consideration.
4. Go to 1 until only two nodes (that means one link) are left.

After termination of the construction rule, we have a Prüfer number with $n - 2$ digits which represents the tree. An efficient implementation of this algorithm uses a priority queue implemented in a heap to hold the nodes of degree 1. The algorithm's time complexity is then $O(n \log n)$.

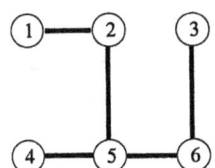

Fig. 6.8. A tree and the corresponding Prüfer number $P = 2565$

Let us demonstrate the construction of the Prüfer number with a brief example. The network in Figure 6.8 has 6 nodes. Therefore, the Prüfer number consists of 4 digits. The lowest numbered node with degree 1 is node 1. This node is connected to node 2 so the Prüfer number starts with a 2. We remove node 1 from further consideration and search for the lowest numbered node with degree 1. We identify node 2 which is connected to node 5. The Prüfer number becomes 25. After removing node 2, node 3 is the lowest numbered node which is eligible (it has degree 1). Node 3 is connected to 6 so we get 256. The node 4 is the lowest eligible node and we add 5 to the Prüfer number. Finally, only two nodes remain in the tree. The algorithm stops and the resulting Prüfer number is 2565.

The Construction of the Tree from the Prüfer Number

The construction of the tree from the Prüfer number follows the construction of the Prüfer number from the tree. It goes as follows:

1. Let P be a Prüfer number with $n - 2$ digits. All node numbers which are not in P can be used for the construction of the tree (are eligible).

2. Let i the lowest numbered eligible node. Let j be the leftmost digit of P.
3. Add the link (i, j) to the tree.
4. Designate i as no longer eligible and remove the leftmost digit j from the Prüfer number.
5. If j does not occur anywhere else in the remaining Prüfer number, designate j as eligible.
6. Go to 2 until no digits remain in the Prüfer number. If no digits are left, then there are exactly two numbers, r and s, which are eligible. Finally, add the link (r, s) to the tree.

We also illustrate this construction rule with a brief example. We want to construct the tree from the Prüfer number $P = 2565$. Eligible nodes are 1, 3 and 4. As 1 is the lowest eligible node, and 2 is the leftmost digit of the Prüfer number, we add the link $(1, 2)$ to the tree. 1 is then no longer eligible, and 2 does not occur anywhere else in the string. Therefore, the nodes 2, 3 and 4 are eligible and P becomes 565. Now 2 is the lowest eligible node and we add the link $(2, 5)$ to the tree. As 5 occurs somewhere else in the string, we do not designate 5 as eligible. Thus, we only remove 2 from our pool of eligible numbers, and then we can add the link $(3, 6)$ to the tree. Now, only the nodes 4 and 6 are eligible and $P = 5$. We continue with adding $(4, 5)$. Finally, all digits are removed from P and the numbers 5 and 6 remain eligible. The link $(5, 6)$ is added to the tree and the algorithm terminates. We have constructed the tree illustrated in Figure 6.8.

6.2.3 Properties

This subsection analyzes the properties of the Prüfer number encoding by using the design issues from Palmer and Kershenbaum (1994a) and Palmer (1994) (see subsection 6.1.6).

Benefits

In the following, we give a brief overview on the benefits of the Prüfer number encoding in relation to genetic and evolutionary algorithms (Palmer & Kershenbaum, 1994a).

The Prüfer number encoding is a very elegant and attractive encoding with some remarkable benefits:

- Every tree can be represented by a Prüfer number.
- Only trees are represented by Prüfer numbers.
- Every Prüfer number represents exactly one tree.
- All trees are represented equally (unbiased).

A look at the construction rule of the Prüfer number shows that the Prüfer number is able to represent all possible trees. Because every tree has at least two nodes with degree 1, the construction rule can be applied to every tree.

The user should notice that the original intent of the Prüfer number was to prove Cayley's theorem (Cayley, 1889) by introducing Prüfer numbers. It was also shown by Prüfer (1918) that Prüfer numbers only represent trees. Therefore, a Prüfer number can be created randomly and it always represents a tree. In contrast to many other representations, no repairing of a randomly chosen individual is necessary. Furthermore, it is also not necessary to repair individuals that are generated by genetic operators in each generation. The first three benefits of the Prüfer numbers can be summarized by denoting the Prüfer number encoding as a one-to-one mapping. The mapping is not only surjective, but also bijective.

One consequence of a one-to-one mapping is that all trees are represented equally. Each tree is represented by exactly one specific Prüfer number. The number of different trees for a graph with n nodes is n^{n-2}, and there are also exactly n^{n-2} different Prüfer numbers for an n node tree. Therefore, GEAs using Prüfer numbers have no problems with redundancy. GEAs using Prüfer numbers can not be affected by the over- or underrepresentation of some individuals.

These advantages make Prüfer numbers an attractive encoding for trees. However, the use of Prüfer numbers is connected to some serious drawbacks.

Drawbacks

The Prüfer number has the disadvantages of

* complex calculation and
* low locality.

In comparison to some other representations, the construction of the Prüfer number is more complex and not straightforward. But, it can be done using the help of a heap in $O(n \log n)$. This seems to be acceptable for most problems.

The most important disadvantage of the Prüfer number is the low locality of the representation. Small changes in the Prüfer number string can lead to large changes in the represented network. This means the mapping from the phenotype to the genotype is not homogeneous. Therefore, the basic mutation operator that searches the local solution space around an individual does not generate offspring that are similar to their parents. A descendant does not inherit the important properties of its parents. Thus, mutation works not as a local search, but more as a random search over the solution space.

A small example illustrates the low locality of the encoding. Changing the last digit in the Prüfer number of Figure 6.9(a) from 3 to 1 yields 2231, which decodes to the links (2,4), (2,5), (3,2), (1,3), and (1,6). Only two of the original tree's five links remain (compare Figure 6.9).

As it is obvious that low locality is the main drawback of the Prüfer number encoding, we focus in greater detail in the following subsection on the problems of low locality.

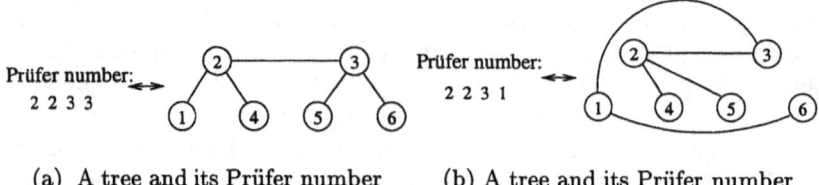

(a) A tree and its Prüfer number
$P = 2233$

(b) A tree and its Prüfer number
$P = 2231$

Fig. 6.9. The low locality of the Prüfer number encoding. A change of one digit changes 3 links in the corresponding tree.

6.2.4 The Low Locality of the Prüfer Number Encoding

As illustrated in the previous subsection, the Prüfer number encoding is affected by low locality. The purpose of this subsection is to investigate the locality of the encoding more closely. The locality of Prüfer numbers is examined by performing two different investigations: Firstly, we perform random walks through the search space and examine the distances between parents and offspring. Secondly, we investigate the neighborhood of the genotypes and phenotypes. We examine the locality of the neighboring individuals, and determine their number. Finally, we present an empirical verification of the theoretical predictions for mutation- and crossover-based evolutionary search algorithms.

Random Walks

In the following, we present computer simulations for evaluating the locality of the Prüfer number encoding. For this purpose we perform random walks through the genotypes/phenotypes and analyze the resulting change in the corresponding phenotypes/genotypes.

Fig. 6.10 shows the encoding of a tree $x_p \in \Phi_p$ as a Prüfer number $x_{g1} \in \Phi_{g1}$ and the encoding of the Prüfer number as a bitstring $x_g \in \Phi_g$. The genetic operators are applied to the genotypes x_g. The Prüfer number itself is a sequence of integers and is represented as a bitstring using the binary encoding. Therefore, the mapping from the bitstring to the Prüfer number $f_g : \Phi_g \to \Phi_{g1}$ is affected by scaling and has the properties discussed in section 3.2. The mapping from the Prüfer numbers to the trees $f_{g1} : \Phi_{g1} \to \Phi_p$ is described in subsection 6.2.2. Notice that for a tree with n nodes, the Prüfer number has $n - 2$ digits, and the bitstring $(n - 2)\lceil \log_2(n) \rceil$ bits.

The locality of the Prüfer number encoding can be measured by performing a random walk through one of the solution spaces Φ_g, Φ_{g1}, or Φ_p, and measuring the distances between parent and offspring in the other two solution spaces. A random walk through a search space Φ is defined by performing iteratively small changes. Therefore, the distance between parent x_p

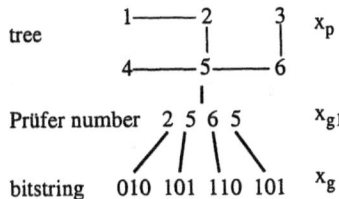

Fig. 6.10. A tree, its Prüfer number and the corresponding bitstring

and offspring x_o in the solution space we are performing our random walk in is $d_{p,o} = 1$. According Figure 6.10 there are three different possibilities:

1. A random walk through Φ_g (one step is a one bit change in the bitstring).
2. A random walk through Φ_{g1} (one step is a change of one digit in the Prüfer number).
3. A random walk through Φ_p (one step is a change of one link in the tree).

A random walk through Φ_g means randomly changing one bit of x_g and examining how many links change in the corresponding tree x_p. Furthermore, the change of one bit in the bitstring x_g results in the change of exactly one digit of the Prüfer number x_{g1}. A random walk through Φ_{g1} means randomly changing one digit of the Prüfer number x_{g1} and measuring how many links are different in the resulting x_p. Notice that the change of one digit in x_{g1} results in up to $\log_2(n)$ different bits in x_g. A random walk through Φ_p means that one link of the tree x_p is replaced by a randomly chosen link, and the difference of bits/digits in the bitstring/Prüfer number is examined.

In Figure 6.11, 6.12, 6.13, and 6.14 we present the results for the random walks. In all our experiments the start individual for the bitstring, the Prüfer number or the tree is chosen randomly. To gain statistically significant information independent of the start individual, 400 steps (mutations) were carried out in each of the 20 runs. Thus, we performed overall 8000 steps in the search space.

The results for a random walk through Φ_g (Fig. 6.11) and Φ_{g1} (Fig. 6.12) shows that only about 40% of the one bit/digit changes lead to a change of one link in the tree. More than 35% (16 nodes) or 50% (32 nodes) of all one bit/digit changes result in networks with at least four different links ($d_{offspring,parent} \geq 4$). The locality of the genotype (bitstring as well as Prüfer number) is low. Low genotypic distances $d_{i,j}^g = 1$ do not correspond to low phenotypic distances $d_{i,j}^p$.

When walking through the phenotypic solution space (trees) the plots in Figure 6.13 show that only about 50% of all one link changes result in a change of less than eight bits (16 nodes), respectively 20 bits (32 nodes) in the bitstring. For the Prüfer number (Fig. 6.14) about 75% of the neighboring individuals are different in more than one digit. The locality of the phenotype is also low.

The random walks through Φ_g and Φ_p have shown that the locality of the Prüfer number representation is low. Most of the small steps in the phenotype

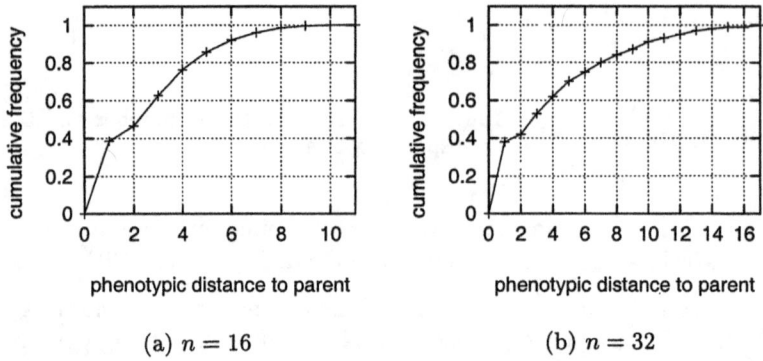

Fig. 6.11. Distribution of phenotypic distances for neighboring bitstrings, on 16 and 32 nodes. We perform a random walk through Φ_g, and the graphs show how many links are different in the tree x_p if one bit in the bitstring x_g is changed.

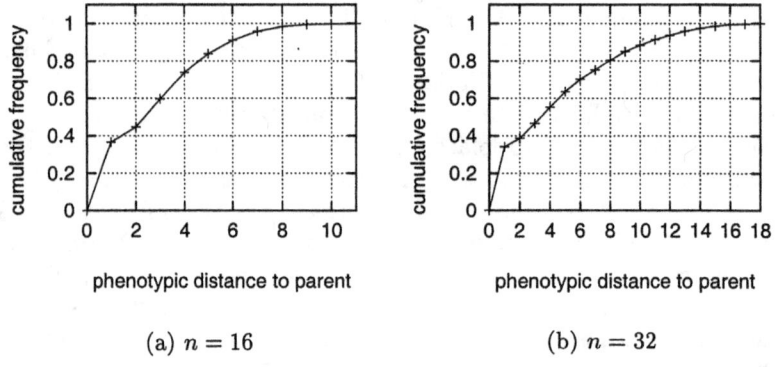

Fig. 6.12. Distribution of phenotypic distances for neighboring Prüfer numbers on 16 and 32 nodes. We perform a random walk through Φ_{g1}, and the graphs show how many links are different in the tree x_p if one digit of the Prüfer number x_{g1} is changed.

respective genotype result in unacceptably high changes in the corresponding genotype and phenotype. In the following, we investigate if the locality is uniformly low everywhere in the search space, or if there are some areas of high locality.

Analysis of the Neighborhood

Performing random walks through the search space has revealed that the locality of the Prüfer number encoding is low. Therefore, we investigate in

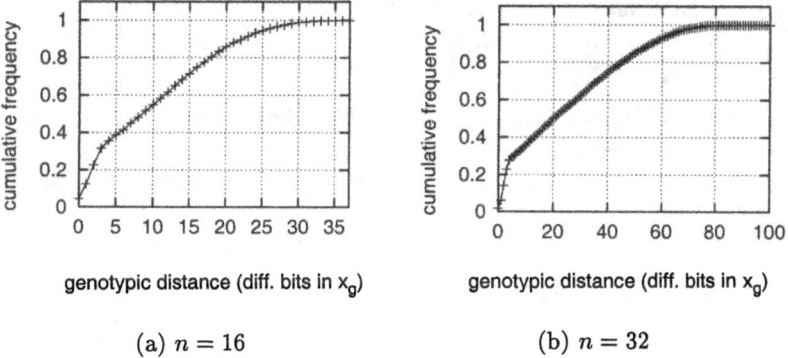

Fig. 6.13. Distribution of genotypic bitstring distances for neighboring trees on 16 and 32 nodes. We perform a random walk through Φ_p, and the graphs show how many bits are different in x_g if one link of the tree x_p is changed.

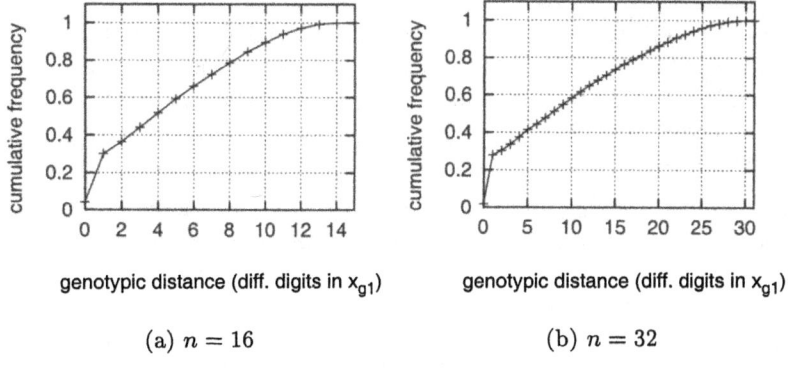

Fig. 6.14. Distribution of genotypic distances for neighboring trees on 16 and 32 nodes. We perform a random walk through Φ_p, and the graphs show how many digits of the Präfer number x_{g1} are different if one link of the tree x_p is changed.

the following whether the locality of the encoding is uniformly low, or if there are differences in locality for different areas of the search space. The search space can be separated into different areas by making assumptions about the structure of the represented graph such as being a star or a list.

To investigate if the locality of the Präfer number encoding is different for different areas of the search space, we choose an individual x_i with specific properties and examine its locality. Doing this we look at all individuals x_j with distance $d_{x_i, x_j} = 1$ and measure the resulting genotypic or phenotypic distance. As an individual is defined to be a neighbor to another individual if the distance between the two individuals is 1, our examination is nothing more than an examination of the neighborhood of specific individuals.

Similarly to the random walks, we investigate the neighborhood of individuals in all three search spaces Φ_g, Φ_{g1} and Φ_p:

- Neighborhood of an individual $x_g \in \Phi_g$ (all neighbors that are different in one bit from the examined individual).
- Neighborhood of a Prüfer number $x_{g1} \in \Phi_{g1}$ (all neighbors that differ in one digit).
- Neighborhood of a tree $x_p \in \Phi_p$ (all neighbors have distance $d_{ij} = 1$).

We examine the complete neighborhood of an individual either in Φ_g, Φ_{g1}, or Φ_p and measure the corresponding distances in the two others. This investigation can be performed for four different types of networks:

(i) Star: One node is of degree $n - 1$ and the rest of the nodes have degree 1.

(ii) Random list: Two nodes are of degree 1 (the first and the last node of the list) and all other nodes have degree 2. The numbering of the nodes is random.

(iii) Ordered list: Like random list, but the nodes in the list are connected in ascending order. Node k is connected to $k + 1$, node $k + 1$ is connected to $k + 2$ and so on. If the highest numbered node n is not a leaf node then it is connected to node 1.

(iv) Tree: An arbitrary tree.

We distinguish between ordered and random lists because the locality of the Prüfer number encoding is slightly different for ordered and random lists.

Figure 6.15, 6.16, 6.17, and 6.18 examine the neighborhood of star, list and arbitrary trees on 16 and 32 node problems. A bitstring representing a tree has length $l = 56$ (16 nodes) respective $l = 150$ (32 nodes); a Prüfer number encoding a tree has either 14 (16 nodes) or 30 (32 nodes) digits. For every problem instance, the complete neighborhood of 1000 randomly chosen individuals is examined. Figure 6.15 and 6.16 show distributions of phenotypic distances for neighboring bitstrings and Prüfer numbers; that is, for bitstrings that differ in one bit, and for Prüfer numbers that differ in one digit. Figure 6.17 and 6.18 show distributions of genotypic distances for neighboring spanning trees; that is, for spanning trees that differ in one link.

Figure 6.15 and 6.16 reveal that the neighborhood of a bitstring, as well as a Prüfer number representing a spanning tree, depends on the structure of the encoded tree. If the bitstring/Prüfer number encodes a star, all genotypic neighbors also have a phenotypic distance of one. This means that the locality of the bitstring/Prüfer number is perfect for stars. If the bitstring/Prüfer number encodes an ordered list, the genotypic neighbors have a maximum phenotypic distance of 4 independent of the number of nodes. However, bitstrings and Prüfer numbers that encode random lists, or random trees, show very low locality, and most of the genotypic neighbors are phenotypically completely different.

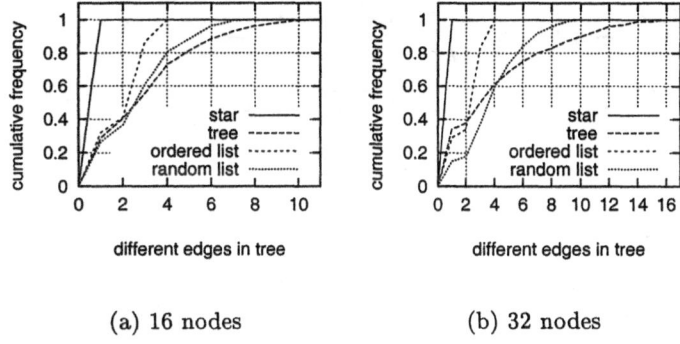

(a) 16 nodes   (b) 32 nodes

Fig. 6.15. Distribution of phenotypic distances for neighboring bitstring genotypes. The graphs illustrate how many links are different in the tree when examining the complete neighborhood of a randomly chosen genotypic bitstring x_g.

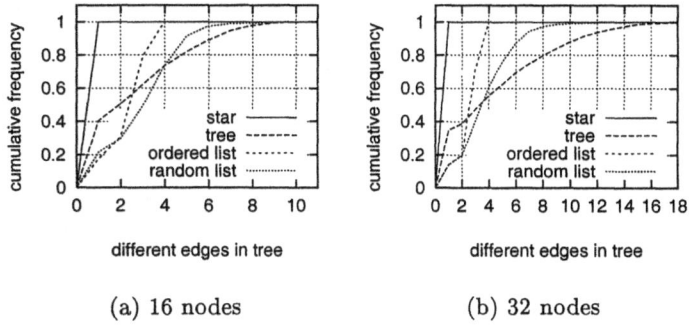

(a) 16 nodes   (b) 32 nodes

Fig. 6.16. Distribution of phenotypic distances for neighboring Prüfer numbers. The graphs illustrate how many links are different in the tree when examining the complete neighborhood of a randomly chosen genotypic Prüfer number x_{g1}.

The neighborhood of a tree is illustrated in Figure 6.17 and 6.18. Similarly to the genotypic neighborhood, all neighbors of a star have a genotypic distance in the Prüfer number space of one (Fig. 6.18). However, up to $\lceil log_2(n) \rceil$ bits are changed in the bitstring that represents the Prüfer number (Fig. 6.17). This is as the bitstring encodes the Prüfer number using a binary encoding. Therefore, the change of one digit in a Prüfer number must

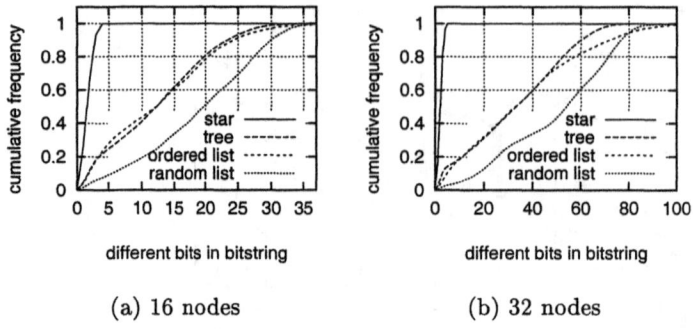

Fig. 6.17. Distribution of genotypic distances for neighboring trees. The graphs illustrate how many bits are different in the bitstring representation x_g of a tree when examining the complete neighborhood of a randomly chosen tree x_p.

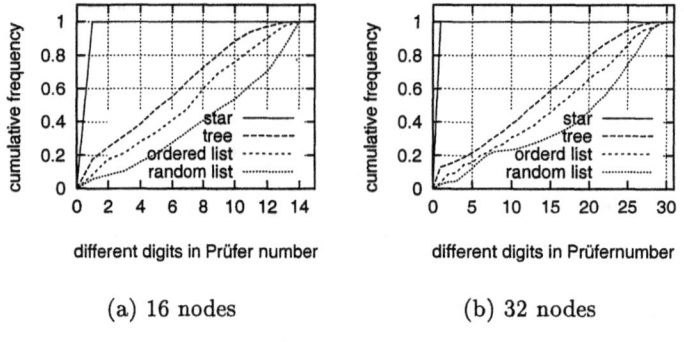

Fig. 6.18. Distribution of genotypic distances for neighboring trees. The graphs illustrate how many digits are different in the Prüfer number x_{g1} when examining the complete neighborhood of a randomly chosen tree x_p.

result in a change of up to $\lceil log_2(n) \rceil$ bits in the bitstring. The locality of the phenotype is perfect for stars. For tree and list networks the change of one link mainly results in a completely different bitstring/Prüfer number. The locality of the phenotype is very low for tree and list networks.

The results show that the locality of the Prüfer number is highly irregular and is dependent on the phenotypic structure of the encoded network. If a Prüfer number encodes a list or arbitrary tree, the locality of the encoding is very low. Most of the genotypic neighbors of the Prüfer number are phenotypically completely different. However, if Prüfer numbers encode stars, the locality of the encoding is perfect. A genotypic neighbor of a star is also a phenotypic neighbor. These results raise two new questions: Why do Prüfer numbers which encode stars have high locality? How large are the areas of high locality? We answer these questions in the following.

Number of Neighbors

We have seen that Prüfer numbers only have high locality if they encode stars. In the following, we want to focus on two main issues: Firstly, we want to find out why Prüfer numbers have high locality especially when encoding stars. And secondly, we want to know how large the areas are of high locality. Finding answers for these questions helps us to more accurately predict the behavior of GEAs for different tree optimization problems. The investigation will show that the number of neighbors has a major impact on the answers to both questions.

The previous investigations have shown that some Prüfer numbers do have high locality. A Prüfer number representing a star has perfect locality ($d_m = 0$) because all phenotypic neighbors of a star are also genotypic neighbors. To shed light on the question of why exactly stars have perfect locality, we calculate the number of neighbors for both Prüfer numbers and trees.

A Prüfer number encodes an n node tree with $n - 2$ digits of base n. Because we can change each of the $n - 2$ digits to $n - 1$ different integers, each Prüfer number has exactly $(n - 1) * (n - 2)$ neighbors. Furthermore, each Prüfer number with $n - 2$ digits is encoded as a bitstring of length $(n-2) * \lceil \log_2(n) \rceil$. So each bitstring has $(n-2) * \lceil \log_2(n) \rceil$ neighbors. A change of one digit in the Prüfer number can result in up to $\lceil \log_2(n) \rceil$ different bits. As we are mainly interested in the Prüfer number encoding, we want to focus in the following only on Prüfer strings and neglect the encoding of Prüfer strings as bitstrings. The reader should notice that the number of neighbors of a Prüfer number is independent of the structure of the encoded tree.

A star on n nodes has $(n - 1) * (n - 2)$ neighbors obtained by replacing one of its links with another feasible link. Therefore, for stars the number of neighbors is the same for the phenotypes and the genotypes. Furthermore, a star's neighbors are represented by the neighbors of its Prüfer number obtained by changing one of the number's symbols; as already mentioned, these neighbors number also $(n - 1)(n - 2)$. For stars, the genotypic and phenotypic neighborhoods coincide, and therefore locality is maximal.

This seems auspicious, but tree localities vary with the shape of the tree. A list is a spanning tree with two leaves and $n - 2$ nodes of degree 2. In a list's Prüfer number, all the symbols are distinct, and each Prüfer number has, as already mentioned, $(n - 1)(n - 2)$ neighbors. However, a list on n nodes has $\sum_{i=1}^{n-1} i(n - i) - 1 = n \sum_{i=1}^{n-1} i - \sum_{i=1}^{n-1} i^2 - (n - 1) = \frac{1}{2} n^2 (n - 1) - \frac{1}{6} n(n - 1)(2n - 1) - n + 1 = \frac{1}{6} n(n - 1)(n + 1) - n + 1$ neighbors (Gerstacker, 1999). Therefore, for lists the number of phenotypic neighbors is much higher than the number of genotypic neighbors. Stars and lists have the smallest and largest phenotypic neighborhoods, respectively. All other spanning trees fall between these extremes which Figure 6.19 plots as a function of the number n of nodes.

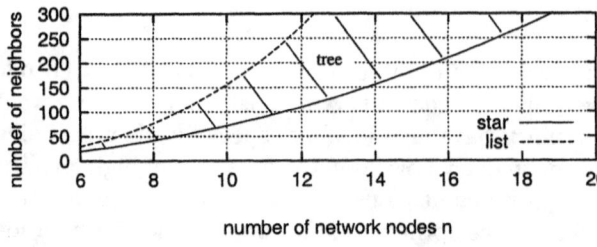

Fig. 6.19. Phenotypic neighborhood sizes for lists and stars, as functions of the number of nodes. The values for all other trees lie between these curves.

We see that the number of neighbors of a phenotype increases when modifying a star towards a list network. However, the number of neighbors of a Prüfer number remains constant and is independent of the structure of the encoded tree. So there is a mismatch between the number of neighbors of non-star trees and of Prüfer numbers. Therefore, the locality of all Prüfer numbers not encoding a star could not be perfect as phenotypes always have a higher number of neighbors than genotypes.

The results concerning the number of neighbors are summarized and illustrated for some example networks in Table 6.2. The number of neighbors for trees is between the number for star and list networks and must be calculated for each network separately. It depends on the degrees of the nodes in the tree.

Table 6.2. Number of neighbors for graph, Prüfer number and bitstring

nodes		graph x_p	Prüfer number x_{g1}	bitstring x_g
n	star	$(n-1)(n-2)$	$(n-1)(n-2)$	$(n-2) * \lceil log_2(n) \rceil$
	list	$\frac{1}{6}n(n-1)(n+1) - n + 1$		
8	star	42	42	18
	list	77		
	tree	62.45 (avg.)		
16	star	210	210	56
	list	665		
	tree	447.2 (avg.)		
32	star	930	930	150
	list	5425		
	tree	2595 (avg.)		

After we have explained why Prüfer numbers can only have high locality when encoding stars, we focus on the question of how large the areas of high

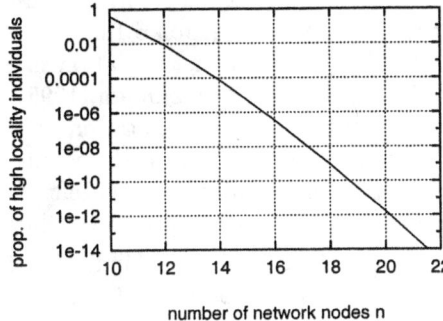

Fig. 6.20. The proportion of spanning trees on n nodes whose Prüfer numbers have high locality, defined as differing from the Prüfer number of a star in no more than $i_{max} = 5$ digits.

locality are. GEAs only search effectively in regions near stars. In these areas the locality is high and the encoding allows a guided search. To approximate the number of individuals with high locality, we extend the definition of neighbors to include trees whose Prüfer numbers differ in at the most i_{max} digits ($i_{max} \ll n$). This means we assume an individual x_i to have high locality if its distance $d_{i,star}$ towards a star is equal or lower than i_{max}. The number of individuals which have the maximum distance i_{max} towards a star ($d_{i,star} \le i_{max}$) can be calculated as $\sum_{i=0}^{i_{max}} \binom{n-2}{i}(n-1)^i$; this value is $O(n^{2i_{max}})$. However, the number of spanning trees on n nodes, and thus the size of the search space, is n^{n-2}. Therefore, as Figure 6.20 illustrates, the proportion of these high-locality individuals is small even for moderate n and diminishes exponentially as n grows. The areas of high locality grow more slowly with increasing problem size n than the overall search space. As a result, we expect GEAs using Prüfer numbers and searching for stars to perform worse with increasing problem size n.

We were able to explain the high locality of Prüfer numbers representing stars by calculating the number of neighbors for different tree structures. The number of neighbors for Prüfer numbers stays constant, whereas for the phenotypes the number of neighbors is different for different network types. For stars, however, there is a one-to-one correspondence and the number of neighbors is the same for genotypes and phenotypes. All other types of trees like lists or arbitrary trees have a higher number of neighbors than stars (or Prüfer numbers). Furthermore, the areas of high locality are very tiny as they grow with $O(n^{2i_{max}})$, where $i_{max} \ll n$, whereas the search space grows with $O(n^{n-2})$. Thus, the locality is in general very low and GEAs searching for optimal networks with structures other than stars must fail.

Performance

In the following, we verify empirically that GEAs using the Prüfer number encoding do not perform well when searching for good solutions in areas where the locality is low. We present results for GEAs only using one-point crossover and for simulated annealing using only mutation. Both search algorithms are applied to the fully easy one-max tree problem from subsection 6.1.5.

Simulated annealing (SA) can be modeled as a GEA with population size 1 and Boltzmann selection (Goldberg, 1990a; Mahfoud & Goldberg, 1995). In each generation, an offspring is created by applying one mutation step to the parent. Therefore, the new individual has distance 1 to its parent. If the offspring has higher fitness than its parent it replaces the parent. If it has lower fitness it replaces the parent with the metropolis probability $P(T) = e^{-\frac{f_{offspring} - f_{parent}}{T}}$, where f denotes the fitness of an individual. The acceptance probability P depends on the actual temperature T which is reduced during the run according to a cooling schedule. With lowering temperature T, the probability of accepting worse solutions decreases. Because the search algorithm uses only mutation, and can in contrast to for example a $(1+1)$ evolution strategy solve difficult multi-modal problems, we use it as a representative of mutation-based evolutionary search algorithms. For further information about simulated annealing the reader is referred to other work (Cavicchio, 1970; Davis, 1987).

In Figure 6.21 we present results for GEAs with $\mu + \lambda$ selection using one-point crossover and no mutation on 16 and 32 node one-max tree problems. $\mu + \lambda$ selection means that we generate λ offspring from μ parents and that we choose the best μ individuals from all $\mu + \lambda$ individuals as parents for the next generation. This selection scheme assures that a once found best individual is preserved during a GA run and not lost again. The structure of the optimal solution is determined to be either a star, list, or an arbitrary tree. For the 16 node problems, we chose $\mu = \lambda = 400$, and for the 32 node problems $\mu = \lambda = 1500$. We performed 250 runs and each run was stopped after the population was fully converged. Figure 6.22 presents results for using simulated annealing. The start temperature $T_{start} = 100$ is reduced in every step by the factor 0.99. Therefore, $T_{t+1} = 0.99 * T_t$. Mutation is defined to randomly change one digit of the Prüfer number. We performed 250 runs and each run was stopped after 5000 iterations.

The results in Figure 6.21 and 6.22 show that if the optimal solution is a randomly chosen star, both search algorithms, the recombination-based GA and the mutation-based SA are able to find the optimal star easily. A search near stars is really a guided search and both algorithms are able to find their way to the optimum. However, if the optimal solution is a random list, an ordered list, or an arbitrary tree, GEAs can never find the optimal solution and are completely misled. Exploring the neighborhood around an individual in an area of low locality results in a blind and random search. Individuals that are created by mutating one individual, or by recombining two individuals, have nothing in common with their parent(s).

The results show that good solutions can not be found if they lie in areas of low locality. A degradation of the evolutionary search process is unavoidable. Evolutionary search using the Prüfer number encoding could only work

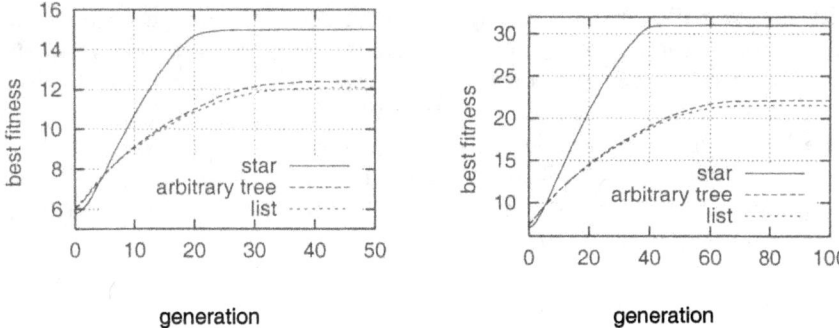

Fig. 6.21. The performance of a GA for a 16 (left) and 32 (right) node one-max tree problem. The plots show the fitness of the best individual over the run. The structure of the best solutions has a large influence on the performance of GAs. If the best solution is a star, GAs perform well. If GAs have to find a best solution that is a list or a tree, they degrades and cannot solve the easy one-max problem.

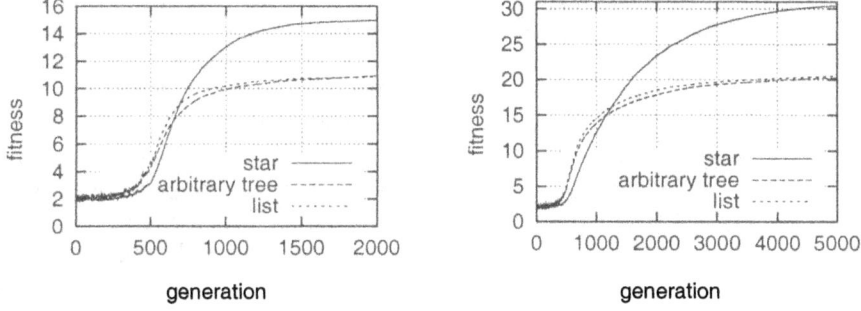

Fig. 6.22. The performance of simulated annealing for a 16 (left) and 32 (right) node one-max tree problem. The plots show the fitness of the best individual over the run. As for recombination-based approaches, the mutation-based simulated annealing fails if the best solution is not a star.

properly if the good solutions are stars. Near stars the locality is high and a guided search is possible. Furthermore, the empirical results confirm the theoretical predictions from subsection 3.3.2 that high locality is a necessary condition for mutation- and recombination-based GEAs. If the locality of an encoding is low, the difficulty of a problem is changed by the representation, and even very simple problems like the one-max tree problem become very difficult and can not be solved any more.

The presented empirical results also shed some light on the contradictive statements about the performance of GEAs using Prüfer numbers. Researchers who investigate problems in which good solution are star-like see acceptable results and favor the use of Prüfer numbers. Other researchers with non-star like optimal solutions, however, observe low performance and advise not to use the encoding. Furthermore, we have seen that the Prüfer number encoding has low locality. Therefore, the distance distortion d_c of the representations is not zero and the difficulty of the problem is changed. As a result, fully easy problems like the one-max tree problem become more difficult, whereas fully difficult problems become more easy. Results about the performance of Prüfer numbers on fully difficult problems are presented later in subsection 8.1.3.

6.2.5 Summary and Conclusions

This section presented an investigation into the locality of Prüfer numbers and its effect on the performance of GEAs using this encoding. We started with a historical review on the use of Prüfer numbers. In subsection 6.2.2, we presented the construction rules for the construction of the Prüfer number. This was followed by a brief overview on the benefits and drawbacks of the encoding. Although the encoding has some remarkable advantages, it is affected by low locality. Some researchers already addressed the low GEA performance of Prüfer numbers due to their low locality. Consequently, we focused in the main part of the section (subsection 6.2.4) on the investigation into the low locality of the Prüfer number encoding. We started investigating the locality of Prüfer numbers more closely by performing random walks through the search space. This was followed by an analysis of the neighborhood of the genotypes and phenotypes. The analysis showed differences in locality. To explain the differences, we calculated the number of neighbors for Prüfer numbers and trees dependent on the structure of the network. Finally, we empirically verified the theoretical predictions by using recombination- as well as mutation-based naturanalogous search algorithms for solving the one-max tree problem.

The historical review showed that there has been a great increase in interest in the Prüfer number encoding over the last two years. However, the suitability of Prüfer numbers for encoding network problems is strongly disputed as some researchers report good results whereas others report failure. By performing random walks through the search space, the low locality of the encoding can be nicely illustrated. A small modification in a genotype mostly results in a completely different phenotype.

The analysis of the neighborhood of individuals answers the question of whether the locality is low everywhere in the search space, and gives an explanation for the contradictive results from different researchers. The results show that the locality of Prüfer numbers representing stars is perfect. However, all other types of networks like lists or arbitrary trees lack locality and the genotypic neighbors of a Prüfer number representing a list or an arbitrary tree have on average not much in common with each other. Therefore, the low locality of the Prüfer numbers does not reduce GEAs performance in all areas of the solution space to the same extent. This can explain the different results using Prüfer numbers existing in the literature.

To answer the questions of why exactly Prüfer numbers encoding stars have high locality, and how large the areas of high locality are, we investigated the number of neighbors a Prüfer number individual has. The analysis shows that for Prüfer numbers, the number of neighbors remains constant. For phenotypes, however, the number of neighbors varies with the structure of the tree. Stars have as many neighbors as the corresponding Prüfer numbers and therefore, the locality around stars is high. When modifying stars towards lists, the number of phenotypic neighbors increases, which makes it impossible to obtain high locality for problems other than stars. Furthermore, the areas of high locality are only of order $O(n^{const})$, whereas the whole search space grows with $O(n^{n-2})$. Thus, the regions of high locality become very small with increasing problem size n, which reduces the performance of GEAs on larger problems.

The results show that Prüfer numbers have low locality. Therefore, the distance distortion $d_c \neq 0$ and the difficulty of problems is changed. Researchers should be careful when using Prüfer numbers on problems of unknown complexity because fully easy problems become more difficult when using the Prüfer number encoding. As a result, GEAs using Prüfer numbers are likely to fail when used on real-world problems of unknown complexity.

6.3 The Link and Node Biased Encoding

When using genetic and evolutionary algorithms for the optimal tree optimization problems, the design of a suitable tree encoding is crucial for finding good solutions. The link and node biased (LNB) encoding represents the structure of a tree using a weighted vector and allows GEAs to distinguish between the importance of the nodes and links in the network.

The purpose of the following section is to examine the link and node biased encoding using the framework from chapter 4. We investigate whether the encoding is uniformly redundant, and how the parameters of the encoding influence redundancy. The investigation reveals that the commonly used simpler version of the encoding using only a node-specific bias is biased towards star networks, and that the initial population is dominated by only a few star-like individuals. The phenotypes are not represented uniformly, but some of them are overrepresented. The more costly link-and-node-biased encoding uses not only a node-specific bias, but also a link-specific bias. Similarly to the node-biased encoding, the link-and-node-biased encoding is also biased towards stars if a node-specific bias is used. However, by increasing the link-specific bias, the bias of the encoding is reduced and the encoding becomes uniformly redundant for a large link-specific bias. Finally, the encoding has great problems if the two biases are too small, because then only minimum spanning tree-like phenotypes can be represented. At the extreme, if both biases are almost zero, only the minimum spanning tree can be represented independently of the values of the LNB vector.

In section 3.1, we have seen that the performance of GEAs using a redundant encoding increases if the good solutions are overrepresented, and decreases if they are underrepresented. Therefore, the performance of GEAs using the LNB encoding is high if the optimal solution is similar to stars or minimum spanning trees. For all other problems, however, a reduction of GEA performance is unavoidable.

6.3.1 Introduction

The link and node biased encoding is a representation from the class of weighted encodings and was developed by Palmer (1994). Additional encoding parameters are necessary to balance the importance of link and node weights. The encoding was proposed to overcome the problems of characteristic vectors, predecessor representations (compare also Raidl and Drexel (2000)) and Prüfer numbers by the introduction of a link- and node-specific bias. Later Abuali, Wainwright, and Schoenefeld (1995) compared different representations for probabilistic minimum spanning tree (PMST) problems and in some cases found the best solutions by using the LNB encoding. Raidl and Julstrom (2000) observed solutions superior to those of several other optimization methods for a similar weighted encoding which was used for the degree-constrained minimum spanning tree (d-MST) problem.

In the following, we want to investigate whether there are any solution candidates preferred in the initial population. This is important because if the encoding prefers some solution candidates, the initial supply of building blocks is affected and GEAs can have problems with redundancy. Degradation of GEAs is inescapable if the optimal solution is underrepresented by the encoding. To get rid of adjusting the encoding parameters, Palmer (1994) presented results in the original paper only using the node-specific bias and neglected the link-specific bias. We start by investigating the limitations of this approach and continue with how the more general link-and-node-biased encoding using both a node- and a link-specific bias influences the bias of the population.

In the following subsection, we give a brief description of the LNB encoding. In subsection 6.3.3, we illustrate that the bias of an encoding has the same effect as non-uniform redundancy (compare section 3.1). Using the notion of redundancy, we can explain some of the properties of biased encodings. This is followed by an investigation into whether the LNB encoding using only node weights is biased, and how the individuals are represented in an initial population. In subsection 6.3.5, we investigate how the setting of both encoding parameters affects the bias of the encoding. Finally, we prove empirically in subsection 6.3.6 the theoretical predictions about the performance of GEAs for the one-max tree problem. The section ends with concluding remarks.

6.3.2 Motivation and Functionality

In this subsection, we want to review the motivation and the resulting properties of the LNB encoding as described in Palmer (1994), Palmer and Kershenbaum (1994a), and Palmer and Kershenbaum (1994b).

As the costs of a communication or transportation network strongly depend on the length of the links, network structures that prefer short distance links often tend to have higher fitness. Furthermore, it is useful to run more traffic over the nodes near the gravity center of an area than over nodes at the edge of this area (Kershenbaum, 1993; Cahn, 1998). Thus, it is desirable to be able to characterize nodes as either interior (some traffic only transits), or leaf nodes (all traffic terminates). As a result, the more important a link is, and the more transit traffic that crosses the node, the higher in general is the degree of the node. Nodes near the gravity center tend to have a higher degree than nodes at the edge of the network. Hence, the basic idea of the LNB encoding is to encode the importance of a node. The more important the node is, the more traffic that should transit over it.

When applying this idea to a network problem, the given distance matrix that defines the distances between any two nodes, is biased according to the importance of the nodes. If a node is not important, the modified distance matrix should increase the length of all links that are connected to this node. Doing this will result with high probability in a leaf node.

When using the node-biased encoding the chromosome b holds the biases for each node, and has length n for an n node network. The values in the distance matrix d_{ij} are modified according to b using the weighting function

$$d'_{ij} = d_{ij} + p(b_i + b_j)d_{max}. \tag{6.2}$$

The bias b_i is a floating number between zero and one, d_{max} is the largest value in the distance matrix and p controls the influence of the biases. In the following, we want to denote this approach as the node-biased encoding.

Using the bias-vector for encoding trees, we get the encoded network structure by calculating the minimum spanning tree (MST) for the modified distance matrix. Prim's algorithm (Prim, 1957) was used in the original work. By running Prim's MST algorithm, nodes that are situated near other nodes will probably be interior nodes of high degree in the network. Nodes that are far away from the other nodes will probably be leaf nodes. Thus, the higher the bias of a node, the higher is the probability that it will be a leaf node. To finally get the tree's fitness, the encoded network is evaluated by using the original distance matrix.

We want to illustrate the functionality of the node-biased encoding with a small example. The vector $b = \{0.7, 0.5, 0.2, 0.8, 0, 1\}$ is a node-biased vector and holds the bias for each node. A distance matrix for the 5-node problem should be defined as

$$D = \begin{pmatrix} - & 2 & 1 & 3 & 4 \\ 2 & - & 5 & 6 & 3 \\ 1 & 5 & - & 4 & 3 \\ 3 & 6 & 4 & - & 10 \\ 4 & 3 & 3 & 10 & - \end{pmatrix} . \tag{6.3}$$

For the construction process, we first have to calculate all values of the modified distance matrix. Using $p = 1$, we get for example for $d'_{0,1} = 2 + (0.7 + 0.5) * 10 = 14$. When calculating all modified distances $d'_{i,j}$ according equation 6.2 we get for the modified distance matrix:

$$D' = \begin{pmatrix} - & 14 & 10 & 18 & 12 \\ 14 & - & 12 & 19 & 11 \\ 10 & 12 & - & 14 & 6 \\ 18 & 19 & 14 & - & 19 \\ 12 & 11 & 6 & 19 & - \end{pmatrix} .$$

Using Prim's algorithm for the modified distance matrix D', we finally get the tree illustrated in Figure 6.23. The represented tree is calculated as the MST using the distance matrix D'. For example, $d'_{0,2} = 10 < d'_{0,i}$, where $i \in \{1,3,4\}$. Because the link between node 0 and node 2 has the shortest distance d', it is used for the represented tree.

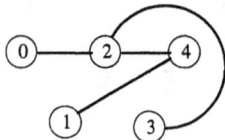

Fig. 6.23. An example tree for the node-biased encoding

Palmer noticed in his original work that each bias b_i modifies a whole row and a whole column in the distance matrix. Thus, not all possible solution candidates can be encoded by the node-biased encoding (Palmer, 1994, pp. 66-67).

To overcome this problem, he introduced in the second, extended version of the representation, an additional link-bias. The chromosome holds biases not only for the n nodes but also for all possible $n(n - 1)/2$ links, and has overall length $l = n(n + 1)/2$. The weighting function for the elements in the distance matrix was extended to

$$d'_{ij} = d_{ij} + P_1 b_{ij} d_{max} + P_2 (b_i + b_j) d_{max} \tag{6.4}$$

with the link-specific bias b_{ij}, the weight of the link-specific bias P_1, and the weight of the node-specific bias P_2. Using this representation the encoding could represent all possible trees. However, the string length is increased from $l = n$ to $l = n(n+1)/2$. In the following, we want to denote this representation as the link-and-node-biased encoding.

In analogy to the node-biased encoding we present a brief example for the link-and-node-biased encoding. The example chromosome holds the node bias $\{0.7, 0.5, 0.2, 0.8, 0, 1\}$ and the link bias

$$
\begin{pmatrix}
- & 0.1 & 0.6 & 0.2 & 0.8 \\
0.1 & - & 0.1 & 0.9 & 0.5 \\
0.6 & 0.1 & - & 0.3 & 0.2 \\
0.2 & 0.9 & 0.3 & - & 0.4 \\
0.8 & 0.5 & 0.2 & 0.4 & -
\end{pmatrix} .
$$

With $P_1 = 1$ and $P_2 = 1$ and using the distance matrix from equation 6.3 we get, for example, for $d'_{0,1} = 2 + 0.1 * 10 + (0.7 + 0.5) * 10 = 15$. Consequently, we get for the modified distance matrix

$$
D' = \begin{pmatrix}
- & 15 & 16 & 20 & 20 \\
15 & - & 13 & 28 & 16 \\
16 & 13 & - & 17 & 8 \\
20 & 28 & 17 & - & 23 \\
20 & 16 & 8 & 23 & -
\end{pmatrix} .
$$

Then we calculate the MST using the modified distance matrix D' and get the tree shown in Figure 6.24.

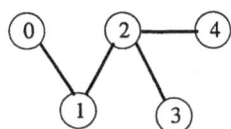

Fig. 6.24. An example tree for the link-and-node-biased encoding

Using the simple node-biased, or the more general link-and-node-biased encoding, makes it necessary to determine the value of one, respectively two, additional encoding parameters P_1 and P_2. In the original work from Palmer, only results for the node-biased encoding and $p = P_2 = 1$ are presented.

Furthermore, we must be aware of the problem that the structure of the represented tree depends on the underlying distance matrix D. The same LNB vector can represent different trees when different distance matrixes are used.

If the setting of the parameters could result in a biased representation of the individuals, a degradation of GEA performance is sometimes unavoidable as illustrated in the following subsection.

6.3.3 Biased Initial Populations and Non-Uniformly Redundant Encodings

The purpose of this subsection is to illustrate that biased representations are affected by the same problems as non-uniformly redundant representations. An investigation into the link and node biased encoding shows that the encoding is redundant. Therefore, the representation can be affected by problems with over- or underrepresentation of individuals.

In the previous subsection, we have seen that each bias b_i in the bias vector b is a floating number between zero and one. By calculating the minimum spanning tree from the modified distance matrix, we get the represented tree. Therefore, there is an infinite number of possibilities for the b_i to represent one of the n^{n-2} trees. This means the LNB encoding is a redundant encoding because each phenotype is represented by an infinite number of different genotypes.

In section 3.1 we have seen that uniform redundancy has no influence on the performance of GEAs. An encoding is uniformly redundant if all phenotypes are represented by the same number of genotypes. However, if some individuals are overrepresented by the encoding the performance of GEAs is influenced. If the optimal solution is similar to the overrepresented individuals, GEA performance increases. If the optimum is similar to underrepresented individuals a degradation of GEAs is unavoidable. As a result, if the encoding is not uniformly redundant, GEA performance depends on the structure of the optimal solution.

In the following we illustrate that the bias of a representation is the same as the redundancy of a representation. Therefore, the results about non-redundant representations from section 3.1 can be also used for biased representations. Palmer described a bias in his thesis (Palmer, 1994, pp. 39) as:

> "It (a representation) should be unbiased in the sense that all trees are equally represented; i.e., all trees should be represented by the same number of encodings. This property allows us to effectively select an unbiased starting population for the GA and gives the GA a fair chance of reaching all parts of the solution space."

When comparing this definition of bias to the definition of redundant encodings (compare section 3.1, p. 33), we see that both definitions are essentially the same: An encoding is biased if some individuals are over-, or underrepresented. Furthermore, Palmer correctly recognized, in agreement with the results about redundant encodings, that a widely usable, robust encoding should be unbiased. However, in contrast to Palmer's statement that only unbiased encodings allow an effective search, we have seen that biased encodings can be helpful if the encoding is biased towards the optimal solution.

The reader should be careful not to confuse the bias of a representation with the ability to represent all individuals. The ability to represent all individuals means that all possible phenotypes can be represented. Palmer and Kershenbaum (1994a) has already shown that the simple node-biased encoding is not able to represent all individuals. However, in contrast, the bias of an encoding describes whether the phenotypes which can be represented are represented uniformly. To the knowledge of the author, this point has not been investigated in the literature before. Therefore, the remainder of this section focuses on the question of whether the LNB encoding overrepresents some individuals.

We have stated that the LNB encoding is a redundant encoding. Therefore, the performance of GEAs goes with $O(r/2^{k_r})$. The question arises regarding whether the encoding is non-uniformly redundant, or not. Palmer developed the LNB encoding with the intent to create a non-biased encoding. Therefore, to be able to judge the performance of GEAs using the LNB encoding, we investigate in the following subsections whether the LNB encoding is really unbiased.

6.3.4 The Node-Biased Encoding

It is known that the node-biased encoding is not capable of representing all possible network structures (Palmer, 1994). The purpose of this subsection is to investigate whether the represented networks are encoded unbiased. The subsection extends prior work (Gaube, 2000; Gaube & Rothlauf, 2001) by unpublished results. We start with a distance matrix where all elements have the same value. This is followed by an investigation where the position of the nodes is chosen randomly.

All Links Have the Same Length

We assume that all values d_{ij} in the unbiased distance matrix are equal. Thus, the values in the biased distance matrix are only determined by b. We denote by b_l the lowest bias in b. It is the bias for the lth node and all other biases are larger. Using this definition, the modified length d' of each link connecting node i and j is always higher than the length of the link connecting either i and l or j and l:

$$d'_{i,l} < d'_{i,j} \text{ for } b_l = \min\{b_1, \ldots b_n\},$$

where $i, j, l \in \{0, \ldots, n\}$, $i \neq l$, $i \neq j$ and $l \neq j$. As the decoding algorithm chooses the shortest $n - 1$ links that do not create a cycle for creating the encoded network, the only structure that could be represented by the node-biased encoding is a star with center l.

For a tree with n nodes, the number of possible stars is n, whereas the number of all possible trees is n^{n-2}. Thus, only a small fraction of trees could be represented by the node-biased encoding. However, at least the represented stars are unbiased as the elements of b are uniformly distributed in the initial population.

Although an empirical proof of a theoretical prediction is redundant, we present an empirical verification of these results for a small 4 node problem in Table 6.3. There are 16 possible networks, and 4 of them are stars with center l, where $l \in \{1, 2, 3, 4\}$. For the experiments, we created 1000 initial populations of size 1000. The distances d_{ij} between all four nodes are equal.

Table 6.3. Average percentage of represented network types for a 4 node problem

non-star	star with center			
	$l=1$	$l=2$	$l=3$	$l=4$
0%	25.01%	24.97%	24.92%	25.10%

We see that it is not possible to create non-stars, and that the stars are represented uniformly. As a result, the node-biased representation is uniformly redundant (it represents the n different stars unbiased) but it can only represent a small portion of the solution space (only stars) if the distances $d_{i,j}$ between the nodes have the same value.

Random length of links

For this investigation we randomly placed the nodes on a two-dimensional quadratic plane of size 1000 x 1000 and randomly created 500 node-biased vectors. The elements of the distance matrix D were calculated using the Euclidean distance (see equation 6.1).

In Figure 6.25 we show the average minimum phenotypic distance $d = \min(d_{x_p, x_{star}})$, with $x_{star} \in \{all\ star\ networks\}$, of a randomly generated node-biased encoded individual x_p towards a star. The distance d measures how similar the phenotype of a randomly created node-biased vector is to one of the n stars. If the distance is low, the phenotype has many edges in common with one of the n stars. We performed experiments for 8, 16, and 32 node problems and positioned the nodes 250 times randomly on the square.

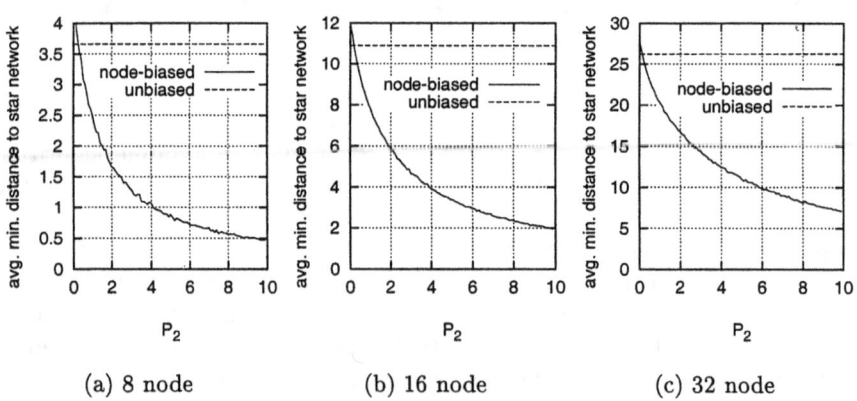

(a) 8 node   (b) 16 node   (c) 32 node

Fig. 6.25. Average minimum phenotypic distance of a randomly generated node-biased individual to a star. With increasing node-bias $p = P_2$ ($P_1 = 0$), the encoding is strongly biased towards stars. If P_2 is large enough, the encoding can represent only one of the n stars.

The plots show that with increasing node-specific bias $p = P_2$, the phenotypic distance between an individual and one of the n stars decreases and a randomly created individual becomes more and more star-like. On the contrary, the distance of a randomly created individual towards list networks increases. We see that the node-biased encoding is strongly biased towards stars. In comparison to an unbiased encoding (like the Prüfer number encoding), where the average distance towards a star stays constant, the average distance decreases with increasing P_2.

This result is not surprising when we take a closer look at equation 6.2 which describes how the distance matrix is biased. The values in the original distance matrix $d_{i,j}$ are modified by adding an additional bias. With increasing node bias $p = P_2$, the influence of $d_{i,j}$ decreases relatively and $d'_{i,j}$ only depends on the node-specific bias b_i and b_j and no longer on the distance d_{ij}. Therefore, with p large enough, $d_{i,j}$ can be completely neglected and the encoding can only encode stars. We have the same situation as when the distances between all nodes are the same. Thus, the results described in paragraph 'All links have the same length' hold true. We see that with $p = P_2$ large enough, every randomly created node-biased individual will be a star, and the minimum distance d of a randomly created vector towards one of the n stars will be zero.

However, not only for very large, but even for reasonable values of $p = P_2$, is the encoding strongly biased. To investigate how often different phenotypic trees are represented in a randomly created population, we ordered the represented trees according to their frequency. In Figure 6.26, we plot the cumulative frequency of the ordered number of copies a specific tree has in a randomly created population for a 4 node problem dependent on the values of $p = P_2$. The frequencies are ordered in ascending order. This means that rank 1. corresponds to the tree structure that is most often encoded (encoded with the highest probability), and rank 16. to the tree that is encoded with the lowest probability. We generated 1000 randomly node-biased vectors and performed 1000 experiments with differently randomly located nodes.

If all individuals are created with the same probability, the cumulative frequency would be linear over all possible tree structures. All possible trees are represented uniformly with probability 100%/16=6.25%. However, for the node-biased encoding, some individuals are created more often. These individuals encode stars and the encoding is biased towards them. For example, when using a node-biased encoding, where $p = P_2 = 0.5$, a randomly generated individual has with a probability of about 50% the same phenotypic structure (rank 1.). The next most frequent tree structure that is represented by a randomly chosen node-biased individual is created with probability of about 75%-50%=25%. The line shows that for $P_2 = 0.5$ about 90% of all randomly generated node-biased individuals only represent three different trees. Therefore, the plots nicely illustrate that a bias results in an overrepresentation of some individuals.

Furthermore, we see that for $p = P_2$ large enough $(P_2 > c)$, the encoding can only represent a maximum of $n = 4$ different individuals which are stars. These 4 stars are encoded unbiased (each of the four individuals is created with 25% probability), similarly to the situation where all links have the same length.

In general, the results show that for medium values for $p = P_2$ some trees are strongly overrepresented, whereas some other tree structures are not represented at all.

For very small values of $p = P_2$, the values of b_i no longer influence the distance matrix and the only tree that can be represented by a node-biased genotype is the minimum spanning tree. Figure 6.26 illustrates that with decreasing P_2, fewer and fewer different networks can be represented. The represented networks, however, become more and more minimum spanning tree-like, and at the extreme that P_2 is zero, only one individual (the minimum spanning tree) can be encoded.

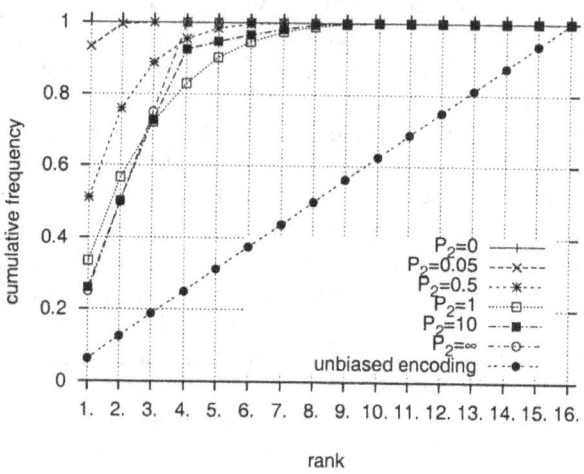

Fig. 6.26. Distribution of network types for a 4 node problem

Figure 6.27 illustrates the frequency of the first, second and fifth most frequent individual over P_2. The plots are based on Figure 6.26. We see that with increasing P_2, the diversity in the population decreases, and for $P_2 > c$ only the four most frequent individuals (they are the four stars) are uniformly represented with 25% frequency. On the other hand with P_2 decreasing, the individuals are biased towards the minimum spanning tree, and for P_2 small enough, only one individual (the minimum spanning tree) can be represented.

The simple node-biased encoding is strongly biased towards either stars or towards the minimum spanning tree. At the extreme, with P_2 large enough, the encoding can only represent stars. But, even with lower P_2, a few star-like individuals dominate a randomly created population, whereas it is impossible to create some solution candidates. For smaller values of P_2, the represented

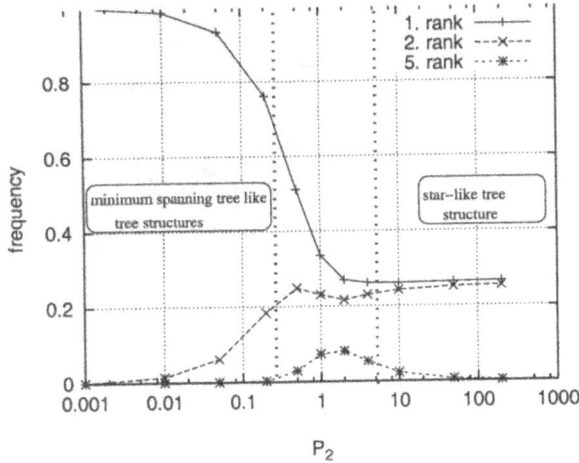

Fig. 6.27. Frequency of how often the first, second and 5th most frequent tree structure is represented by the node-biased encoding for a 4 node problem.

networks become more and more minimum spanning tree-like, and for the extreme that P_2 is small enough, only the minimum spanning tree can be encoded. In the following subsection, we want to investigate how the bias is influenced by the setting of both parameters when using the link-and-node-biased encoding.

6.3.5 The Link-and-Node-Biased Encoding

Palmer (1994) proposed the link-and-node-biased encoding using a node- and a link-bias as a way to overcome some of the problems with the node-biased (NB) encoding. In the following, we investigate how the non-uniform redundancy which we have noticed for the NB encoding is influenced by the choice of the two parameters P_1 and P_2.

Similarly to the previous subsection, we investigate the bias of the LNB encoding by randomly creating link-and-node-biased vectors and measuring their distance towards one of the n stars. The more links an individual has in common with one of the stars, the more star-like it is and the lower is the distance. In Figure 6.28 we present results for randomly created 8, 16 and 32 node-problems. The average minimum distance towards one of the n stars is plotted over P_1 and P_2 and compared to an unbiased encoding (Prüfer number). The parameters P_1 and P_2 vary between 0 and 1, and we generated 1000 individuals for randomly positioning the nodes 250 times.

The results show for all three problem instances that the bias of the LNB encoding strongly depends on the node-specific bias P_2. With increasing values of P_2 the individuals are strongly biased towards stars. With P_2 dominating P_1, we notice the same behavior as for the node-biased encoding. With increasing node-specific bias P_2, the encoding can only represent star-like structures, and for P_2 large enough, the distance of an individual towards one of the n stars will become zero.

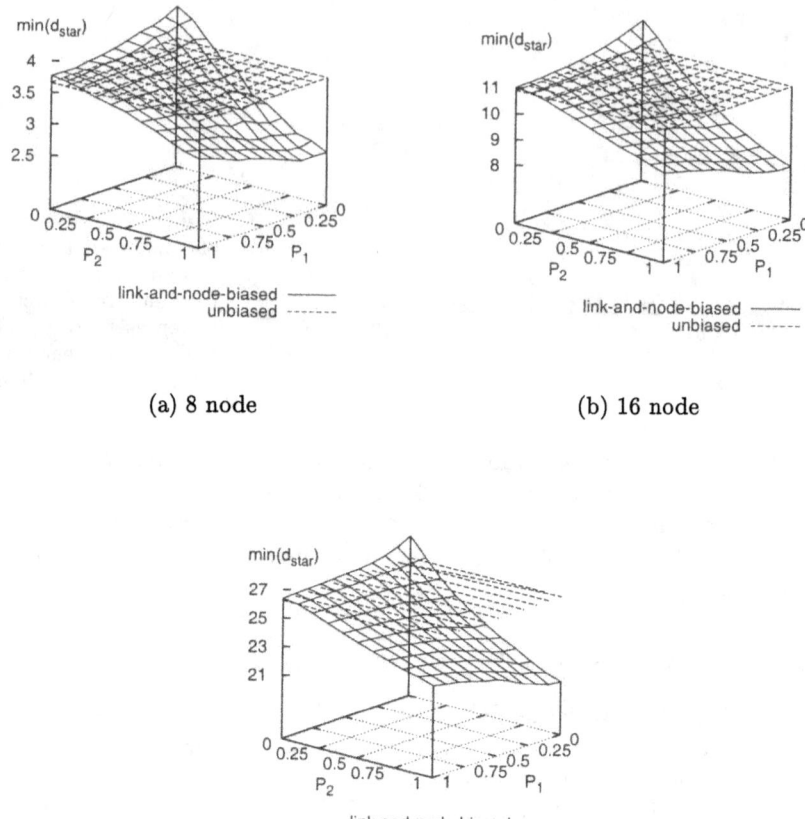

(a) 8 node   (b) 16 node

(c) 32 node

Fig. 6.28. Average minimum phenotypic distance of a randomly generated link-and-node-biased individual with 8 or 16 nodes to a star. By increasing the node-specific bias P_2, an individual is strongly biased towards star structures. Higher values for the link-specific bias P_1 result in a more unbiased representation. Small values of P_1 and P_2 result in a bias towards the minimum spanning tree.

However, the results also raise some hope that with increasing link-specific bias P_1 the bias of the representation can be reduced. To more closely investigate the dependency of the bias on P_1, Figure 6.29 presents how the minimum distance towards stars depends on P_1 for $P_2 = 0$. The plots show that with increasing P_1 the minimum distance towards stars stays constant, and the encoding becomes unbiased. In comparison to unbiased representations there is still a small bias. However, as it is very small, we want to disregard it.

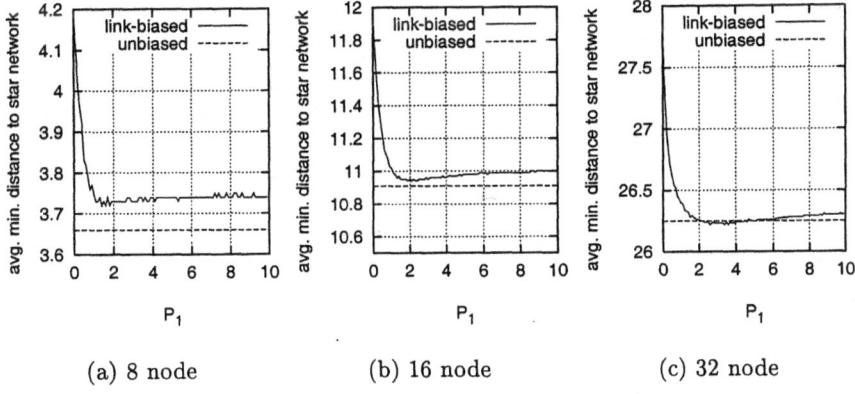

(a) 8 node   (b) 16 node   (c) 32 node

Fig. 6.29. Average minimum phenotypic distance of a randomly generated link and node biased individual to a star over the link-specific bias P_1. The node-specific bias $P_2 = 0$. For P_1 large enough, the encoding is almost completely unbiased. For P_1 very small, only the minimum spanning tree can be encoded.

This behavior that the LNB encoding becomes unbiased with increasing P_1 can be explained by looking more closely at equation 6.4. With $P_2 = 0$, P_1 large enough, and $b_{ij} \in \{0, 1\}$, each element $d_{i,j}$ of the distance matrix is modified separately by the bias $b_{i,j}$. To each distance $d_{i,j}$ a random value with mean $0.5 * P_1 d_{max}$ is added. Therefore, no links are preferred and using only a link-bias results in an almost unbiased encoding. The small rest bias is probably a result of the construction of the MST from the modified distance matrix.

Finally, we want to emphasize that small values of P_1 and P_2 reduce the performance of the encoding dramatically, as the bias then has no influence on the structure of the represented network. The encoding can only represent networks that are similar to the minimum spanning tree.

The results show, in agreement with the node-biased encoding, that especially with increasing P_2, a randomly created link-and-node-biased individual is strongly biased towards stars. However, by increasing the link-specific bias P_1 the population becomes less biased. For P_1 large enough, the encoding is unbiased as the node-bias can be neglected and the link-bias alone does not prefer specific network structures.

6.3.6 Empirical Results

In the following, we prove theoretical predictions about GEA performance by empirical results.

The previous subsections have shown that both LNB encodings, the node-biased encoding as well as the link-and-node-biased encoding are biased towards stars if they use a node-bias. With increasing node-bias P_2, star-like structures are strongly overrepresented in a randomly generated population. We know from section 3.1 that redundancy favors genetic search if the optimal solutions are overrepresented by the encoding, and hurts genetic search if the optimal solutions are underrepresented. Therefore, we expect good GEA performance if the optimum is a star, and low GEA performance if the optimum is a non-star such as a random list.

Furthermore, we have seen that with the link-bias P_1 large enough, the encoding becomes unbiased and is able to represent all individuals uniformly. This means the performance of GEAs should be independent of the structure of the optimal solution with P_1 large enough. Finally, we know that the representation can not work for very small values of P_1 and P_2 because then the bias has no influence on the structure of the network and the encoding can only represent minimum spanning tree-like network structures. At the extreme, if $P_1 = P_2 = 0$, the biases have no influence at all, and the only network that could be represented is the minimum spanning tree.

To investigate how the performance of GEAs using the LNB encoding depends on P_1 and P_2, we use the one-max tree problem from subsection 6.1.5. We define the best solution to either be a star or a random list and present the performance of GEAs in Figure 6.30, 6.31, 6.32, and 6.33. We use a simple GA on a $n = 16$ node problem with only one-point crossover, no mutation, tournament selection of size 3, a population size of 300, and terminate the run after the population is fully converged. For each parameter setting we perform 100 runs with different randomly chosen positions of the 16 nodes. The distances between the nodes are calculated according the Euclidean metric (compare equation 6.1).

Figure 6.30(a) and 6.30(b) present the average fitness of the best individual over the run dependent on different values of the node-specific bias P_2 ($P_1 = 0$) if the optimal solution is either a star or a list. We see that with increasing P_2, GEAs find the optimal star much faster, whereas GEAs fail completely when searching for the optimal list. The reader should also notice that with increasing node bias P_2 the initial population becomes more and more star-like, and the average fitness of the best individual in the initial population becomes higher if the optimum is a star. If the node-bias is very small ($P_2 \to 0$) only the MST can be encoded and the only individual a GA can find is therefore the MST. As a result, GAs fail for small values of P_2.

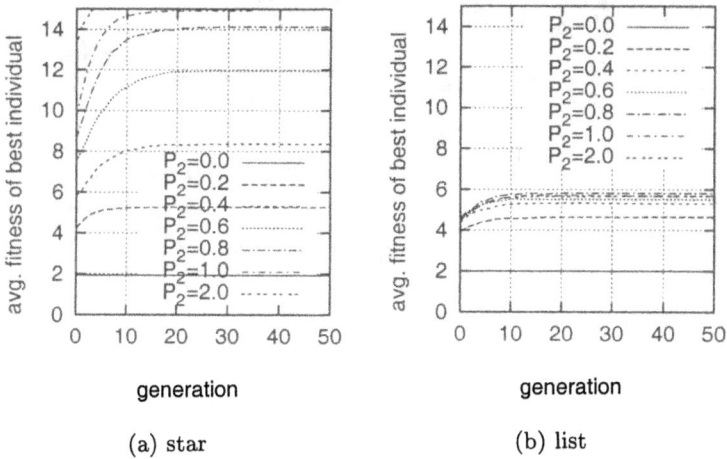

(a) star   (b) list

Fig. 6.30. Average fitness of the best individual dependent on different values of the node-specific bias P_2. The link-specific bias is set to $P_1 = 0$. We use the 16 node one-max tree problem and the best solution is either a star or a list. The results reveal that GEAs using the node-biased encoding perform better with increasing P_2 if the optimal solution is a star. If the optimal solution is a list, GEAs fail. We see that the performance of GEAs using only a node-specific bias strongly depends on the structure of the optimal solution.

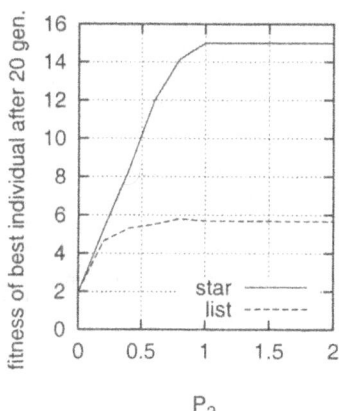

Fig. 6.31. We compare for a 16 node one-max tree problem how the fitness of the best individual after 20 generations depends on the structure of the optimal solution. If the optimal solution is a star the fitness increases with increasing node-specific bias P_2. If the optimal solution is a list, GAs fail and the fitness at the end of the run is independent of P_2.

The problem with the node-bias P_2 becomes more obvious when looking at the best solution at the end of the run dependent on P_2 as illustrated in Figure 6.31. If the optimum is a star, GEAs perform better and better with increasing P_2. However, if the optimum solution is a list, GEAs are not able to find the optimal solution.

The situation is different when investigating the influence of the link-specific bias P_1 ($P_2 = 0$) on the performance of GEAs as illustrated in Figure

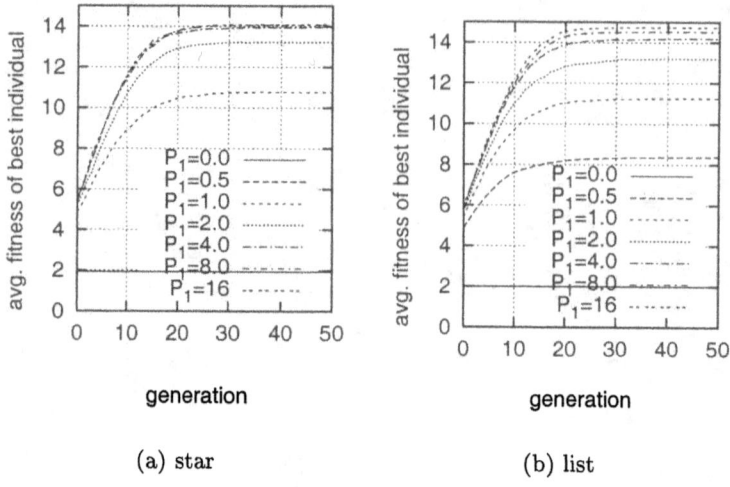

(a) star   (b) list

Fig. 6.32. Average fitness of the best individual dependent on different values of the link-specific bias P_1. The node-specific bias is set to $P_2 = 0$. We use the one-max tree problem and the best solution is either a star or a list. The results reveal that with P_1 large enough, the LNB encoding is unbiased, and the performance of GEAs is independent of the structure of the optimal solution.

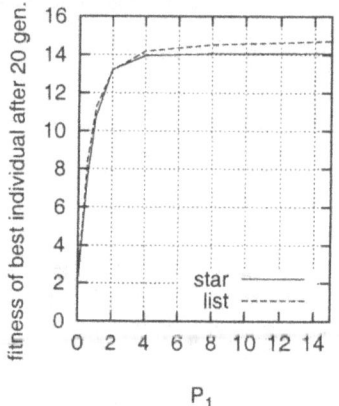

Fig. 6.33. We compare for a 16 node one-max tree problem how the fitness of the best individual after 20 generations depends on the structure of the optimal solution. We only use a link-specific bias P_1. We see that the performance of GEAs is independent of the structure of the optimal solution. Therefore, if the link-bias P_1 is large enough, the encoding is unbiased.

6.32 and 6.33. For both problems, GEAs work properly with P_1 large enough. GEAs searching for a star perform as well as when searching for a list. If the link-bias is large enough, the encoding is (almost) unbiased and GEA performance is independent of the structure of the optimal solution. As we have seen before the encoding is slightly biased towards lists and GAs perform slightly better when searching for optimal lists. However, as the effect is very small we want to neglect it.

The empirical results prove the theoretical predictions from the previous subsections. With a large value of the node-specific bias P_2, the encoding is strongly biased towards stars, and GEAs fail if the optimal solution is a non-star structure like a list. With the link-specific bias P_1 large enough the encoding becomes unbiased and GEAs perform well independently of the structure of the optimal solution. If both biases are very small, only the minimum spanning tree can be represented and GEAs not searching for the MST fail.

6.3.7 Conclusions

After a brief introduction, we reviewed the link and node biased encoding (LNB) as described in Palmer (1994). We described the encoding and illustrated that when setting the link-specific bias to zero, we get the simplified node-biased encoding which is mostly used in the literature. For both types of encodings, the node-biased and the link-and-node-biased encoding, we presented a small example. In section 6.3.3, we showed that the LNB encoding is redundant, and that an encoding with a bias is the same as a non-uniformly redundant encoding. Then we focused in subsection 6.3.4 on the node-biased encoding and illustrated that it is biased towards either stars or to the minimum spanning tree. This is followed by an investigation into the more general link-and-node-biased encoding which revealed that with increasing link-bias, the encoding becomes less biased. Finally, subsection 6.3.5 proved the predictions about GEA performance by empirical results.

Analyzing the notion of biased encodings as given by Palmer (1994) we recognize that a biased encoding is the same as a non-uniformly redundant encoding. Therefore, we can use the framework from chapter 4 for the analysis of the performance of GEAs using the redundant LNB encoding. The performance of GEAs using redundant representations goes with $O(r/2^{k_r})$, where r denotes the number of copies that are given to the best phenotypic BB and k_r denotes the order of redundancy. Palmer, who introduced the LNB encoding drew the conclusion at the end of his thesis that the

> "... new Link and Node Bias (LNB) encoding was shown to have all the desirable properties ..." (Palmer, 1994, pp. 90)

illustrated in subsection 6.1.6 including those to be unbiased that means uniformly redundant. However, we have shown in this section that this claim is not true. With increasing node-specific bias, the encoding becomes more and more biased towards stars. At the extreme for the node-specific bias large enough, the LNB encoding is not able to represent anything other than stars. Therefore, GEAs using a large node-specific bias can not work properly at all.

Fortunately, the encoding becomes uniformly redundant with increasing link-specific bias. If the node-specific bias can be ignored, and the link-specific bias is large enough, the encoding represents all phenotypes uniformly, and GEAs perform independently of the structure of the optimal solution.

Finally, the encoding has problems if both biases are very small because then the encoding can only represent structures that are similar to the minimum spanning tree. At the extreme, when P_1 and P_2 are very small, the genotype has no influence on the phenotype and the only network that can be represented using the LNB encoding is the minimum spanning tree.

Because optimal solutions for the optimal communication spanning tree problem (see section 8.2) often tend to be star- or MST-like, the LNB encoding could be a good choice for this problem. In general, however, the encoding has some serious problems, especially when using the simplified node-biased encoding. Researchers should therefore be careful when using this encoding for other problems because some network structures are not encoded at all, and a randomly generated individual is biased towards either star or minimum spanning trees.

As a result, we strongly encourage users to use, as long as they have no idea about the structure of the optimal solution, higher values for the link-specific bias, and to not use the node-specific bias. Otherwise, GEAs are likely to have great problems in finding optimal non-star structures, and a reduction of GEA performance is unavoidable.

6.4 The Characteristic Vector Encoding

The characteristic vector is one of the most common approaches for encoding the structure of a network (Davis, Orvosh, Cox, & Qiu, 1993; Berry, Murtagh, McMahon, & Sugden, 1997; Ko, Tang, Chan, Man, & Kwong, 1997; Dengiz, Altiparmak, & Smith, 1997c; Dengiz, Altiparmak, & Smith, 1997b; Dengiz, Altiparmak, & Smith, 1997a; Berry, Murtagh, McMahon, Sugden, & Welling, 1999; Premkumar, Chu, & Chou, 2001). Examples for the use of the characteristic vector encoding can be found in Tang et al. (1997) and Sinclair (1995).

The purpose of this section is to use the framework from section 4 for the analysis of the characteristic vector encoding. The investigation shows that characteristic vectors are able to represent invalid solutions. Therefore, the encoding is redundant and a repair mechanism is necessary that constructs valid trees from invalid genotypes. In contrast to the link and node biased encoding, characteristic vectors are uniformly redundant but affected by stealth mutation. Stealth mutation is a result of the repairing process and brings already extinguished schemata back into the population. Repairing invalid solutions works like additional mutation and increases the run duration t_{conv}.

In the following subsection, we describe the functionality of the characteristic vector encoding. This is followed in subsection 6.4.2 by a discussion about what to do with representations that are able to represent invalid solutions. We illustrate that the characteristic vector encoding can represent invalid solutions and we propose a repair mechanism for the encoding. In section 6.4.3, we investigate the effects of the repair mechanism. We show that characteristic vectors are uniformly redundant because the proposed repair mechanism is unbiased. However, repairing invalid solutions results in stealth mutation which increases the run duration t_{conv}. The section ends with a brief summary.

6.4.1 Encoding Trees with the Characteristic Vector

In the following, we briefly describe the characteristic vector (CV) encoding and review some of its important properties.

The characteristic vector encoding can be used for the encoding of trees. Further information and examples for its use can be found in Berry, Murtagh, and Sugden (1994) and Palmer (1994).

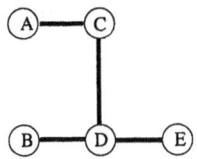

Fig. 6.34. A five node tree

A characteristic vector is a binary vector that indicates if a possible link is used or not in the network. For an n-node network there exist $n(n-1)/2$ possible links, and a characteristic vector of length $l = n(n-1)/2$ is necessary for encoding the structure of an n-node network. All possible links must be numbered, and each link must be assigned to a position in the vector. In Table 6.4, we give an example of a characteristic vector for a 5 node tree. The nodes are labeled from A to E. The link from node A to B is assigned to the first position in the string, the link from A to C is assigned to the second position, and so on. To indicate if the ith link is established, the value at position i is set to one. If no link is established, the value of the allele is set to zero. The tree that is represented by Table 6.4 is shown in Figure 6.34.

Table 6.4. The characteristic vector for the tree in Figure 6.34

0	1	0	0	0	1	0	1	0	1
A-B	A-C	A-D	A-E	B-C	B-D	B-E	C-D	C-E	D-E

The encoding has interesting properties. Firstly, it is able to represent all possible trees. The encoding can also represent non-trees, but we will discuss this problem in the next subsection. Furthermore, all alleles of a characteristic vector have the same contribution to the construction of the phenotype. Therefore, we expect no problems due to non-uniformly scaled alleles, domino convergence, or genetic drift.

Finally, the locality of the encoding is perfect and the distance distortion d_c is low. Neighboring genotypes correspond to neighboring phenotypes. Two neighboring phenotypes x_i and x_j that have the phenotypic distance $d_{i,j}^p = 1$ differ in exactly two positions in the genotype. In general, we can calculate the phenotypic distance d^p between two individuals x_i and x_j as

$$d_{i,j}^p = 0.5 d_{i,j}^g,$$

where d^g denotes the genotypic Hamming distance and d^p denotes the distance between trees as defined in subsection 6.1.2. Therefore, the distance distortion $d_c = 0$. In contrast to other encodings such as the Prüfer number encoding (compare section 6.2), the problem difficulty remains unchanged when assigning the genotypes to the phenotypes. The BBs have the same complexity for the genotypes and phenotypes.

However, as already mentioned, the characteristic vector encoding can also represent non-trees. We discuss this problem in the following subsection.

6.4.2 Repairing Invalid Solutions

We describe in the following how to handle invalid solutions (non-trees) which can be represented by the characteristic vector encoding. Most common is the repairing of invalid genotypes.

Every characteristic vector that represents a tree must have exactly $n-1$ ones, the represented graph must be connected, and there are no cycles allowed. This makes the construction of trees from randomly chosen characteristic vectors demanding as most of the randomly generated characteristic vectors are invalid, and not trees. For an n-node network, there are $2^{n(n-1)/2}$ possible characteristic vectors, but only n^{n-2} valid trees (Prüfer, 1918). The probability of randomly getting a tree is $\frac{n^{n-2}}{2^{n(n-1)/2}} < \frac{2\ln(n)}{\ln(2)n} < 3\ln(n)/n$. Therefore, the chance of randomly creating a CV that represents a tree decreases exponentially with the problem size n (Palmer, 1994).

Randomly chosen characteristic vectors which should represent a tree can be invalid in two different ways:

- There are cycles in the represented graph.
- The graph is not connected.

We get a cycle for the example in Table 6.4 if we set the first allele (A-B) to one. This characteristic vector does not represent a tree any more because there is a cycle of A-C-D-B-A. On the other side, if we alter any of the ones in the characteristic vector to zero we get two disconnected trees. If A-C is switched to zero, this will be the case. One subtree consists of the node A and the other consists of a star with center D.

If we do not want to remove invalid solutions from the population, there are two different possibilities to handle invalid solutions: Firstly, we can ignore the invalid solutions and leave them in the population, or secondly, we repair them. Some GEA approaches report to some extent good results when accepting invalid solutions (Orvosh & Davis, 1993; Davis, Orvosh, Cox, & Qiu, 1993). However, when leaving invalid solutions in the population we must ensure that a valid individual is created at the end of the run. Furthermore, we must find a way to evaluate invalid solutions. This can be difficult as for tree problems a fitness function exists for trees, but not for non-trees. Finally, the largest problem is that for tree problems, the probability of generating a valid individual drops exponentially $O(\exp(-n))$ and therefore, only a very small fraction of the individuals are valid at all. Due to these problems, most of the traditional GEAs choose the second possibility and repair all infeasible solutions by some kind of repair mechanism.

The repairing of invalid solutions is mainly carried out in two steps (Berry, Murtagh, & Sugden, 1994):

1. Remove links that cause cycles.
2. Add links to obtain a connected network.

When repairing a characteristic vector that should represent a tree, the cycles in the graph must be identified. If we randomly choose the characteristic vector $c = 1100010100$ for a 5 node network, we are faced with the cycle A-C-D-B-A in the graph. We could choose one of the links A-C, C-D, D-B or B-A, and remove it randomly. When we choose the link A-C we get $c = 1000010100$. As there are no more loops we can stop removing links and continue checking whether the graph is fully connected. As there are only three ones in c and we have a tree with 5 nodes, the graph could not be connected and we have to add one link. As the node E is separated from the rest of the tree, a link from E to a randomly chosen node A, B, C or D, has to be added. The link C-E is chosen and we finally get the $c = 1000010110$. A closer look at the repair mechanism shows that the order of the repair steps does not matter.

After reviewing how to handle invalid solutions, we want to examine some important properties of the encodings.

6.4.3 Bias and Stealth Mutation

In this subsection we examine the characteristic vector representation more closely. We have already seen that the characteristic vector encoding is redundant. Consequently, we have to investigate if the representation is uniformly redundant. Otherwise, GEA performance depends on the structure of the optimal solution. Furthermore, a repair mechanism for repairing invalid solutions is necessary when using characteristic vectors. We illustrate that the unbiased repair mechanism from the previous subsection results in mutation effects, even if no mutation is used. This effect is denoted as stealth mutation.

A Redundant, but Unbiased Encoding

We illustrate in the following that the characteristic vector encoding with repair mechanism can be described as a redundant encoding. Furthermore, we show that when using the repair mechanism from the previous subsection that the encoding is unbiased that means uniformly redundant.

The genotype-phenotype mapping f_g constructs a valid tree $x_p \in \Phi_p$ with the help of the repair mechanism from every possible valid or invalid characteristic vector $x_g \in \Phi_g$. This means that n^{n-2} phenotypes are represented by $2^{n(n-1)/2}$ genotypes. Therefore, we see that the characteristic vector encoding is a redundant encoding independently of whether the invalid individuals are repaired or remain unrepaired in the population. If they remain unrepaired in the population they must be evaluated using the fitness function defined on the phenotypes (that means on trees). Therefore, the invalid unrepaired individuals must be assigned in some way to the feasible trees and we have the same situation as when the individuals are repaired.

Recognizing that the characteristic vector is a redundant encoding, we can use the insights into the effects of redundant encodings from section 3.1. Therefore, we are especially interested as to whether the characteristic vector encoding is biased that means non-uniformly redundant. A closer look at the repair mechanism from subsection 6.4.2 shows that the removed, respective added links are chosen randomly. Furthermore, we know that an encoding is unbiased if every phenotype is represented on average by the same number of genotypes. The random repair process shows exactly this behavior as it does not favor any particular genotype. Therefore, the characteristic vector encoding is unbiased, that means uniformly redundant.

To investigate empirically whether the characteristic vector encoding is unbiased, we randomly create a characteristic vector and measure the average minimum distance towards a star and the average distance to the minimum spanning tree. In Table 6.5, we present the mean μ and the standard deviation σ of the two distances. We show the average minimum distance to one of the n stars and the average distance to the minimum spanning tree. We randomly created 10 000 characteristic vector individuals and compare the characteristic vector representation to an unbiased representation like the Prüfer number encoding. The numbers indicate that the characteristic vector is, in contrast to the LNB encoding, unbiased.

Because the characteristic vector encoding is uniformly redundant, GEA performance is independent of the structure of the optimal solution. We present in Figure 6.35 results for the performance of GEAs for a 16 node one-max tree problem using only uniform crossover, no mutation, and tournament selection of size 3. The plots show the probability of success $1 - \alpha$ (finding the optimal solution) and the fitness of the best individual after the population is completely converged. We performed 250 runs for either a star, a list or an arbitrary tree is the optimal solution. Before evaluating an invalid individual in each generation, we repair it according to the algorithm

Table 6.5. Average minimum distance to stars and average distance to the minimum spanning tree. The numbers reveal that the characteristic vector representation is uniformly redundant and is not biased to some solutions.

| n | minimum distance to star | | | | distance to MST | | | |
| | unbiased | | CV | | unbiased | | CV | |
	μ	σ	μ	σ	μ	σ	μ	σ
8	3.67	0.643	3.66	0.645	5.16	0.993	5.19	0.991
16	10.91	0.783	10.91	0.787	13.08	1.072	13.08	1.087
32	26.25	0.818	26.23	0.827	29.08	1.311	29.05	1.310

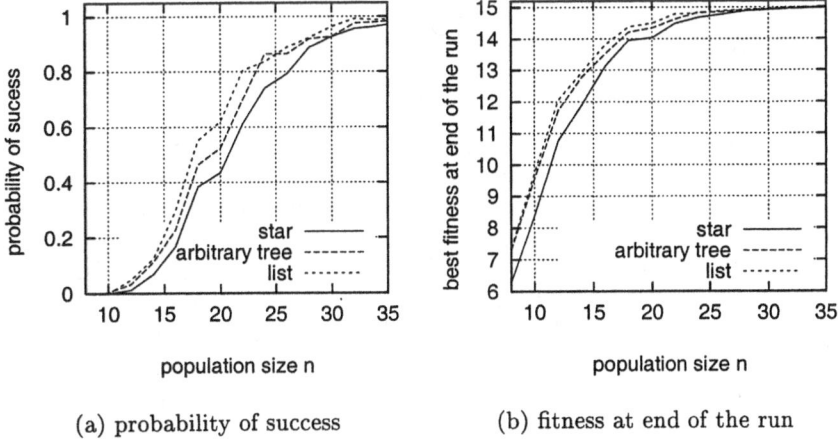

(a) probability of success   (b) fitness at end of the run

Fig. 6.35. Performance of GEAs using the CV encoding for optimizing a 16 node one-max tree problem with the best solution either a star, a list, or an arbitrary tree. Invalid solutions are repaired. GEA performance is approximately independent of the structure of the optimal solution. Therefore, the redundant characteristic vector encoding using the repair mechanism from subsection 6.4.2 is unbiased, and has no problem with over- or underrepresentation of individuals.

outlined in subsection 6.4.2, and the repaired characteristic vector replaces the invalid solution. The plots show that GEA performance is approximately independent of the structure of the optimal solution.

Finally, we briefly discuss the possibility of using specific mutation and recombination operators that always create only valid solutions. Then, no repairing is necessary and we do not have to worry any more about a bias. Every individual that is created during the GEA's run would be valid. However, one problem is the creation of the initial, necessarily valid generation if the fraction of valid solutions is tiny and only a genotype-phenotype mapping, but not the inverse, exists. Furthermore, the creation of thus-like "intelligent" crossover and mutation operators leads to a direct encoding (compare section 7.2). Then, the genetic operators are, in contrast to our used standard x-point crossover or bit-flipping mutation, not based on the Hamming distance between individuals any more (compare subsection 3.3.2). Instead, the operators are problem-specific and there exists no explicit representation ($\Phi_g = \Phi_p$).

We have illustrated that the characteristic vector encoding is redundant. However, as the proposed repairing of invalid solutions does not prefer some phenotypes, the encoding is unbiased and GEA performance is independent of the structure of the optimal solution.

Stealth Mutation

To repair invalid characteristic vector individuals, the insertion of links is necessary if the tree is underspecified. This results in mutation-like effects even if no mutation is used. We denote this as stealth mutation and discuss in the following its effect on GEA performance.

When recombining two valid parents which are encoded using the CV encoding, the offspring are often underspecified. The represented tree is not connected and the repair mechanism we presented in subsection 6.4.2 inserts links randomly to construct a valid individual. Therefore, links which do not exist in both parents could be used for the construction of the offspring. It could even happen that a link that does not not exist in any of the individuals in the population can find its way back into the population by the repair mechanism. This means, although we only use recombination, we still get some kind of mutation when we use characteristic vectors. This effect caused by the repair process should be called *stealth mutation*.

When using the notion of BBs, stealth mutation results in a continuous supply of new BBs during a run. New BBs are created randomly in the population during GEA-run, even if they are not present in the start population, or not properly mixed and lost. This effect makes a comparison of this encoding to other representations difficult as stealth mutation muddies the results. GEAs need longer to converge but can perform better on easy problems.

In Mühlenbein and Schlierkamp-Voosen (1993) and Thierens and Goldberg (1994), the time until convergence was found for the one-max problem as $t_{conv} = \pi\sqrt{l}/2I$ with the selection intensity I and the string length l. I depends only on the used selection scheme and for tournament selection of size 2 we get $I = 1/\sqrt{\pi}$. With $l = n(n-1)/2$, we get $t_{conv} \approx \pi\sqrt{\frac{\pi}{2}}n$. n denotes the size of the problem (number of nodes).

We compare in Figure 6.36 for the one-max tree problem the theoretical prediction for the run duration t_{conv} with the empirical results for the characteristic vector encoding. For the empirical analysis, we use a simple GA with no mutation, tournament selection of size $s = 2$, and uniform crossover. The population size is large enough to reliably find the optimal solution.

The results show a non-linear dependency of the run duration t_{conv} over the problem size n. The plots indicate that GEAs using CVs struggle because of more repair operations and stealth mutation. The search for good solutions depends more on the random effects of mutation than on recombination. We

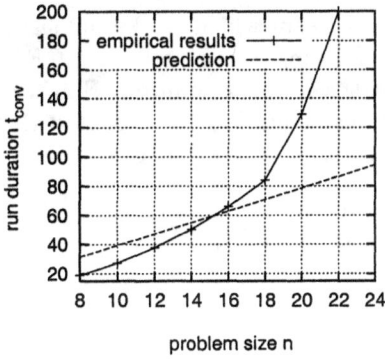

Fig. 6.36. Run duration t_{conv} over problem size n for the one-max tree problem using tournament selection of size 2 and uniform crossover. For selectorecombinative GAs t_{conv} should grow linearly with the problem size n. However, if GEAs use the CV encoding t_{conv} grows due to problems with stealth mutation approximately exponentially with increasing number of nodes n.

can assume that for simple problems, the stealth mutation of the CV helps GEAs to regain lost links randomly. With higher problem size, the probability of randomly finding the correct link decreases, and the run duration of GEAs using CVs is increased.

6.4.4 Summary

This section examined important properties of the characteristic vector encoding in the context of trees. We started with a description of the encoding and briefly reviewed its important properties. In subsection 6.4.2, we illustrated the problem of invalid solutions and how invalid solutions can be repaired. This was followed in subsection 6.4.3 by an investigation into the properties of the encoding. We recognized that the characteristic vector encoding is redundant and, in contrast to the LNB encoding, unbiased. Therefore, GEA performance is independent of the structure of the optimal solution. Furthermore, it was shown that the repair process results in stealth mutation which increases run duration. We proved the theoretical predictions about the increase of the run duration by empirical results for the one-max tree problem.

We have seen in this section that an encoding which can represent not only valid, but also invalid, solutions encodes a valid phenotype by more than one genotype. Therefore, such an encoding is redundant and the results about redundant encodings from section 3.1 can be used. Redundancy is independent on whether the invalid solutions are repaired, or if they remain untouched in the population. In both situations, it is necessary to evaluate the invalid genotypes and to assign a fitness value to every invalid solution. Furthermore, we have seen that an encoding which can represent invalid solutions is unbiased if the construction of a valid phenotype from an invalid genotype is unbiased and does not favor some geno- or phenotypes.

Our investigation in the characteristic vector encoding has shown that the encoding is redundant if it is used for encoding trees. Only n^{n-2} valid trees are encoded by n^{n-2} valid and $2^{n(n-1)/2} - n^{n-2}$ invalid solutions. To repair invalid solutions, we presented a repair mechanism which works unbiased. Therefore, the whole characteristic vector encoding is unbiased, and GEAs perform independently of the structure of the optimal solution. However, repairing invalid solutions results in additional mutation. We denoted this effect as stealth mutation and showed that it increases the run duration t_{conv}.

6.5 Conclusions

In the previous sections, we used the framework from chapter 4 for an analysis of some of the most common tree representations. By doing this, we were able to illustrate the relevance of the basic design principles more clearly and to understand the influence of common tree representations on the performance of GEAs.

We started in section 6.1 by providing the necessities for analyzing tree representations. We defined the network design problem and presented the used metric for graphs. This was followed by the schema analysis for graphs which we used in the following for measuring phenotypic problem difficulty. Based on the schema analysis for graphs we presented in subsection 6.1.5 scalable test problems for trees (one-max tree and deceptive trap tree problems). The section ended with a review of design criteria for trees as given by Palmer (1994).

Section 6.2 presented an investigation into the properties of the Prüfer number encoding. After an historical review, the construction rule, and known properties of the encoding, we focused in subsection 6.2.4 on the low locality of the encoding. We performed random walks through the search spaces and showed that the locality of the representation is low. This was followed by an investigation into the locality of neighboring individuals. The section ended with empirical proof of the theoretical predictions about GEA performance derived from section 3.3 concerning the locality of the encoding.

In section 6.3, we examined the link and node biased (LNB) encoding which was developed by Palmer (1994). We started by illustrating the motivation for developing the encoding and described its different variants. This was followed by illustrating that the link and node biased encoding is redundant, and that the bias of an encoding is the same as non-uniform redundancy. The investigation revealed that the LNB encoding is strongly biased towards stars as long as a node-specific bias is used. Therefore, stars are overrepresented in a randomly generated population, and the performance of GEAs depends on the structure of the optimal solution. Finally, we proved the theoretical predictions about GEA performance by empirical results.

Finally, in section 6.4, we focused on the characteristic vector encoding as an example for a redundant, but unbiased encoding. We described how trees can be represented by the characteristic vector encoding, and how invalid solutions can be handled by repairing them. Furthermore, we illustrated that repairing invalid solutions is the same as redundancy. The investigation into the properties of the characteristic vector encoding revealed that the encoding is uniformly redundant, but affected by stealth mutation which increases the time to convergence t_{conv}.

In this section we applied the framework about the influence of representations on GEA performance to the analysis of existing tree representations. The framework allowed us to predict how the performance of GEAs, measured by run duration and solution quality, is affected by the use of different representations for trees. We were able to compare representations in a theory-based manner, to predict the performance of GEAs using different representations, and to analyze representations guided by theory. The analysis showed that the proposed elements of the framework – redundancy, BB-scaling, and locality/distance distortion – can be used properly for analyzing representations. We want to briefly summarize the insights our analysis revealed:

Our investigation into the locality of the Prüfer number encoding has shown that the locality is different in different areas of the search space. For networks that are similar to stars, the encoding has high locality, and the BB complexity is the same for the genotypes and phenotypes. However, for non-stars, the encoding modifies problem difficulty which makes fully easy problems more difficult. These insights explain the inconsistent statements about Prüfer number performance in the literature. If the optimal solution was accidentally star-like, the encoding shows an acceptable performance; if it was non-star-like, GEAs fail.

The investigation into the LNB encoding illustrated the possible effects of redundant encodings. The LNB encoding is not uniformly redundant but strongly biased towards stars as soon as a node-specific bias is used. Therefore, GEAs using the LNB encoding have great problems in finding non-star-like good solutions. Only for the link-bias large enough and neglectable node-bias, does the encoding become unbiased, the redundancy uniform, and GEAs do work well independently of the structure of the optimal solution.

Finally, we presented the characteristic vector encoding as an example of a representation that is also redundant but unbiased. We recognized that an encoding that allows the representation of invalid solutions, like the CV encoding, can be described as a redundant encoding. Such an encoding is unbiased if the repair process is unbiased, that means it does not favor some of the valid genotypes. However, the use of a repair process results in stealth mutation which increases the run duration t_{conv}.

Last but not least, we presented in this chapter a schema analysis for graph problems. Using it we were able to measure the phenotypic problem complexity of a graph problem and to classify problems to be easy or difficult. Furthermore, it can help us to judge if encodings preserve problem difficulty because we can measure if the problem complexity remains constant when mapping the phenotypes on the genotypes. Based on the schema analysis for graphs, we provided a fully difficult deceptive trap and a fully easy one-max tree optimization problem. Because standard test problems for network problems are not widely used, both scalable test problems are helpful for comparing the performance of different tree encodings. Furthermore, the test problems allow users to easily examine if GEA performance depends on the structure of the optimal solution. This is important for investigating effects that can be caused by non-uniformly redundant encodings.

This chapter has applied the principles of representations from chapter 3 to common tree encodings. By identifying Prüfer numbers to have low locality, the LNB encoding to be redundant but biased, and the CV encoding to be redundant but unbiased, we were able to predict the behavior and performance of genetic and evolutionary algorithms using these representations. In general, by applying the presented theory of representations to other not mentioned, or new representations, the behavior of GEAs using these encodings can be much better predicted. Therefore, we want to encourage researchers to use the presented theory about representations from chapter 3 and 4 for analyzing other representations. By following this advice, the behavior and performance of GEAs could be predicted much better and GEAs can be used much more efficiently.

7. Design of Tree Representations

In the previous chapters, we have gained new insights into the theory of representations. As a result, we are able to predict, using the framework presented in chapter 4, the influence of redundant, exponentially scaled, or low locality representations on GEA performance. However, the presented framework can not only be used to analyze the performance of existing representations, but also to develop representations in a theory-guided manner.

The purpose of this chapter is to use the framework about representations for developing tree representations. According to the new insights into representations, there are several requirements representations should fulfill: A representation should be robust according to the location of the optimal solution in the search space. Furthermore, it should allow genetic operators to work properly and to propagate the high-quality BBs from the parents to the offspring. Finally, the representation should not increase the difficulty of the problem but allow GEAs to perform the best they can. Based on these assumptions, we design two tree representations.

The network random key (NetKey) representation combines the advantageous elements from the LNB and CV representation. The information about the tree is stored, in a similar way to random keys, as an ordered permutation of links. The NetKey representation allows GEAs to distinguish between high and low quality links, it is uniformly redundant, it has high locality, and it allows the efficient use of standard recombination operators.

For the direct tree (NetDir) representation we will choose a completely different approach and construct a direct representation for trees. No explicit genotype-phenotype mapping exists and the genotypes are the same as the phenotypes. The "representation" does not change problem difficulty and GEAs using it perform independently of the structure of the optimal solution. However, the standard recombination operators can not be applied to a direct tree representation. It is necessary to develop specific operators for trees with respect to the processing of building blocks. The NetDir representation illustrates that when using direct representations, the design issue of finding good representations is transformed into the black art of finding good operators.

The chapter is divided into two parts. In section 7.1 we present the network random keys, and in section 7.2 the direct representation.

7.1 Network Random Keys (NetKeys)

We have learned in the previous chapters that high quality representations should be robust, allow genetic operators to work properly, and should not increase problem difficulty. However, Prüfer numbers and LNB encoding have shown to have difficulties with some of these issues. Prüfer numbers make easy problems more difficult, and the LNB encoding is not uniformly redundant. This means that the performance of GEAs depends on the location of the optimal solution in the search space. Only the characteristic vector encoding promises good GEA performance, but it has some problems with invalid solutions and stealth mutation.

Therefore, we combine the advantageous properties of the characteristic vector and the LNB encoding to create a more powerful representation: the *network random keys* (NetKeys). The NetKey encoding belongs to the class of weighted encodings. In contrast to other representations, for example the binary encoding which can only indicate whether a link is established or not, weighted encodings allow us to encode the importance of a link. Consequently, an additional construction algorithm is necessary which constructs a valid tree from the weighted representation (the random key sequence) of length $l = n(n-1)/2$ considering the importance of the links.

The section is structured as follows. We start by illustrating how the Net-Keys can be created by combining the characteristic vector encoding with some elements of the link and node biased encoding. This is followed in subsection 7.1.2 by the functionality of the NetKeys. We illustrate the random keys which store the importance of the links as weighted vectors, and the construction algorithm which constructs a valid tree from a random key sequence according to the importance of the links. Subsection 7.1.3 summarizes the benefits of the NetKey encoding. This is followed by an investigation into whether the NetKey representation is uniformly redundant, or not. We measure the distance towards star and minimum spanning tree networks, and provide empirical verification that GEAs using NetKeys perform independently of the structure of the optimal solution. Before closing the section with concluding remarks, in subsection 7.1.5 we present a model for the population sizing and run duration of GEAs using NetKeys for the one-max tree problem.

7.1.1 Motivation

Chapter 6 has examined some problems with existing tree representations. We illustrate in this subsection how we can define the principial functionality of the new NetKey encoding by combining the characteristic vector encoding with some interesting elements of the LNB encoding.

We have seen in subsection 6.4.3 that the characteristic vector encoding is a redundant, but unbiased encoding. However, the genetic operators mutation and crossover produce invalid solutions which can be under- or over-specified.

Thus, a repair mechanism is necessary that restores valid solutions. Because an allele only indicates if a link is established or not, the repair mechanism must rely on random link insertion or deletion. The repair mechanism works randomly and has no information on how to repair invalid solutions. Nevertheless, the CV encoding has advantageous properties and we want to use it as the basis for the construction of the NetKeys.

When trying to improve the CV encoding, we have to overcome the problem that the repair mechanism can not distinguish between the importance of the links. Furthermore, we must ensure that the construction of the phenotype from the genotype remains unbiased. Therefore, we replace the binary alleles, which only indicate if a link is established or not, by continuous alleles which encode the importance of a link by a weighted value (a randomly chosen number). Then, we can design an unbiased construction mechanism which is able to distinguish between the importance of links and results only in valid solutions.

By using a weighted instead of a binary encoding, the NetKey encoding inherits some properties of the LNB encoding when using only a link-specific bias. As we have seen, the phenotypes are constructed in an unbiased way if the weight of the link-specific bias is large enough. However, in contrast to the LNB encoding, the alleles of the NetKeys directly encode the importance of a link and not the bias of the distance matrix. Furthermore, we do not use Prim's algorithm (Prim, 1957) which constructs the minimum spanning tree from the modified distance matrix, but we have developed a new algorithm which allows us to use the information about the importance of a link and which is unbiased.

7.1.2 Functionality

For describing the functionality of the NetKey encoding, we have to separate the representation into two parts: Firstly, the sequence of random keys which stores the importance of the links as a weighted vector of length $l = n(n - 1)/2$. Secondly, the construction algorithm which constructs a tree from the random key sequence. In the following, we illustrate the functionality of both elements of the NetKey encoding.

Random Keys

By substituting the zeros and ones in the characteristic vector encoding by continuous values that can describe the importance of the links, the first part of NetKey functionality is defined. However, the idea to use a weight for describing the importance of an allele is not new, and has already been presented in a different context as the so called random key encoding. For other work about weighted encodings in the context of tree representations the reader is referred to Palmer (1994) or Raidl and Julstrom (2000). In the following, we summarize the history and properties of random keys (RKs).

The random key representation for representing permutations was first presented by Bean (1992). Later, the encoding was also proposed for single and multiple machine scheduling, vehicle routing, resource allocation, quadratic assignments, and traveling salesperson problems (Bean, 1994). Norman and Bean (1994) refined this approach (Norman & Bean, 2000) and applied it to multiple machine scheduling problems (Norman & Bean, 1997). An overview of using random keys for scheduling problems can be found in Norman (1995). In Norman and Smith (1997) and Norman, Smith, and Arapoglu (1998), random keys were used for facility layout problems. In Knjazew (2000) and Knjazew and Goldberg (2000), a representative of the class of competent GAs (fast messy GA (Goldberg, Deb, Kargupta, & Harik, 1993)) was used for solving ordering problems with random keys.

The random key representation uses random numbers for the encoding of a solution. A key sequence of length l is a sequence of l distinct real numbers (keys). The values are initially chosen at random, are floating numbers between zero and one, and are only subsequently modified by mutation and crossover. An example for a key sequence is $r = (0.07, 0.75, 0.56, 0.67)$. Of importance for the interpretation of the key sequence is the position and value of the keys in the sequence. If we assume that $Z_l = \{0, \ldots, l-1\}$ then a permutation σ can be .defined as a surjective function $\sigma : Z_l \to Z_l$. For any key sequence $r = r_0, \ldots, r_{l-1}$, the permutation σr of r is defined as the sequence with elements $(\sigma r)_i = r_{\sigma(i)}$. The permutation r^s corresponding to a key sequence r of length l is the permutation σ such that σr is decreasing (i.e., $i < j \Rightarrow (\sigma r)_i > (\sigma r)_j$). The ordering corresponding to a key sequence r of length l is the sequence $\sigma(0), \ldots, \sigma(l-1)$, where σ is the permutation corresponding to r. This mathematical definitions describes that the positions of the keys in the key sequence r are ordered according to the values of the keys in descending order. In our example we have to identify the position of the highest value in the key sequence (0.75 at position 2). The next highest value is 0.67 at position 4. We continue ordering the complete sequence and get the permutation $r^s = 2 \to 4 \to 3 \to 1$. In the context of scheduling problems this permutation can be interpreted as a list of jobs that are executed on one machine (We start with job 2, then continue with job 4, job 3, and job 1). From a key sequence of length l, we can always construct a permutation of l numbers. Every number between 1 and l (resp. 0 and $l-1$) appears in the permutation only once as the position of each key is unique. Here are some properties of the encoding.

- A valid permutation r^s of l numbers can be created from all possible key sequences as long as there are no two keys r_i that have the same value.[1] Therefore, every random key sequence can be interpreted as a permutation of l numbers.
- There are many possibilities for the construction of a key sequence r from a permutation of numbers r^s. Every element r_i of the sequence can be scaled

---

[1] $r_i \neq r_j$ for $i \neq j$ and $i, j \in [1, l]$

up by some factor and r still represents exactly the same permutation. As long as the relative ordering of the keys in the key sequence is the same, different key sequences always represent the same permutation. It is necessary that r^s is a permutation of l numbers, otherwise no key sequence r can be constructed from r^s.

- RKs encode both the relative position of a number in the permutation r^s (encoded by the value of the key at position i in comparison to all other keys), and the absolute position of i in r^s. The relative position of a number i in the permutation r^s is determined by the numbers that precede and follow i. It is determined directly by the weights of the keys r_i. All numbers j in the sequence r^s that follow i correspond to lower-valued keys $(r_j < r_i)$, whereas all numbers j that precede i correspond to higher-valued keys $(r_j > r_i)$. In the context of scheduling problems all jobs where the corresponding key has a higher value than the ith key are executed before job i, and all jobs with a corresponding key with lower value are executed after i. In contrast, the absolute position of a number i in the permutation r^s cannot be encoded directly, but is only indirectly determined by the value of the ith key. The absolute position describes at which position in the permutation r^s a number i appears. A large value at the ith position leads us to a position at the beginning of the permutation, and a low value leads to a position at the end.

- The distinction between relative and absolute position of a number in the permutation r^s is important for the locality of RKs. The locality of an encoding describes how well the genotypic neighbors correspond to the phenotypic neighbors. A coding has high locality if mutating a genotype changes the corresponding phenotype only slightly. A look at RKs shows that the locality of RKs is high for ordering problems. A small change in the genotype (the key sequence r) leads to a small change in the phenotype (the permutation r^s). The change of one key changes the relative position of exactly one number. However, one must be careful with the definition of the neighborhood. If the absolute position of the numbers in r^s is important, a change of one key is disastrous. If the value of the key r_i with the highest value is modified, only the number i changes its relative position in the permutation r^s, but up to l numbers change their absolute position in the permutation. However, as we use RKs to represent a permutation of numbers, only the relative, and not the absolute positions of the numbers in the permutation must be considered. And for problems where the relative positions of numbers are important, the locality of RKs is high.

- When using genetic and evolutionary algorithms with RKs, standard crossover and mutation operators can be used and are expected to work well. No repair mechanism, or problem-specific operators, are necessary when using this encoding for ordering problems. The standard one- or multi-point crossover schemes work well (Bean, 1994) because the relative ordering of the positions in the parents is preserved and transferred to the offspring (Fox & McMahon, 1991). Due to the high locality of the encoding we expect standard mutation operators to work well and to construct offspring that are similar to their parents.

We have seen that random keys have interesting properties. When using them for the encoding of trees, we still have to define exactly how a tree can be constructed from them. We want to present the construction of a tree from a random key sequence in the following.

Construction of the Tree from the Random Keys

After we have presented random keys as the basis for the NetKey encoding, we still have to define a construction algorithm which creates a valid tree from a random key sequence. Both elements, the random keys and the construction algorithm are necessary for the new NetKey encoding. To get an unbiased and uniformly redundant encoding, we demand the construction algorithm not to favor some phenotypes but to work uniformly.

We have seen that we are able to give priority to the objects in the permutation when using random keys. As NetKeys use continuous variables that could be interpreted as the importance of the link, it is possible to distinguish between more and less important links. The higher the value of the allele, the higher the probability that the link is used for the tree.

When constructing the tree, the positions of the keys in the key sequence r are interpreted in the same way as for the characteristic vector. The positions are labeled and each position represents one possible link in the tree. From a key sequence r of length $l = n(n-1)/2$, a permutation r^s of l numbers can be constructed. Then the tree is constructed from the permutation r^s as follows:

1. Let $i = 0$, G be an empty graph with n nodes, and r^s the permutation of length $l = n(n-1)/2$ that can be constructed from the key sequence r. All possible links of G are numbered from 1 to l.
2. Let j be the number at the ith position of the permutation r^s.
3. If the insertion of the link with number j in G would not create a cycle, then insert the link with number j in G.
4. Stop, if there are $n-1$ links in G.
5. Increment i and continue with step 2.

With this calculation rule, we can construct a unique, valid tree from every possible random key sequence. Thus, the NetKey encoding is now completely described: The new encoding uses random keys which allows us to give priority to some links, and the construction rule uses this information and gradually builds a valid tree.

We want to illustrate the functionality of the NetKey encoding with an example. We use the key sequence from Table 7.1. The permutation $r^s = 10 \rightarrow 8 \rightarrow 6 \rightarrow 9 \rightarrow 2 \rightarrow 7 \rightarrow 1 \rightarrow 5 \rightarrow 4 \rightarrow 3$ can be constructed from the

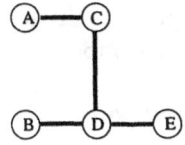

Fig. 7.1. A five node tree

random key sequence r. We start constructing the graph G by adding the link D-E (position 10) to the tree. This is followed by adding C-D (position 8) and B-D (position 6). If we add the link C-E (position 9) to the graph, the cycle C-E-D-C would be created, so we skip C-E and continue by adding A-C (position 2). Now we have a tree with four edges and terminate the construction algorithm. We have constructed the tree shown in Figure 7.1.

Table 7.1. A key sequence r for the five node tree in Figure 7.1

position	1	2	3	4	5	6	7	8	9	10
value	0.55	0.73	0.09	0.23	0.40	0.82	0.65	0.85	0.75	0.90
link	A-B	A-C	A-D	A-E	B-C	B-D	B-E	C-D	C-E	D-E

The computational effort for constructing the phenotype from the genotype is similar for the NetKey and the characteristic vector representation. The calculation of the permutation from the key sequence r can be done in $O(l \log(l))$ (sorting an array of l numbers). The process of constructing the graph from the permutation r^s is comparable to repairing an invalid graph that is constructed from a characteristic vector and its effort depends on the specific structure of the graph.

Similarly to the random keys, the NetKey encoding has high locality. A mutation (changing the value of one key) results either in no change of the corresponding phenotype if the relative ordering is not changed, or the change of two edges if the relative position is changed. Therefore, the maximum distance $d_{i,j}$ between two neighboring NetKey individuals is one (compare subsection 6.1.2 about the definition of distance). The reader should notice, that a mutation of one key of the genotype often dramatically changes the absolute positions of the numbers in the permutation r^s. But the construction rule we defined is only based on the relative ordering of r^s. Therefore, we do not have to worry about the change of the absolute positions.

We have illustrated in this subsection how the NetKey encoding can be used for representing trees. In the following, we summarize the advantages of the encoding.

7.1.3 Advantages

By combining the random key representation which allows GEAs to give priority to some links with a construction algorithm which constructs a valid tree from the random key sequence, we have defined all components of the NetKey encoding. In the following subsection, we summarize the important benefits of this new encoding.

The use of the NetKey encoding has some remarkable advantages:

- Standard crossover and mutation operators work properly.
- The encoding allows a distinction between important and unimportant links.
- There is no over- or under-specification of a tree possible.

In the following, we briefly discuss these benefits. The NetKey encoding allows standard genetic operators to work properly. In subsection 3.3.2 we have demanded mutation operators to create a child which is similar to its parent. Therefore, a small genotypic distance between two individuals should result in a small phenotypic distance. Then, the encoding has high locality. A look at the NetKey encoding shows that the mutation of one key results either in the same or in a neighboring tree which makes it a high locality encoding. Furthermore, we have demanded recombination operators to create an offspring which inherits the properties of its parents. In terms of metric, the distance of an individual to its parents should be smaller than the distance between both parents. In terms of links, an offspring should inherit the links from its parents. Standard recombination operators, like x-point or uniform crossover, show this behavior when used for NetKeys: If a link exists in a parent, the value of the corresponding key is high in comparison to the other keys. After recombination, the corresponding key in the child has the same, high value and is therefore also used with high probability for the construction of the child. This means, the NetKey encoding has high heritability concerning standard recombination operators. As a result, both types of operators, mutation and recombination, work well when used for the NetKey encoding.

A further benefit of the NetKey encoding is that GEAs are able to distinguish between important and unimportant links. In contrast to the characteristic vector encoding which only allows us to store information about whether a link is established or not, we are able to identify the important links in the network when using NetKeys. The algorithm which constructs a tree from the keys uses the high-quality links and ensures that they are not lost during the GEA run. In contrast, low quality links are not used for the construction of the tree. The importance of being able to distinguish between important and unimportant becomes clearer when comparing NetKeys to CVs. The CV encoding can not store information about the importance of a link. Therefore, the repair process must delete or insert links randomly. High quality links can be accidentally removed, or low quality links can find their way back into the population.

Finally, NetKeys always encode valid trees. No over- or underspecification is possible. The construction process which builds a tree from a random key ensures that NetKeys encode valid solutions only. Thus, we do not need an additional repair mechanism.

We see that the NetKey encoding has some remarkable benefits. However, we have not yet investigated whether the encoding is biased. We want to do this in the following subsection.

7.1.4 Bias

Section 7.1.2 has revealed that NetKeys are a redundant encoding. To ensure that GEAs perform independently of the structure of the optimal solution, NetKeys should be uniformly redundant, that means unbiased. In the following subsection, we examine the bias of the NetKey encoding. We measure, in analogy to subsection 6.4.3, the minimum distance of randomly generated NetKey individuals towards one of the n star networks and the average distance towards the minimum spanning tree. This is followed by empirical evidence of the uniform redundancy of the encoding.

We know that NetKeys, similarly to the LNB encoding, are redundant because they encode the importance of a link as a continuous value. Therefore, there is an infinite number of genotypes that represent one specific phenotype. We know from subsection 4.4.1 that GEA performance depends on the location of the optimal solution in the search space if an encoding is non-uniformly redundant. GEAs searching for the optimal solution only perform well if the encoding is not biased towards the low-quality solutions.

The investigation into the bias of the characteristic vector encoding (see subsection 6.4.3) has shown that the CV encoding is unbiased. Furthermore, we know from subsection 7.1.1 that we can interpret NetKeys as an improved version of the characteristic vector encoding. For both encodings we need a repair/construction algorithm which constructs a valid phenotype from the genotypic vector. We want to examine if the construction algorithm of the NetKeys is also, similarly to the characteristic vector, unbiased.

Table 7.2. Average minimum distance to star networks and average distance to the minimum spanning tree. The numbers reveal that the NetKey encoding is uniformly redundant and not biased towards some solutions.

| | minimum distance to star | | | | distance to MST | | | |
| | unbiased | | NetKey | | unbiased | | NetKey | |
n	μ	σ	μ	σ	μ	σ	μ	σ
8	3.67	0.643	3.75	0.602	5.16	0.993	5.24	0.961
16	10.91	0.783	11.00	0.759	13.08	1.072	13.13	1.041
32	26.25	0.818	26.34	0.800	29.08	1.311	29.07	1.319

In Table 7.2, we present for randomly created NetKey individuals the average minimum distance to a star and the average distance to the minimum spanning tree (MST). We randomly create 10 000 individuals for each problem instance and show the mean μ and the standard deviation σ of the distance. The numbers indicate that although there is a small bias we

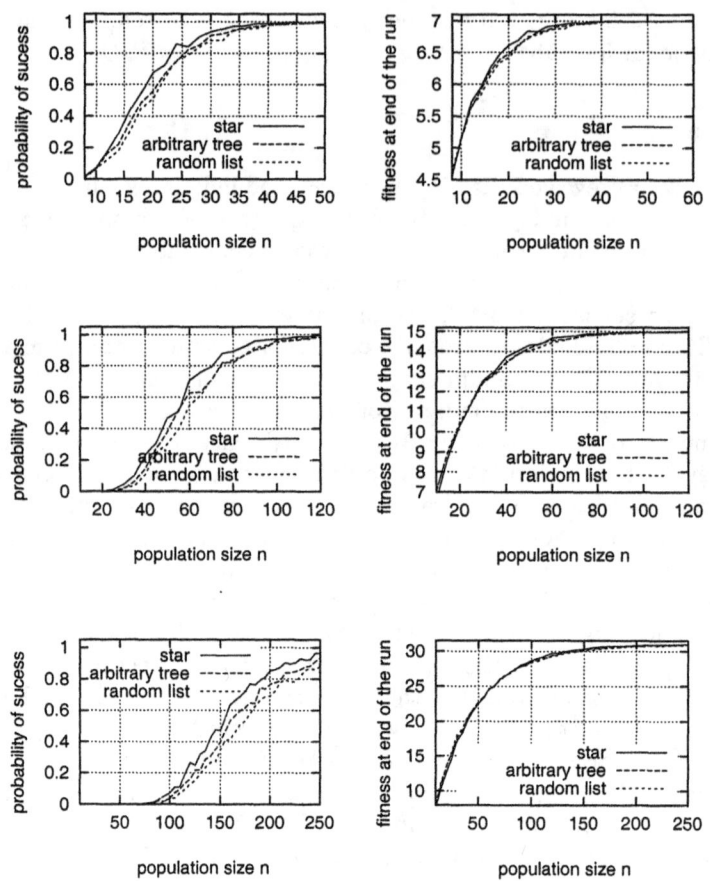

Fig. 7.2. Performance of GAs using NetKeys for 8 (top), 16 (middle) and 32 (bottom) one-max tree problems. The plots show either the probability of success (left) or the fitness at the end of a run (right). The GAs search for the optimal star, list, or arbitrary tree. The plots indicate that GA performance is independent of the structure of the optimal solution. Therefore, NetKeys are uniformly redundant.

can treat the NetKey encoding as unbiased and assume that all phenotypes are represented uniformly. The small bias is expected to be a result of the construction process.

In the remainder of this subsection, we prove the uniform redundancy of the NetKeys by an empirical investigation into the performance of GEAs searching for different types of optimal solutions.

For examining how GEA performance depends on the structure of the optimal solution, we use the one-max tree problem from section 6.1.5. Our GA uses uniform crossover only, no mutation, tournament selection without replacement of size 3, and stops after the population is fully converged. On the left of Figure 7.2, we present the probability of finding the optimal solution (a randomly chosen star, list, or arbitrary tree) over the population size n. The right side shows the fitness of the best individual at the end of the run over the population size n. Both plots confirm that GEAs perform almost independently of the structure of the optimal solution. The reader should notice that the figures indicate a slightly better performance for star networks. We believe that the omitting of invalid links during the construction process results in this very small, marginal, and difficult to predict influence on GA performance. Therefore, we neglect this small bias and assume that NetKeys are unbiased.

Our investigation has revealed that the NetKey encoding is unbiased. GEAs using the encoding perform independently of the structure of the optimal solution.

7.1.5 Population Sizing and Run Duration for the One-Max Tree Problem

Finally, we investigate the necessary population size and run duration of GAs using NetKeys. We present theoretical models for the one-max tree problem which we derived from existing theory and provide empirical verification.

Population Sizing

When extending the population sizing equation of Harik, Cantú-Paz, Goldberg, and Miller (1997) from a binary alphabet to a χ-ary alphabet, we get:

$$N_{min} = -\frac{\chi^k}{2} \ln(\alpha) \frac{\sigma_f}{d} \sqrt{\pi},$$

where χ is the cardinality of the alphabet, α is the probability of failure, σ_f is the overall variance of the function, and d is the signal difference between the best and second best BB. For calculating σ_f we have to investigate how to decide among the competing BBs. For the one-max tree problem we have to find these $n - 1$ links the optimal solution is constructed from. The key sequence r that represents a tree with n nodes consists of $l = n(n-1)/2$ different keys. For the construction of the tree these $n - 1$ keys r_i are used that have the highest value. Therefore, we can split the $n(n-1)/2$ different keys r_i in $n/2$ different groups of size $(n - 1)$. Finding the optimal solution means that all keys r_i with the links i contained in the optimal solution can be found in one group which is considered for the construction of the tree and thus contains the keys with the $n - 1$ highest values of r_i. A good decision among competing BBs means deciding between the $n/2$ different groups of size $n - 1$ and identifying the correct one. A key k_i can belong either to the one group that is considered for the construction of the tree (the key has a high value), or to one of the $n - 2$ groups that are not considered. This is similar to the needle in a haystack model and the standard deviation for such a case is (Goldberg, Deb, & Clark, 1992)

$$\sigma_f = \frac{\sqrt{2l(n-2)}}{n} = \sqrt{\frac{(n-1)(n-2)}{n}} \approx \sqrt{n}.$$

As we have $n/2$ different partitions, the cardinality χ of the alphabet is $n/2$. Using these results and with $k = 1$ (the one-max tree problem is fully easy, and there are no interdependencies between the alleles), and $d = 1$, we get an approximation for the population size N_{min}:

$$N_{min} = -\frac{\sqrt{\pi}}{4} \ln(\alpha) \sqrt{n(n-1)(n-2)} \approx -\frac{\sqrt{\pi}}{4} \ln(\alpha) n^{1.5}.$$

The necessary population size N_{min} goes with $O(n^{1.5})$.

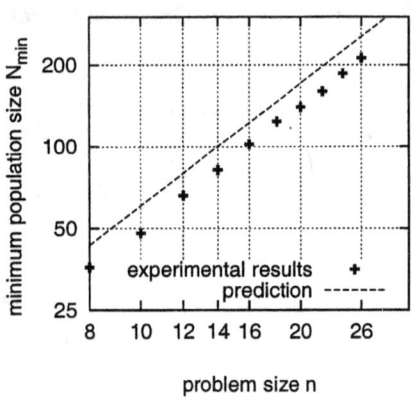

Fig. 7.3. Minimum pop size N_{min} for NetKeys over the problem size n for the one-max tree problem. The probability of finding the optimal solution is $P_n = 0.95$. The population size goes with $O(n^{1.5})$.

In Figure 7.3, the minimum necessary population size N_{min} that is necessary for solving the one-max tree problem with probability $P_n = 1 - \alpha = 0.95$ is shown over the problem size n. We use a simple GA with tournament selection without replacement of size 3, uniform crossover and the NetKey encoding. We perform 500 runs for each population size and for $N > N_{min}$ the GA is able to find the optimum with probability $p = 0.95$ ($\alpha = 0.05$). Although, we have to make some assumptions in our derivation, and the exact influence of the construction algorithm of the phenotype from the encoding is difficult to describe theoretically, the population sizing model gives us a good approximation of the expected population size N which goes with $O(n^{1.5})$.

Run Duration

In Mühlenbein and Schlierkamp-Voosen (1993) and Thierens and Goldberg (1994), the time until convergence is defined as $t_{conv} = \pi\sqrt{l}/2I$ with the selection intensity I and the string length l. I depends only on the used selection scheme and is $I = 3/(2\sqrt{\pi})$ for a tournament size of 3 (Bäck, Fogel, & Michalewicz, 1997, C 2.3). With $l = n(n-1)/2$, we get $t_{conv} \approx const * n$. The run duration t_{conv} should go linearly with the problem size n.

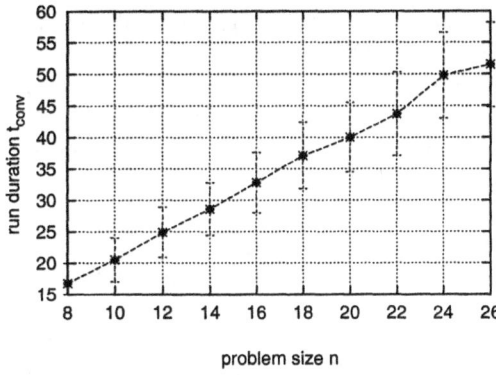

Fig. 7.4. Run duration t_{conv} over problem size n for the one-max tree problem using tournament selection without replacement and uniform crossover

In Figure 7.4, we show the run duration t_{conv} over the problem size n for tournament selection without replacement of size 3 and uniform crossover. t_{conv} measures the number of generations until the population is completely converged. The population size was chosen as $N = 2 * N_{min}$ and N_{min} is from Figure 7.3. The population size N is large enough to ensure that the optimal solution was found in all 500 runs which we performed for every problem instance. The results show that t_{conv} grows, as predicted, linearly with increasing n.

7.1.6 Conclusions

This section presented the NetKey encoding. We started by illustrating how we can combine the characteristic vector representation with some elements of the link and node biased encoding to get the NetKey encoding. This was followed in subsection 7.1.2 by the functionality of the NetKey encoding. We explained the principles of random keys and illustrated how we can construct a valid tree from a random key sequence. After all components of the NetKey encoding were defined, we summarized in subsection 7.1.3 the important advantages of the new encoding. Because NetKeys are a redundant encoding, subsection 7.1.4 presented an investigation into the bias of the encoding. Finally, based on existing theory, we developed in subsection 7.1.5 a population sizing and run duration model for GEAs using NetKeys and optimizing the one-max tree problem.

The sections showed that using the framework outlined in chapter 4 allows theory-guided design of high-quality representations. Based on the new insights into the principles of representations, we were able to develop the new

NetKey encoding. NetKeys are based on the characteristic vector encoding, but use continuous variables for encoding information about the represented tree. The investigation into the properties of the NetKeys revealed that the encoding is uniformly redundant, that standard crossover and mutation operators work properly, and that the representation does not change problem difficulty.

Moreover, the NetKey encoding overcomes some problems of the characteristic vector encoding. In contrast to CVs, trees can not be over- or under-specified when using NetKeys. Furthermore, NetKeys allow a distinction between the importance of the links that should be used for constructing a tree. This is possible because NetKeys encode the importance of the links as a weighted vector, whereas the characteristic vector encoding only encodes whether a link is established or not.

Based on the presented results we encourage further study of NetKeys for encoding both trees and other networks. The use of existing theory for formulating a population sizing model as well as a time to convergence model illustrated the benefits we can get from using existing theory. We encourage users to use existing theory for predicting GEA behavior more frequently. Finally, even though more work is needed, we believe that the properties presented are sufficiently compelling to immediately recommend increased application of the NetKey encoding.

7.2 A Direct Tree Representation (NetDir)

The previous section about the development of the NetKey encoding has shown that a better knowledge about representations can be advantageously used for the design of efficient representations. However, even with theory about representations at hand, the design of new representations is still difficult. A lot of intuition, knowledge of theory, and also luck is necessary for developing good, high-quality, representations. One solution to overcome the problems of finding good representations could be the use of direct representations. When using a direct representation, no explicit genotype-phenotype mapping exists and direct representations are defined by the optimization problem itself. The genotypic representation of the problem is just the phenotype. Therefore, the direct representation does not change problem difficulty and directly encodes the problem.

The purpose of this section is to develop a direct representation for trees (NetDir) and to illustrate that when using direct representations, the design task of finding proper representations is substituted by the search for good crossover and mutation operators. When using GEAs based on the notion of schemata, these problem-specific operators must obey the linkage in the phenotypes and process BBs properly. Therefore, it is not possible by using direct representations to get rid of the difficulties in designing efficient optimization methods.

This section starts with a brief historical review of direct representations for trees. In subsection 7.2.2, we discuss the properties of direct representations. We demonstrate the benefits and drawbacks of using direct representations for GEAs. Because the NetDir representation represents trees as graph structures and not as a list of alleles, standard genetic operators can not be used any more. Therefore, problem-specific operators are necessary. Consequently, in subsection 7.2.3 we develop mutation and crossover operators for the NetDir representation. The section ends with a short summary.

7.2.1 Historical Review

In the following subsection, we give a brief historical review of the use of direct representations for trees.

One of the first approaches to direct representations for trees was presented by Piggott and Suraweera (1993). Offspring individuals are created by randomly copying $n-1$ edges from both parents to the offspring. However, the creation of an offspring does not ensure that the offspring represents a fully connected tree. Therefore, a penalty for invalid solutions is necessary.

Li and Bouchebaba (1999) overcame the problem of invalid solutions and designed more advanced operators such as path crossover and mutation. These operators always generate feasible new solutions. Although, Li and Bouchebaba did not compare their new representation with other representations, the results presented were promising.

Raidl (2000) introduced edge crossover and edge insertion mutation for a degree constrained tree problem. New offspring are created by edge crossover in three steps. Firstly, a child inherits all edges which exist in both parents. Then, the offspring gets the edges which exist only in one parent. Finally, the tree is completed with randomly chosen edges concerning the degree constraints. A direct comparison of this approach to other existing approaches for solving the degree-constrained MST problem is difficult because an additional heuristic for generating good initial solutions was used.

Recently, Li (2001) presented an implementation of a direct encoding. The implementation is based on predecessor vectors and the effort for crossover and mutation goes with $O(d)$, where d is the length of the path in a tree. The work illustrates that even a direct representation of trees needs to be represented on a computer system.

7.2.2 Properties of Direct Representations

The purpose of this subsection is to take a closer look at the properties of direct representations.

We have already seen in subsection 2.1.2 and 2.1.3 that GEAs using direct representations do not use a specific genotype-phenotype mapping $f_g : \Phi_g \to \Phi_p$ any more but work directly on the phenotypes. In contrast to the so called indirect representations, where the genotypic space is different from the phenotypic space, $\Phi_g \neq \Phi_p$, the operators are directly applied to the phenotypes $x_p \in \Phi_p$. This situation is illustrated in Figure 7.5. Therefore, differences between different implementations of direct encodings are not the used representation (all use trees) but how the genetic operators crossover and mutation are applied to the phenotypes.

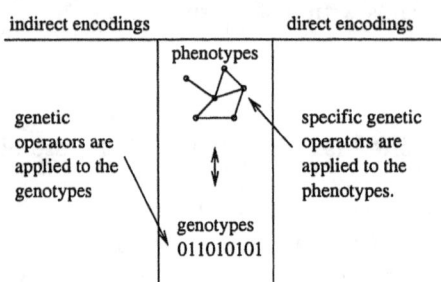

indirect encodings

direct encodings

phenotypes

genetic operators are applied to the genotypes

specific genetic operators are applied to the phenotypes.

genotypes
011010101

Fig. 7.5. Difference direct versus indirect representations

When using direct representations, the genotypes are the same as the phenotypes. Therefore, the representation is directly determined by the structure of the problem and the user does not need to define a proper representation. At a first glance, it seems that the use of direct representations releases us from the pain of designing efficient representations. However, when using direct representations, we are confronted with two other, serious problems:

- No standard mutation and recombination operators can be used.
- It is difficult to design proper problem-specific operators.

In the following, we briefly discuss these drawbacks of direct representations.

For traditional, indirect representations, a large variety of different genetic operators with known properties are available. These standard operators are well investigated and well understood. However, when using direct representations, standard operators like n-point or uniform crossover can no longer be used. For each direct representation, problem-specific operators must be developed. Therefore, most of the theory that predicts behavior and performance of GEAs using standard representations and standard operators is useless.

Furthermore, the development of proper problem-specific mutation and crossover operators is a difficult task. High quality operators must be able to detect the BBs and propagate them properly. Because even the design of proper simple crossover and mutation operators is demanding when using direct representations, the use of more advanced recombination methods as they are used in probabilistic model building GAs becomes almost impossible. These so called competent GAs no longer use standard genetic operators but build new generations according to a probabilistic model of the parent generations (Harik, 1999; Pelikan, Goldberg, & Cantú-Paz, 1999; Pelikan,

Goldberg, & Lobo, 1999). Competent GAs are developed for a few standard representations (mainly binary encoding) and result in better performance than traditional simple GAs. However, because direct representations and problem-specific genetic operators can hardly be implemented in competent GAs, direct representations can not benefit from their increased performance.

We know that it is difficult to design high-quality representations when using an indirect representation and standard operators. However, the task of creating efficient GEAs does not become easier when using direct representations because standard GEA operators can not be used any more and the design of problem-specific operators working on direct representations is difficult.

7.2.3 Operators for NetDir

When using direct representations for trees, problem-specific operators must be developed. In the following, we present the mutation and crossover operators for the NetDir representation.

Mutation

Subsection 3.3.2 illustrated that, in general, mutation operators should create offspring which are similar to the parent. Therefore, most mutation operators create offspring with a minimal distance to the parent.

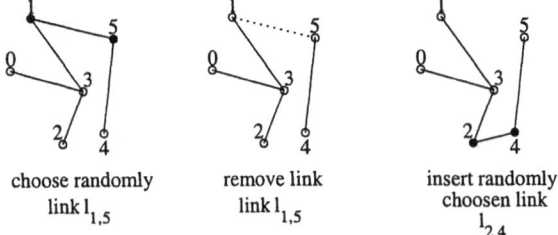

choose randomly   remove link   insert randomly
link $l_{1,5}$ link $l_{1,5}$ choosen link
                                                    $l_{2,4}$

Fig. 7.6. The mutation operator for the NetDir representation. The phenotypic distance between parent (left) and offspring (right) is $d = 1$.

Applying the mutation operator of the NetDir representation to an individual results in a neighboring phenotype. The mutation operator is applied directly to a phenotype $x_1^p \in \Phi_p$ and results in an offspring $x_2^p \in \Phi_p$ with phenotypic distance $d_{x_1,x_2} = 1$. Φ_p denotes the phenotypic space of all trees. Mutation randomly changes one link in the tree. We illustrate the mutation operator in Figure 7.6. The link $l_{1,5}$ is randomly chosen for deletion. After deleting this link we have two unconnected subtrees. Finally, a node is chosen from each of the two unconnected subtrees and the link connecting the two nodes is inserted ($l_{2,4}$).

When using linear, indirect, representations and applying mutation to individuals, an allele of the string is mutated with mutation probability p_m. Therefore, the probability for an individual to remain unchanged by mutation is $P = (1 - p_m)^l$, where l denotes the length of the string. The situation is different for the NetDir representation because no linear representation exists. Therefore, mutation for the NetDir representation is defined as randomly mutating an individual n times with probability p_m, where n is the number of nodes in the graph. Therefore, the probability that the individual remains unchanged is $P = (1 - p_m)^n$.

Crossover

The situation becomes slightly more complicated for the crossover operator. In subsection 3.3.2, we wanted crossover operators to create offspring that are similar to the parents. The offspring should inherit the high-quality substructures of their parents. In terms of metric, crossover operators should ensure that the distances between an offspring and its parents are smaller than the distance between both parents. In terms of schemata, high-quality crossover operators should be able to detect the linkage between the alleles in the string (Harik & Goldberg, 1996) and offspring should inherit the high-quality schemata from their parents. Consequently, the crossover operator of the NetDir representation only uses links that exist in the parents for the creation of the offspring. Therefore, the offspring have similar properties than the parents and the schemata are propagated properly.

In the following, we describe the functionality of the crossover operator. We denote a complete undirected graph as $G = (V, E)$, where $v \in V$ denotes the n different nodes and $l_{i,j} \in E$ denotes the link between node $i \in V$ and $j \in V$. Two parents are denoted as $G_1 = (V, E_1)$ and $G_2 = (V, E_2)$. The two offspring are denoted as $G_{o1} = (V, E_{o1})$ and $G_{o2} = (V, E_{o2})$. The crossover goes with the following scheme:

1. The set of all nodes V is randomly separated into two subsets V_1 and V_2, where $V_1 \cap V_2 = \{\}$ and $V_1 \cup V_2 = V$.
2. All links $l_{i,j} \in E_1$, where $i, j \in V_1$ are added to G_{o1}. All links $l_{i,j} \in E_1$, where $i, j \in V_2$ are added to the second offspring G_{o2}.
3. All links $l_{i,j} \in E_2$, where $i, j \in V_2$ are added to G_{o1}. All links $l_{i,j} \in E_2$, where $i, j \in V_1$ are added to the second offspring G_{o2}.
4. Do the following steps for each offspring individual separately.
5. There are at least two unconnected subtrees $G_{s1} = (V_{s1}, E_{s1})$ and $G_{s2} = (V_{s2}, E_{s2})$, where $G_{s1} \neq G_{s2}$. Add randomly an edge $e_{i,j} \in (E_1 \cup E_2)$ to the offspring, where either $i \in V_{s1} \wedge j \in V_{s2}$ or $i \in V_{s2} \wedge j \in V_{s1}$.
6. If the offspring is not fully connected, go to 5.

The crossover operator consists of two parts. At first, complete substructures are passed from the parent to the offspring (item 1-3). Then, the yet unconnected substructures are connected by adding links that exist in one of the two parents (item 4-6).

There are several choices for dividing the set of all nodes V into two subsets V_1 and V_2 (item 1). If we assume that the n nodes are numbered, we can use uniform, one-point, or n-point crossover. For uniform crossover the probability that each node belongs to either V_1 or V_2 is 0.5. For one-point crossover we have to choose a crossing point $c \in \{1, 2, \ldots, n-1\}$. The nodes with numbers smaller than c belong to V_1; the nodes with numbers equal or larger than c belong to V_2.

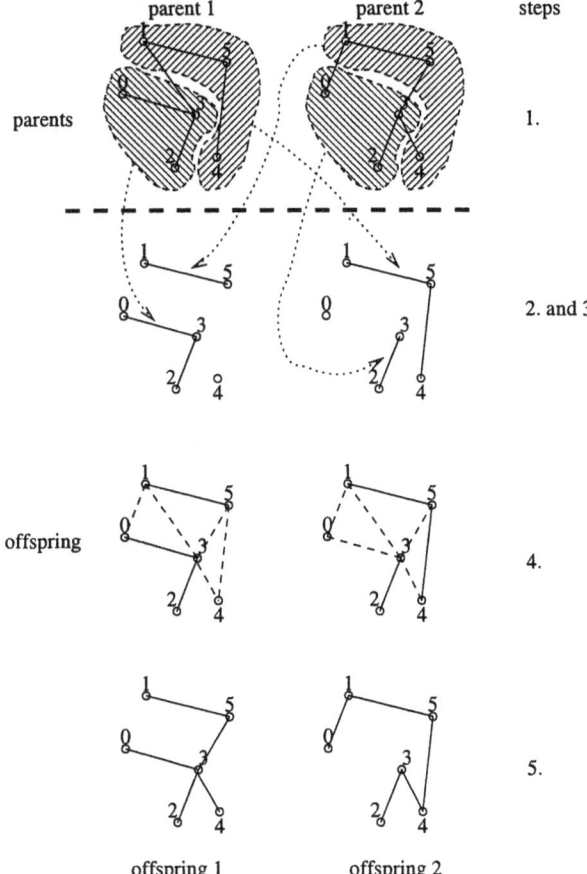

Fig. 7.7. The crossover operator of the NetDir representation. The offspring only inherit information from their parents. No randomly created links are used for the construction of the offspring.

Figure 7.7 illustrates the crossover operator with a small 6-node example. In a first step, the 6 nodes are separated according to uniform crossover into two subsets $V_1 = \{0, 2, 3\}$ and $V_2 = \{1, 4, 5\}$. Then, the links $l_{0,3}$

and $l_{2,3}$ from parent 1 and the link $l_{1,5}$ from parent 2 are added to off-spring 1. Analogously, $l_{1,5}$ and $l_{4,5}$ from parent 1 and $l_{2,3}$ from parent 2 are used for the construction of offspring 2. After copying the substructures from the parents to the offspring, the remaining separated subtrees must be connected. We do this by randomly copying edges which are able to connect the separated subtrees from the parents to the offspring until the offspring are completely connected. Offspring 1 has three unconnected subtrees $(G_1 = (\{0, 2, 3\}, \{(0, 3), (2, 3)\}), G_2 = (\{1, 5\}, \{(1, 5)\})$, and $G_3 = (\{4\}, \{\}))$. Therefore, the links $l_{0,1}$, $l_{1,3}$, $l_{3,5}$, $l_{3,4}$, and $l_{4,5}$ can be used for completion of offspring 1. After randomly choosing $l_{3,5}$ and $l_{3,4}$, offspring 1 is fully connected and we can stop. Offspring 2 also has three unconnected subtrees $(G_1 = (\{1, 4, 5\}, \{(1, 5), (4, 5)\}), G_2 = (\{2, 3\}, \{(2, 3)\})$, and $G_3 = (\{0\}, \{\}))$. For offspring 2, the links $l_{0,1}$, $l_{0,3}$, $l_{1,3}$, $l_{3,4}$, and $l_{3,5}$ can be used for completion. With choosing $l_{0,1}$ and $l_{3,4}$, offspring 2 is fully connected and we can terminate the algorithm.

When measuring the distances between the individuals, the distance between the parents is $d_{p1,p2} = 3$. The distance of offspring 1 to parent 1 is $d_{p1,o1} = 2$, and to parent 2 is $d_{p2,o1} = 1$. The distance between offspring 2 and parent 1 is $d_{p1,o2} = 2$, and to parent 2 is $d_{p2,o2} = 2$. We see that the distance between the offspring and their parents is smaller or equal to the distance between the parents. The offspring exist mostly of substructures of their parents.

7.2.4 Summary

This section presented the direct NetDir representation. After a short historical review of direct tree representations, in subsection 7.2.2 we discussed the properties of direct representations. Because direct representations directly encode the structure of the problem, standard mutation and crossover operators can not be used any more. Therefore, we presented in subsection 7.2.3 tree-specific mutation and crossover operators for the NetDir representation.

The purpose of this section was not just to present another, new representation but to illustrate that the design of efficient GEAs does not become easier when using direct representations. When using direct representations, an engineers' intuition and knowledge is not a prerequisite for the design of representations – these are determined a priori by the structure of the problem – but for the design of proper problem-specific genetic operators. Therefore, direct representations do not provide efficient GEAs for free, but in comparison to indirect representations, the overall difficulty of designing efficient GEAs remains the same or even increases.

In this section, we presented the NetDir representation as an example of a direct representation for trees. The NetDir representation encodes trees directly. Therefore, standard crossover and mutation operators can not be used any more. However, existing theory about GEAs is based on standard operators and standard representations. As a result, the existing theory does not hold any more for direct representations, and it is very difficult to predict GEA performance. Furthermore, the presented framework for representations can also not be used any more because no explicite genotype-phenotype mapping exists.

In the previous chapters, we mostly neglected the interdependencies between operators and representations. We simply defined a proper metric on the genotypes and assumed that the used standard genetic operators perform well. This section brought us to the edge of representation theory and led us into the field of genetic operators. The section showed us that there is a tradeoff between efficient operators and efficient representations and that we must bear both in mind when designing high-quality GEAs. However, as we focus on representations and not on operators, this section marks the furthest end of our journey through the theory of representations. It is the last piece in the puzzle of representations we want to solve and marks the end of our theoretical investigations into representations. The only thing that is left is to verify our theoretical insights into tree representations with empirical results.

8. Performance of Genetic and Evolutionary Algorithms on Tree Problems

In the previous chapters, the presented theory about representations was mainly used for analysis and design of representations. The investigations into the properties of representations were based on theory and helped us to understand what happens when GEAs use a specific representation. However, in practice, GEA users are often less interested in theory about representations but want simple instruments for a fast and rough prediction of the expected performance of a representation. They have several representations at hand and want to know which representation they should choose for their problem. We do not want to leave them alone with their problems, but illustrate how they can advantageously use the proposed theory.

This chapter illustrates for scalable test and real-world tree problems how the performance of GEAs using different types of representations can be predicted by using the provided framework about representations. Based on the framework, we give qualitative predictions of solution quality and time to convergence for different types of tree representations. Doing this, this chapter also provides an exhaustive comparison of the performance of different tree representations. For our comparison we choose the Prüfer number (see section 6.2), characteristic vector (see section 6.4), NetKey (see section 7.1), NetDir (see section 7.2), and the link and node biased representation (see section 6.3). These representations are used on scalable test problems like the one-max tree and deceptive tree problem and on various test instances of the optimal communication spanning tree problem from the literature. The results show that using the outlined theory makes the use of representations in the real world easier.

The test instances of the optimal communication spanning tree problem are chosen because the exact specifications of the problems are either easily available (Raidl, 2001; Rothlauf, Goldberg, & Heinzl, 2002) or published (Palmer, 1994; Berry, Murtagh, & McMahon, 1995). For summarizing purposes the exact distance and demand data for the test instances of the optimal communication spanning tree problems are listed in appendix A. To make results comparable, other test instances which are not available, or which are not sufficiently described in the literature are not considered in this chapter.

The following section 8.1 provides a comparison of GEA performance for scalable test problems. After a brief analysis of representations in subsection 8.1.1, we present in subsection 8.1.2 results for the fully easy one-max tree problem and in subsection 8.1.3 results for the the fully difficult deceptive tree problem. For each of the two test problems, we provide theoretical predictions and empirical evidence. Then, we focus in section 8.2 on the optimal communication spanning tree problem. This problem is defined on trees and researchers have proposed some test instances in the literature (Palmer, 1994; Berry, Murtagh, & McMahon, 1995; Raidl, 2001; Rothlauf, Goldberg, & Heinzl, 2002). For each problem, we deduce predictions of GEA performance and present empirical results. The chapter ends with a brief summary.

8.1 GEA Performance on Scalable Test Tree Problems

This section compares the performance of different types of representations for the one-max tree and the deceptive tree problem. After a brief analysis of tree representations based on the framework from chapter 4, we present results for the one-max tree problem in subsection 8.1.2 and for the deceptive tree problem in subsection 8.1.3. For both problems, we provide brief descriptions, theoretical predictions about GEA performance, and empirical evidence.

8.1.1 Analysis of Representations

This subsection briefly summarizes the most important properties of the different tree representations from the previous chapters. We summarize the results concerning redundancy, bias, scaling and distance distortion for the Prüfer number, characteristic vector, NetKey, NetDir, and LNB encoding.

The investigation into the redundancy of representations has revealed that only the Prüfer number and the NetDir encoding is not redundant. The characteristic vector encoding, the NetKeys, and the LNB encoding with large link-specific bias P_1 are uniformly redundant. Therefore, GEA performance is independent of the structure of the optimal solution. In contrast, the LNB encoding is biased if the link-specific bias is not large enough. For a large node-specific bias P_2, the encoding is biased towards stars and if P_1 and P_2 are small, the encoding is biased towards the minimum spanning tree. Therefore, GEA performance depends on the structure of the optimal solution.

To investigate and verify the bias of the representations, we can measure the average distance of a randomly generated individual to star networks or to the MST. In Table 8.1, we present for a randomly generated individual the average minimum phenotypic distance towards a star, $\min(d_{star})$, and the average phenotypic distance towards the MST, d_{MST}. We randomly generate 10 000 individuals and show the mean and the standard deviation of the distance. Because the encoded structures depend for the LNB encoding on the distances between the n nodes, we randomly place the nodes on a 1000 x 1000

Table 8.1. Average minimum phenotypic distance to star networks, $min(d_{star})$, and average phenotypic distance to the minimum spanning tree, d_{MST}. The numbers reveal that Prüfer number, characteristic vector, NetKey, NetDir, and LNB ($P_1 = 20, P_2 = 0$) encoding are unbiased. The LNB encoding is biased towards stars for $P_1 = 0/P_2 = 1$ and towards the minimums spanning tree for P_1 and P_2 small.

problem size n		8 node		16 node		32 node	
distance d		μ	σ	μ	σ	μ	σ
Prüfer number	$min(d_{star})$	3.66	0.65	10.91	0.79	26.25	0.83
	d_{MST}	5.18	1.06	13.09	1.23	29.09	1.31
CV	$min(d_{star})$	3.67	0.65	10.92	0.78	26.26	0.81
	d_{MST}	5.18	1.05	13.09	1.23	29.05	1.31
NetKey	$min(d_{star})$	3.74	0.61	11.0	0.77	26.35	0.792
	d_{MST}	5.25	1.06	13.13	1.22	29.06	1.31
NetDir	$min(d_{star})$	3.66	0.63	10.90	0.78	26.26	0.82
	d_{MST}	5.18	1.05	13.08	1.24	29.04	1.31
LNB ($P_1 = 0, P_2 = 1$)	$min(d_{star})$	2.43	1.11	7.74	1.96	20.19	2.87
	d_{MST}	3.30	1.05	8.78	1.56	20.65	2.23
LNB ($P_1 = 1, P_2 = 0$)	$min(d_{star})$	3.75	0.61	11.00	0.77	26.41	0.77
	d_{MST}	3.57	1.14	9.52	1.67	22.36	2.28
LNB ($P_1 = 1, P_2 = 1$)	$min(d_{star})$	3.05	0.90	9.22	1.32	23.02	1.77
	d_{MST}	3.91	1.08	10.45	1.53	24.51	2.03
LNB ($P_1 = 20, P_2 = 0$)	$min(d_{star})$	3.75	0.62	10.99	0.78	26.34	0.80
	d_{MST}	5.15	1.06	12.87	1.28	28.56	1.47

square. The results for the Prüfer number and the NetDir representations show the distances of a randomly, unbiased individual. The figures confirm the predictions from above. Characteristic vector, NetKey, and LNB encoding with large P_1 are uniformly redundant because they have the same distances as non-redundant representations (Prüfer numbers and NetDir).

The investigation into the scaling of the BBs has revealed that all examined tree representations have uniformly scaled BBs. There are no BBs that have a higher contribution to the fitness of a tree. Therefore, the dynamics of genetic search are not changed and all alleles are solved implicitly in parallel.

Investigating the locality and distance distortion of tree representations has shown that the Prüfer number representation has in general low locality. However, the locality of Prüfer numbers is not low everywhere. When encoding star networks the locality of Prüfer numbers is perfect. As a result, Prüfer numbers change BB-complexity and problem difficulty for optimal solutions others than stars. If a problem is fully easy, GEAs have great problem in finding solutions other than star-like. If the problem is fully difficult, problems where the optimal solution is not a star become easier to solve. In contrast to Prüfer numbers, the other tree representations have high locality. Small changes in the phenotype result in small changes in the genotype. Furthermore, the distance distortion d_c is low for the CV, NetKey, NetDir, and LNB

Table 8.2. Summary of important properties of tree representations

	redundancy	bias	locality/dist. distortion	BB-scaling	additional comments
Prüfer number	not redundant	-	High locality around stars, Low locality elsewhere	uniformly scaled	
CV	uniformly redundant	no bias	high locality	uniformly scaled	stealth mutation
NetKey	uniformly redundant	no bias	high locality	uniformly scaled	
NetDir	not redundant	-	high locality	uniformly scaled	direct representation
LNB $(P_1 = 0, P_2 = 1)$	non-uniformly redundant	bias towards star and MST	high locality	uniformly scaled	
LNB $(P_1 = 1, P_2 = 0)$	non-uniformly redundant	bias towards MST	high locality	uniformly scaled	
LNB $(P_1 = 1, P_2 = 1)$	non-uniformly redundant	bias towards star and MST	high locality	uniformly scaled	
LNB $(P_1 = 20, P_2 = 0)$	uniformly redundant	no bias	high locality	uniformly scaled	same behavior as NetKeys

encodings because the phenotypic distances between individuals correspond to the genotypic distances.

We summarize the properties of the representations in Table 8.2. Using these results, we are able to predict GEA performance in the following subsections.

8.1.2 One-Max Tree Problem

In this subsection we compare the performance of different representations for the one-max tree problem.

Problem Description

We give a brief description of the *one-max tree problem*.

The one-max tree problem was defined in subsection 6.1.5 as a fully easy tree problem. The problem is fully easy for GEAs as the phenotypic size of the BBs is 1. All tree schemata that contain the optimal solution have higher fitness than their competitors.

For the one-max tree problem, an optimal solution x^p_{opt} is chosen a priori either randomly or by hand. The fitness of an individual x^p_1 is defined as the phenotypic distance $d_{x^p_{opt}, x^p_1}$ to the optimal solution. The higher the fitness, the more links an individual has in common with the optimal solution.

More information about the one-max tree problem is provided in subsection 6.1.5 and 6.1.4.

Theoretical Predictions

We give predictions on GEA performance for the one-max tree problem. The predictions are based on the results about the redundancy and the locality/distance distortion of an encoding.

We have seen that some representations do not uniformly represent the phenotypes. Therefore, GEA performance depends on the structure of the optimal solution. Using the results from subsection 8.1.1, we expect the CV, NetKey, NetDir and LNB ($P_1 = 20$, $P_2 = 0$) encoding to perform independently of the structure of the optimal tree. GEAs using LNB encodings with small biases P_1 and P_2 will perform better if the optimal solution is similar to the MST. However, GEAs using these encodings need more generations and find less BBs if the optimal solution is not similar to the MST. GEAs using the LNB encoding with $P_1 = 0$ and $P_2 = 1$ are biased towards stars and show high performance for when the best solution is a star. When the best solution is an arbitrary tree or a list, GEAs will fail when using the LNB encoding.

After we have examined the effects of redundant encodings, we focus on locality and distance distortion. We have seen that Prüfer numbers have high locality around stars, but low locality elsewhere. This means, the genotypic size of the BBs k_g is larger than the phenotypic size of the BBs $k_p = 1$ if the optimal solution is not a star. Therefore, the performance of GEAs using Prüfer numbers is low if the optimal solution is not a star. In contrast, the other representations (NetKey, NetDir, CV and LNB) have high locality and low distance distortion, and the problem difficulty is not changed.

Finally, we want to predict the influence of stealth mutation on the characteristic vector encoding. Stealth mutation works like additional mutation. As we only use crossover and no mutation in our empirical comparison, BBs can not come back into the population once they are extinct. However, stealth mutation brings back lost BBs. Therefore, we expect GEAs using the CV encoding to be able to find more BBs than when using other encodings, but to need much more time to do this. The run duration and the number of found BBs at the end of the run increases when using CVs. This effect of stealth mutation is amplified by the low difficulty of the one-max problem. This problem is expecially easy for mutation-based GEAs because the landscape leads GEAs to the correct solution. Therefore, mutation, as well as stealth mutation, increases GEA performance.

Empirical Results

We provide empirical results for the different representations.

For our experiment, we use a simple genetic algorithm without mutation. We use tournament selection without replacement of size 2 and uniform crossover. The run is stopped after the population is fully converged or a maximum of 100 generations (8 node and 16 node problems) or 200 generations (32 node problems) is reached. We perform 100 runs for every population size and problem instance. We present results for the one-max tree problem with 8, 16, and 32 nodes. The optimal solution is either a star, a list, or an arbitrary tree, and is chosen randomly. The performance of GEAs is determined by the number of correct BBs at the end of the run and the number of generations until the population is fully converged.

The performance of GEAs using different representations for the one-max tree problem is shown in Figure 8.1 (8 node and best solution is a star), 8.2 (8 node and best solution is a list), 8.3 (8 node and best solution is an arbitrary tree), 8.4 (16 node and best solution is a star), 8.5 (16 node and best solution is a list), 8.6 (16 node and best solution is an arbitrary tree), 8.7 (32 node and best solution is a star), 8.8 (32 node and best solution is a list), and 8.9 (32 node and best solution is an arbitrary tree). The plots on the left show the number of BBs found at the end of the run over the population size n. The plots on the right show the number of generations t_{conv} until the population is completely converged. We draw results for Prüfer number, characteristic vector, NetKey, NetDir, and the LNB encoding. For the LNB encoding, we present results for $P_1 = 0/P_2 = 1$, $P_1 = 1/P_2 = 0$, $P_1 = 1/P_2 = 1$, and $P_1 = 20/P_2 = 0$. The figures show that the performance of GEAs searching for the optimal list is about the same as when searching for the optimal arbitrary tree.

The performance of GEAs using characteristic vector, NetKey, NetDir, or LNB ($P_1 = 20/P_2 = 0$) representation is independent of the structure of the optimal solution. The number of correct BBs and the run duration is not affected by the optimal solution being a star, list, or arbitrary tree. In contrast, GEAs using Prüfer numbers perform better when searching for optimal stars than when searching for optimal trees or lists because the locality of Prüfer numbers is higher around stars then around trees or lists. With increasing problem size n GEAs perform worse for the optimal solution to be a star because the areas of high locality (stars) are surrounded by larger areas of low locality and GEAs can not find their way to high locality areas. Furthermore, the number of high locality individuals exponentially decreases with increasing problem size n (compare 6.2.4).

LNB encodings with $P_1 = 0/P_2 = 1$ are biased towards stars and GEAs perform very well when searching for the optimal solution is a star, but

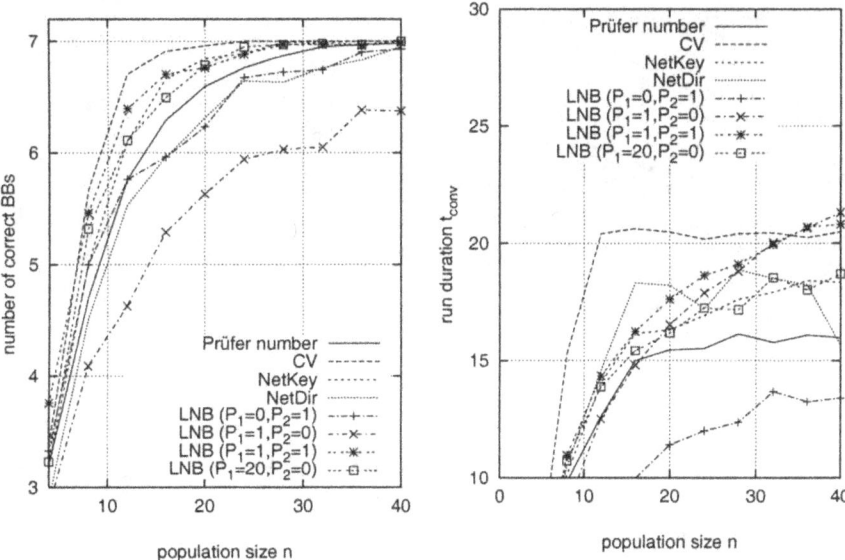

Fig. 8.1. Performance of GAs for star 8 node one-max tree problems. The plots show either the number of correct BBs at the end of the run (left) or the number of generations until the population is fully converged (right).

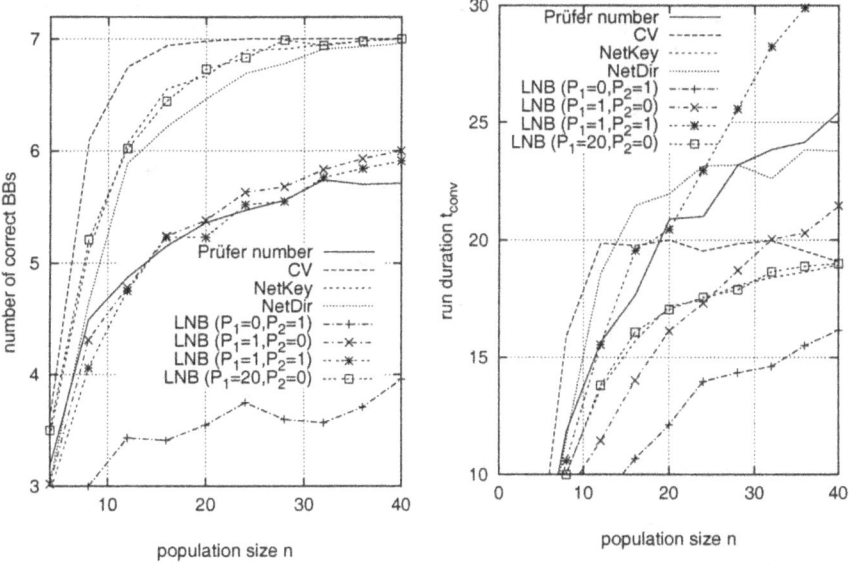

Fig. 8.2. Performance of GAs for list 8 node one-max tree problems. The plots show either the number of correct BBs at the end of the run (left) or the number of generations until the population is fully converged (right).

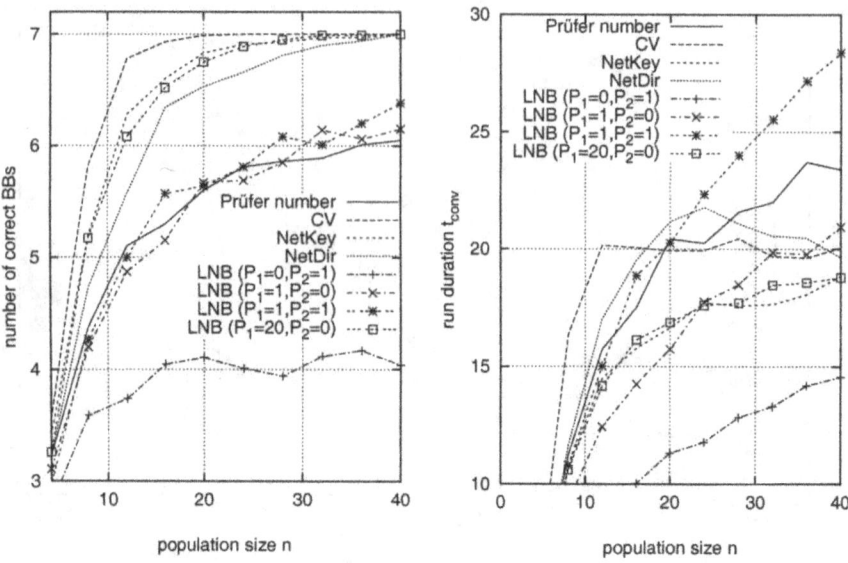

Fig. 8.3. Performance of GAs for tree 8 node one-max tree problems. The plots show either the number of correct BBs at the end of the run (left) or the number of generations until the population is fully converged (right).

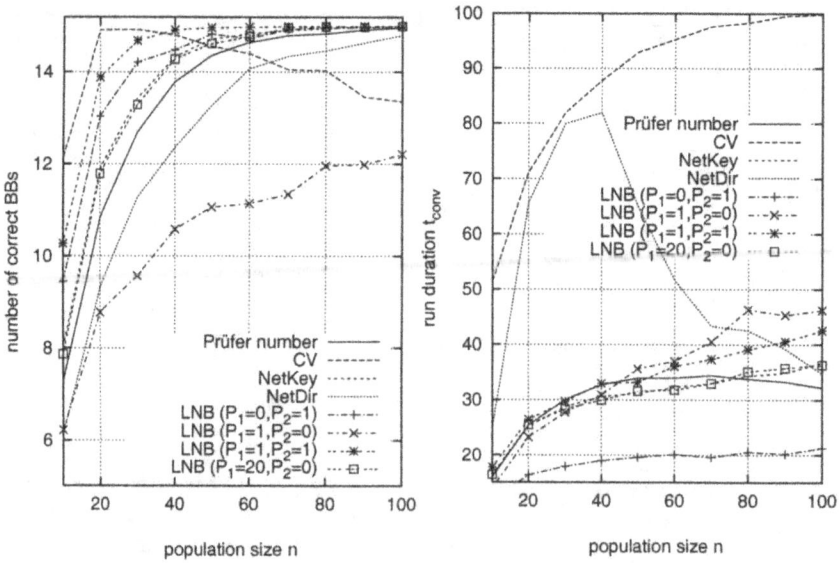

Fig. 8.4. Performance of GAs for star 16 node one-max tree problems. The plots show either the number of correct BBs at the end of the run (left) or the number of generations until the population is fully converged (right).

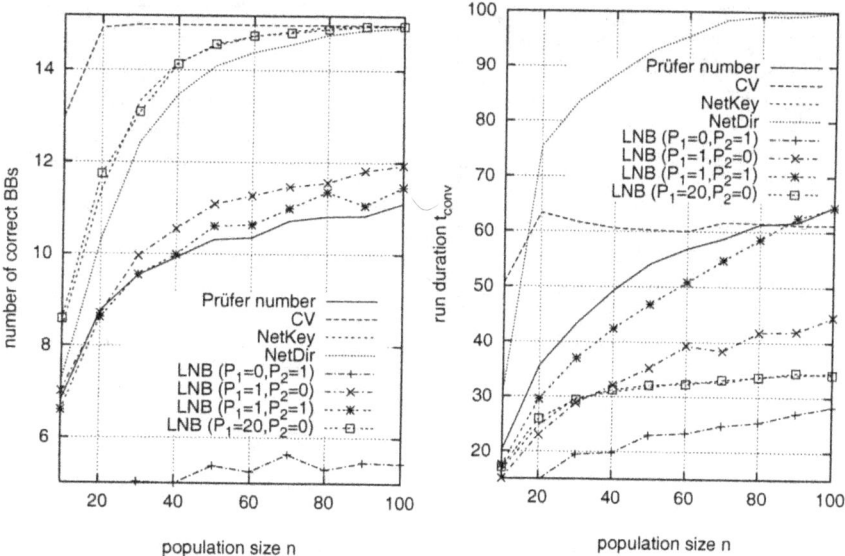

Fig. 8.5. Performance of GAs for list 16 node one-max tree problems. The plots show either the number of correct BBs at the end of the run (left) or the number of generations until the population is fully converged (right).

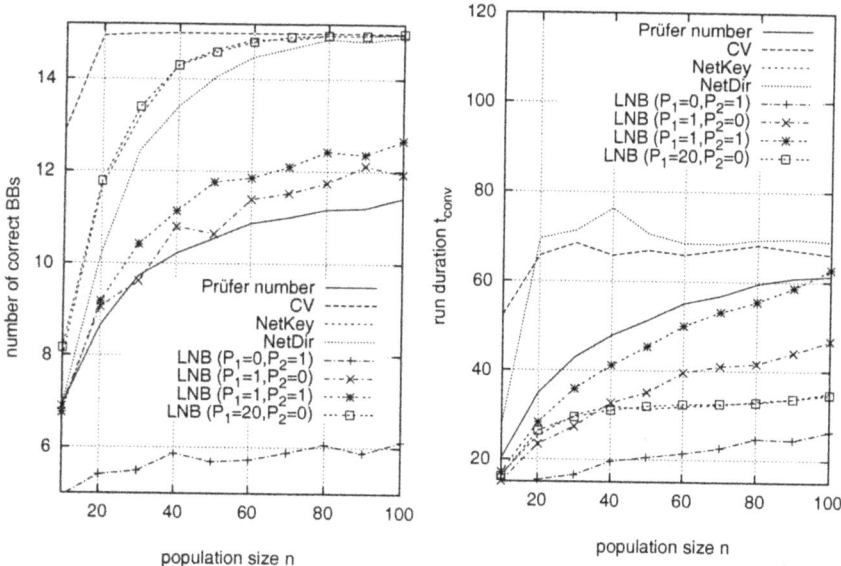

Fig. 8.6. Performance of GAs for tree 16 node one-max tree problems. The plots show either the number of correct BBs at the end of the run (left) or the number of generations until the population is fully converged (right).

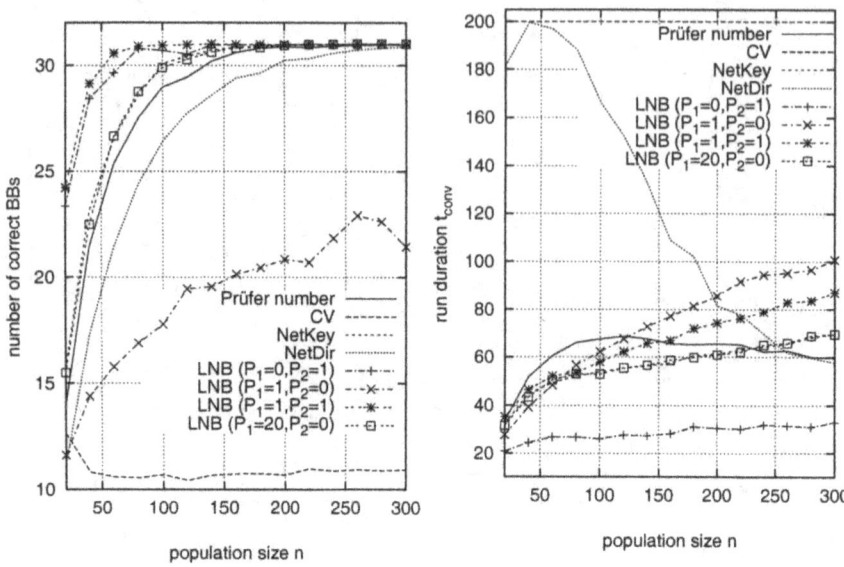

Fig. 8.7. Performance of GAs for star 32 node one-max tree problems. The plots show either the number of correct BBs at the end of the run (left) or the number of generations until the population is fully converged (right).

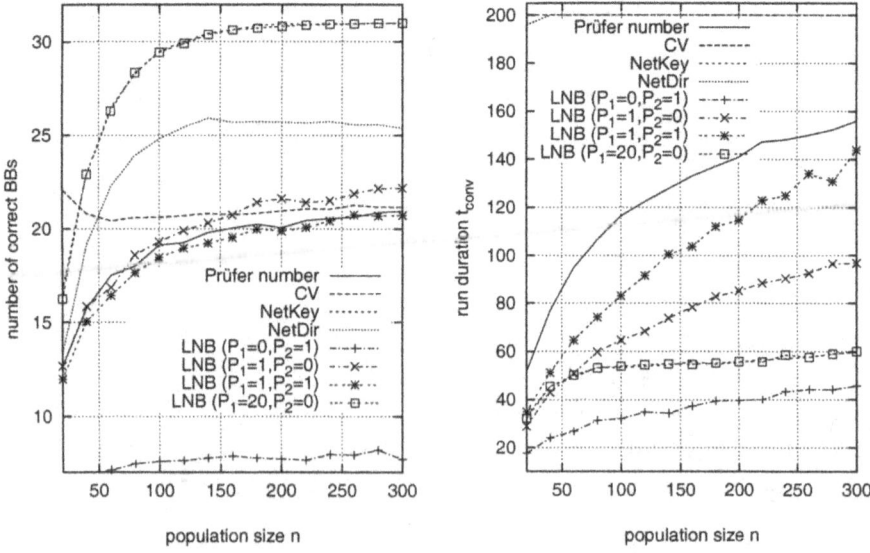

Fig. 8.8. Performance of GAs for list 32 node one-max tree problems. The plots show either the number of correct BBs at the end of the run (left) or the number of generations until the population is fully converged (right).

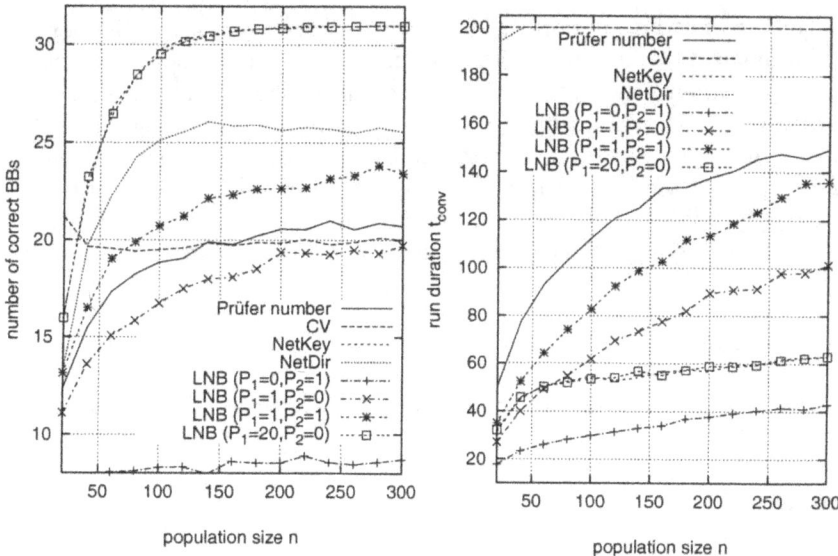

Fig. 8.9. Performance of GAs for tree 32 node one-max tree problems. The plots show either the number of correct BBs at the end of the run (left) or the number of generations until the population is fully converged (right).

fail completely when searching for lists or arbitrary trees. If LNB encodings only use a small link-specific bias P_1, the encoding is biased towards MSTs. Therefore, GEAs using this encoding show low performance when searching for stars, lists, or arbitrary trees. When adding a small node-specific bias ($P_1 = 1/P_2 = 1$), LNB encodings are also biased to stars and GEAs show a high performance when searching for optimal star networks. However, for optimal lists and trees, GEA performance is low for $P_1 = 1/P_2 = 1$ and $P_1 = 1/P_2 = 0$.

We developed the NetKey encoding to allow GEAs to distinguish between important and unimportant BBs. The results show that GEAs using NetKeys are able to find a high number of correct BBs after a few generations. As predicted, the LNB encoding with large link-specific bias ($P_1 = 20/P_2 = 0$) shows similar performance to the NetKey encoding. Only GEAs using the CV encoding are able to identify a higher number of correct BBs for small problems due to the stealth mutation, but need a greater number of generations. For the one-max tree problem, GEAs using CVs use stealth mutation and not crossover as the main search operator.

We see that the empirical results confirm the theoretical predictions. However, we can not predict the performance of GEAs using the NetDir representation. This direct representation uses a problem-specific crossover operator and we can not use the framework about representations for it. GEAs using the NetDir encoding perform slightly worse than when using NetKeys but need a much larger number of generations. The large number of generations is unexpected and can not be predicted by using the existing theory.

The results show that only NetKeys and the LNB encoding with a large link-specific bias show high performance for the one-max tree problem and different types of optimal solutions. GEAs using the characteristic vector encoding are able to find more correct BBs for small problem instances due to stealth mutation at the end of the run, but need a greater number of generations.

8.1.3 Deceptive Tree Problem

This subsection compares the performance of different representations for the deceptive tree problem.

Problem Definition

We give a brief description of the deceptive tree problem.

The deceptive tree problem was defined in subsection 6.1.5. The problem is fully difficult for crossover-based GEAs as all tree schemata with $k < n - 1$ that contain the optimal solution have lower fitness than their competitors.

For the deceptive tree problem, an optimal solution x_{opt}^p is chosen a priori either randomly or by hand. The fitness of the optimal solution is $n - 1$. The fitness f_i of other individuals $x_i^p \neq x_{opt}^p$ is defined as $d_i = d_{x_{opt}, x_i} - 1$. The higher the fitness of an individual, the less links it has in common with the optimal solution.

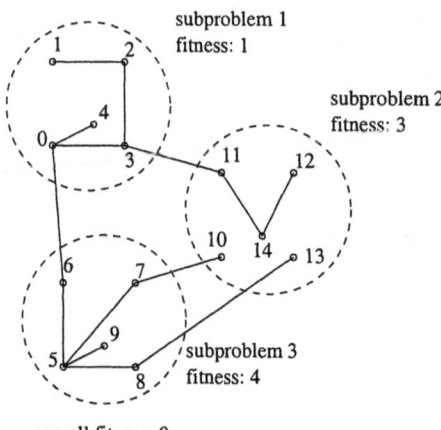

subproblem 1
fitness: 1

subproblem 2
fitness: 3

subproblem 3
fitness: 4

overall fitness: 9
number of correct BBs: 1

Fig. 8.10. Calculation of the fitness for three concatenated, deceptive traps of size $n = 5$. Although, a tree with 15 nodes has 14 links, only a maximum of 12 links is considered for the calculation of the fitness.

Even small instances of the deceptive tree problem can not be solved by GEAs with reasonable problem sizes. Therefore, to be able to construct larger problem instances, we concatenate m deceptive traps of size n. The size of a deceptive trap is denoted by the number of nodes that are part of this sub-problem. As a result, we can construct problems with $m*n$ nodes. When concatenating deceptive traps to larger problems, we have to consider that the subproblems (BBs) of size n containing $n-1$ links must be connected to form a fully connected tree with $mn-1$ links. Therefore, at least $(mn-1)-m(n-1) = m - 1$ links are not considered for the fitness of an individual. We illustrate in 8.10 how the fitness of an individual is calculated if we concatenate 3 deceptive sub-problems of size $n = 5$. The optimal solution should be the star with center $5i$, where $i \in \{0, 1, 2\}$. If we have three concatenated subproblems then there are three groups of nodes, and each group consists of 5 nodes. For example, subproblem 3 consists of the nodes 5, 6, 7, 8, and 9. Subproblem 3 in Figure 8.10 shows the optimal solution to the subproblem which is a star with center 5. The fitness of an individual is the sum of the fitness of the deceptive sub-problems. Therefore, the maximum fitness of an individual is 12 (all three BBs are correct), and the minimum fitness is 0. When concatenating deceptive sub-problems to a larger problem, it can happen that an individual has no links in common with the optimal solution of a sub-problem. In this case, the contribution of this sub-problem to the overall fitness of the individual is $n - 2$. In our example, sub-problem 2 has no links in common with the optimal solution and its contribution to the overall fitness is n-2=3.

For more information about the deceptive tree problem the reader is referred to subsection 6.1.5.

Theoretical Predictions

We predict the performance of GEAs for the deceptive tree problem based on the results about redundancy and locality/distance distortion.

Due to uniform redundancy, we expect the performance of GEAs using the CV, NetKey, NetDir, and LNB ($P_1 = 20, P_2 = 0$) encoding to be independent of the structure of the optimal solution. The LNB encoding using a small link- or node-specific bias is either biased towards stars or the minimum spanning tree. Therefore, GEA performance would be high if we only had one BB, and the optimal solution is a star or the MST. However, when concatenating m sub-problems to a larger problem, the overall optimal solution has nothing in common with either a star or a MST. Thus, GEAs using LNB encodings with small bias will fail.

We have learned that low locality encodings make fully difficult problems easier and fully easy problems more difficult. Therefore, GEAs using Prüfer numbers will perform well when used on the fully deceptive tree problem where the optimal solution is a non-star, although, concatenating m difficult deceptive traps, results in an overall problem which is no longer fully difficult (only the sub-problems are fully difficult). In comparison to the one-max tree problem where GEAs using Prüfer numbers failed completely, we expect a

better performance for the deceptive tree problems when using Prüfer numbers. Because relative problem difficulty increases with decreasing number of
sub-problems m and increasing size of the sub-problems, GEA performance
using Prüfer numbers will also increase. Furthermore, because Prüfer numbers have higher locality around stars, GEAs have more difficulty in finding
optimal stars than optimal lists when solving the deceptive tree problem.
Consequently, GEA performance will be higher when searching for optimal
non-stars than when searching for optimal stars.

Finally, for the characteristic vector encoding, we expect the same effects
as for the one-max tree problem. Due to stealth mutation, we expect GEAs
to need more generations and to be able to find more correct BBs. However,
the overall number of fitness calls that are necessary to find the optimum
strongly increases.

Empirical Results

We present empirical results for the deceptive tree problem.

For the empirical comparison, we concatenate 4 instances of a deceptive
tree problem of size 3 and 4. Therefore, our problems have either 12 (size
3) or 16 nodes (size 4). The minimum fitness of an individual is 0 and the
maximum fitness is either $4*2 = 8$ (4 instances of the 3 node trap) or $4*3 = 12$
(four instances of the 4 node trap).

In our experiments, we use a simple genetic algorithm without mutation.
Because uniform crossover would result in high BB-disruption, we use two-
point crossover in all runs. Furthermore, we use tournament selection without
replacement of size 3. The runs are stopped after the population is fully
converged or the number of generations exceeds 200 generations. We perform
100 runs for every population size and problem instance. We present the
number of correct BBs at the end of the run and the number of generations.
Because we have seen for the one-max tree problem that GEAs show the
same performance when searching for lists as when searching for arbitrary
trees, we do not consider the case that the optimal solution is a list.

The performance of GEAs where the optimal solution is either a star (top)
or an arbitrary tree (bottom) is shown in Figure 8.11 (four concatenated 3-
node problems) and Figure 8.12 (four concatenated 4-node problems). As
before, the plots on the left show the number of correct BBs at the end of
the run, and the plots on the right show the number of generations until
the population is fully converged. We present results for the Prüfer number,
characteristic vector, NetKey, NetDir, and the LNB representation ($P_1 =
0/P_2 = 1$, $P_1 = 1/P_2 = 0$, $P_1 = 1/P_2 = 1$, and $P_1 = 20/P_2 = 0$).

As predicted, the performance of the uniformly redundant encodings (CV,
NetKey, NetDir and LNB ($P_1 = 20/P_2 = 0$)) is independent of the structure
of the optimal solution. Because LNB encodings with small biases P_1 and P_2
are biased towards stars and the minimum spanning tree, the performance
of GEAs using these types of encodings is low and the same for stars and

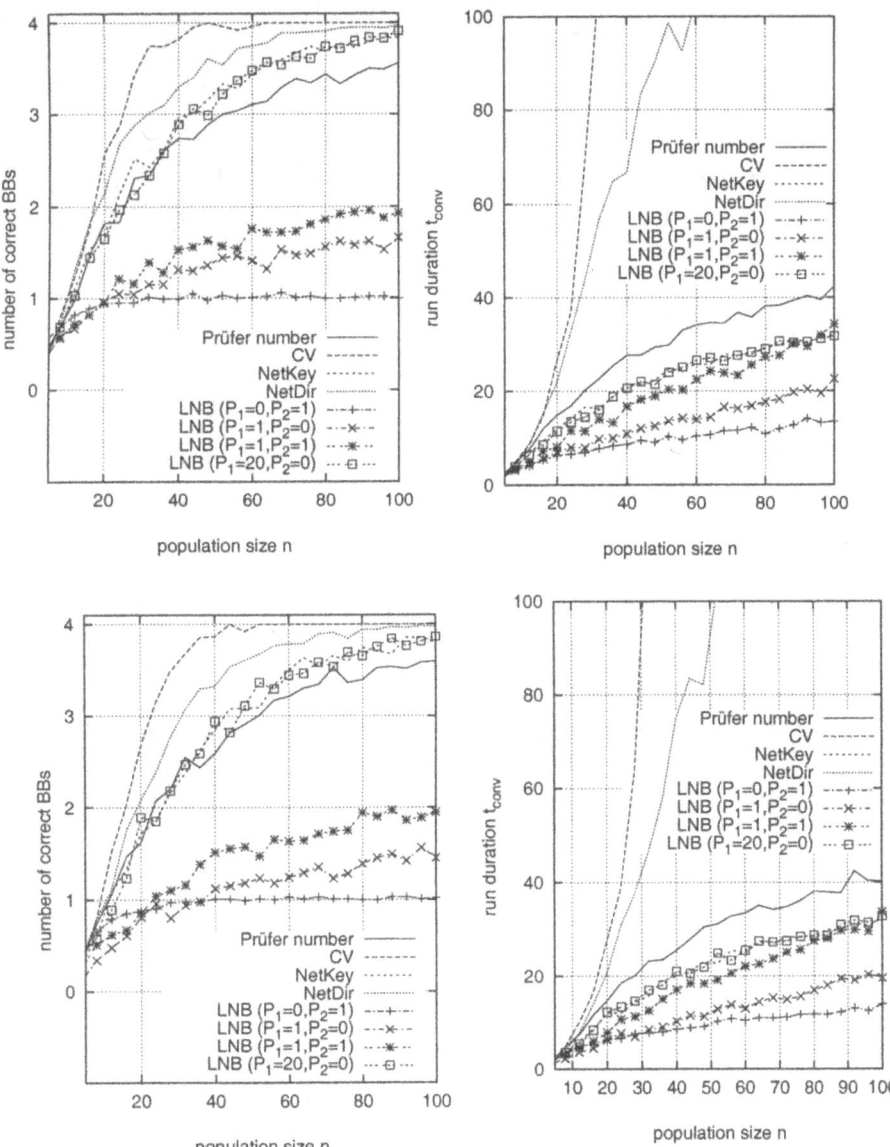

Fig. 8.11. Performance of GAs on four concatenated deceptive sub-problems of size three. The optimal solution to the three node sub-problems are either stars (top) or arbitrary trees (bottom). The plots on the left show the number of correct BBs at the end of the run and on the right the number of generations until the population is fully converged. In analogy to the one-max tree problem, GEAs using the NetKey or LNB encoding with large link-specific bias show high performance. In contrast, GEAs using Prüfer numbers do not fail as the low locality of the encoding reduces problem difficulty and makes it easier for GEAs to find their way to the optimal solution.

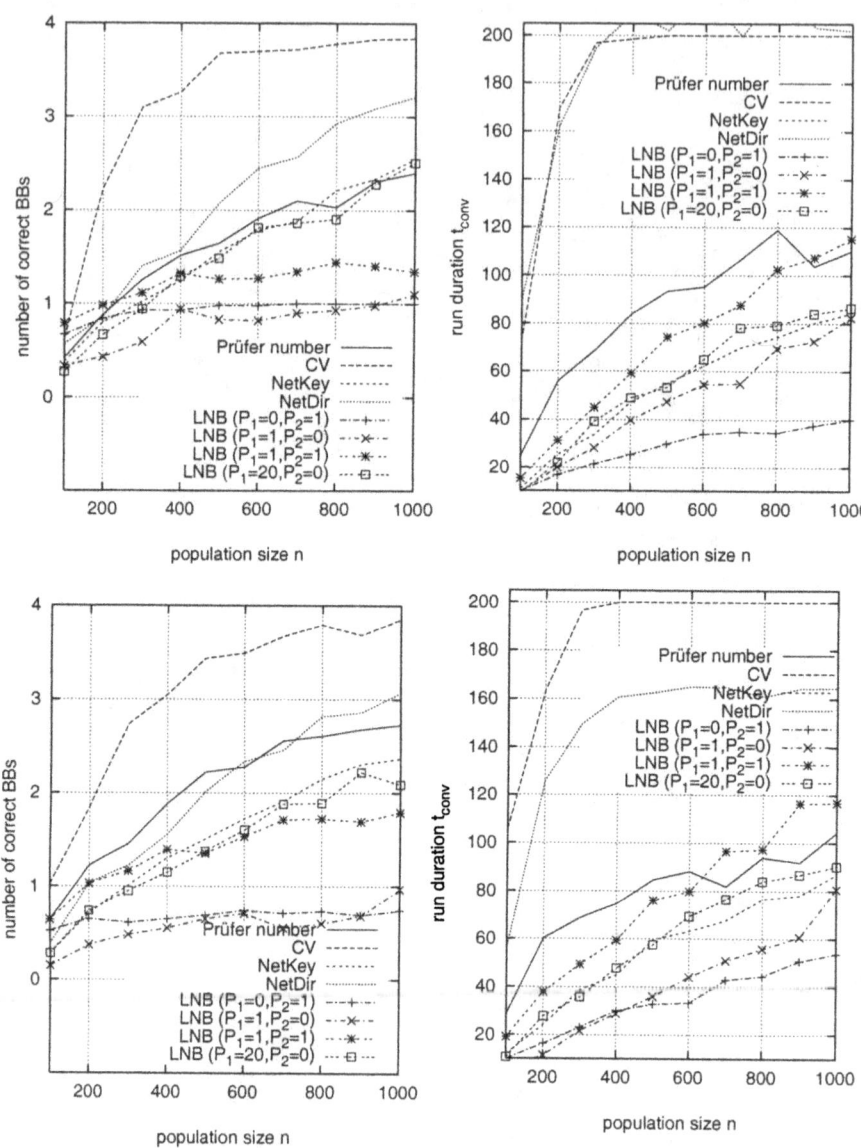

Fig. 8.12. Performance of GAs on four concatenated deceptive sub-problems of size four. The optimal solution to the four node sub-problems are either stars (top) or arbitrary trees (bottom). The plots on the left show the number of correct BBs at the end of the run and on the right the number of generations until the population is fully converged. With increasing problem size, GEAs using Prüfer numbers perform better as the low locality of the encoding reduces problem difficulty and makes it easier for GEAs to find their way to the optimal solution.

arbitrary trees. Because the optimal solutions have nothing in common with a star or the MST, GEAs must fail.

In comparison to the fully easy one-max problems, GEAs using Prüfer numbers perform much better when searching for optimal non-stars. The low locality of the encoding helps GEAs to find their way to the optimal solution. As expected, the solution quality is lower when searching for optimal stars because the locality around star-like structures is higher.

As we have already seen for the one-max tree problem, the LNB encoding with a large link-specific bias $P_1 = 20$ results in the same GEA performance as when using the NetKey encoding. GEAs using one of these two encodings are able to detect a high proportion of correct BBs and only need a few generations. Due to stealth mutation, GEAs using the CV representation perform well for the difficult deceptive tree problems. However, stealth mutation strongly increases the number of generations. The results for the NetDir representation are surprising. GEAs using this encoding are able to find a high proportion of BBs but need many more generations.

Coincidentially with the results for the one-max tree problem, only Net-Keys and LNB encodings with a large link-specific bias allow GEAs to reliably detect a high proportion of correct BBs after a few generations. In contrast to the one-max tree problem, GEAs using Prüfer numbers show a comparable, high performance for this difficult problem as the low locality of Prüfer numbers makes difficult problems easier to solve.

8.2 GEA Performance on the Optimal Communication Spanning Tree Problem

This section illustrates how we can use the framework about representations for predicting and verifying GEA performance on some instances of the optimal communication spanning tree (OCST) problem. Consequently, this section also provides a comprehensive comparison for the performance of different tree representations on OCST problems. We present results for test instances from Palmer (1994), Raidl (2001), Berry, Murtagh, and McMahon (1995), and Rothlauf, Goldberg, and Heinzl (2002).

In the following subsection, we give a brief description of the OCST problem. This is followed in subsection 8.2.2 by a brief discussion of the expected performance of GEAs for OCST problems. In subsection 8.2.3, we provide results for the 6, 12 and 24 node problems from Palmer (1994). Subsection 8.2.4 presents results for two 10 and 20 node test instances from Raidl (2001). In subsection 8.2.5, we examine one 6 node and two 35 node test instances from Berry, Murtagh, and McMahon (1995). Finally, subsection 8.2.6 presents four real-world OCST problems. We provide a brief description of the problem and empirical results for all test instances.

8.2.1 Problem Definition

This subsection gives a brief definition of the optimal communication spanning tree problem.

The optimal communication spanning tree (OCST) problem (Hu, 1974) finds a tree that connects all given nodes and satisfies their communication requirements for a minimum total cost. The number and positions of the network nodes are given a priori and the cost of the network is determined by the cost of the links. A link's flow is the sum of the communication demands between all pairs of nodes communicating directly, or indirectly, over the link. The cost for each link is not fixed a priori but depends on the length and capacity of the link. A link's capacity must satisfy the link's flow and this flow depends on the entire tree structure. For more information about tree design problems the reader is referred to subsection 6.1.1.

Fig. 8.13. A communication spanning tree on 15 nodes where the path connecting nodes 3 and 14 is emphasized.

Like other constrained spanning tree problems, the OCST problem is NP-hard (Garey & Johnson, 1979, p. 207). Thus, several genetic and evolutionary based algorithms were proposed for solving the problem (Davis, Orvosh, Cox, & Qiu, 1993; Berry, Murtagh, & Sugden, 1994; Kim & Gen, 1999). Researchers proposed encodings such as characteristic vectors (Davis, Orvosh, Cox, & Qiu, 1993; Tang, Man, & Ko, 1997; Sinclair, 1995; Berry, Murtagh, & Sugden, 1994), Prüfer numbers (Kim & Gen, 1999), NetKeys (Rothlauf, Goldberg, & Heinzl, 2002), Blob code (Julstrom, 2001), weighted encodings (Raidl & Julstrom, 2000), and other types of representations for encoding the problem.

Figure 8.13 shows a communication spanning tree on 15 nodes and emphasizes the path connecting nodes 3 and 14.

8.2.2 Theoretical Predictions

In the following subsection, we briefly discuss which theoretical predictions of GEA performance we can make for OCST problems.

It is difficult to predict GEA performance on a problem of unknown complexity. If we have no information about the structure of the optimal solution and the difficulty of the problem, we are not able to make any predictions regarding GEA performance. However, we know from experience that many real-world problems are easy. Therefore, the problems are often decomposable and the BBs are not fully deceptive, but of lower order.

Therefore, due to the easiness of OCST problems we expect GEAs using Prüfer numbers to fail when used on these problems. The low locality of the encoding will increase the difficulty of easy problems and GEA search will result more in a random than in a guided search. Furthermore, we know that the performance of GEAs using uniformly redundant encodings like the NetKey, NetDir, CV, or LNB ($P_1 = 20, P_2 = 0$) encoding is independent of the structure of the optimal solution. Neglecting other effects such as stealth mutation, we expect that these types of encodings result in similar GEA performance. When taking stealth mutation into account and focusing on small problem instances, we believe that GEAs using CVs are able to find the good solutions very reliably, but need more generations. For large problem instances, we expect GEAs using CVs to fail. Stealth mutation becomes stronger with increasing problem size and there is no guided, but only random search possible.

We have seen that the performance of GEAs using LNB encodings with small biases strongly depends on the structure of the optimal solution. If the optimal solution is similar to a star or to the MST, we expect GEAs using these types of encodings to perform very well. Former work indicated that real-world OCST problems often tend to have star- or MST-like optimal solutions (Kershenbaum, 1993; Cahn, 1998). If this assumption is true, we expect GEAs using LNB encodings with small biases to have a high performance on OCST problems. If the good solutions are not similar to stars or to the MST, we expect GEAs to fail.

8.2.3 Palmer's Test Instances

This subsection presents results for the OCSTP test instances from Palmer (1994).

Problem Description

Palmer described OCST problems with 6, 12, 24, 47, and 98 nodes in his thesis. The inter-node traffic requirements were inversely proportional to the distances between the nodes. The nodes correspond to cities in the United States and the distances between the nodes (link cost) $d_{i,j}$ were obtained from a tariff database. Thus, the cost C of a network is defined as

$$C = \sum_{l_{i,j}} d_{i,j} * t_{i,j},$$

where $l_{i,j}$ denotes the used links, $d_{i,j}$ the distance between the nodes i and j, and $t_{i,j}$ the amount of traffic on that link. For the exact distance and requirement matrix for the 6, 12 and 24 node problem the reader is referred to Palmer (1994) or to appendix A.1. Unfortunately, the data for the 47 and 98 node problems is no longer available[1].

Empirical Results

To estimate the performance of GEAs using different types of representations, we show in Table 8.3 the minimum phenotypic distance of a randomly generated individual towards a star, $\min(d_{star})$, and the minimum phenotypic distance towards the minimum spanning tree, d_{MST}. For the investigation, we randomly created 10 000 individuals for each representation and each problem. The results show that the CV, NetKey, NetDir, and LNB ($P_1 = 20, P_2 = 0$) encoding have about the same distances $\min(d_{star})$ and d_{MST} as the Prüfer number encoding which is unbiased. Therefore, these encodings are also unbiased that means uniformly redundant. LNB encodings with small biases have smaller distances $\min(d_{star})$ and d_{MST} and therefore are biased towards the minimum spanning tree and towards stars.

In Table 8.4, we present the properties of the best solutions we found for the three test problems. The results show that the optimal solutions have about the same minimal distances towards a star than a randomly created unbiased individual. However, the optimal solutions have a smaller distance towards the MSTs than a randomly generated individual. This means that the good solutions are biased towards the MST. Therefore, we expect GEAs using LNB encodings with small biases to have high performance. For these types of encodings, MST-like individuals are overrepresented.

In Figure 8.14 (6 and 12 node) and Figure 8.15 (24 node), we present empirical results for the 3 available test problems. Due to illustrative purposes the two plots showing the cost of the best individual over the population size in Figure 8.15 use different scaling for the same data. For our experiments, we use a simple genetic algorithm without mutation. We use uniform crossover and tournament selection without replacement of size 2. The runs are stopped after the population is fully converged or the number of generations exceeds 200. We perform 100 runs for every population size. GEA performance is indicated by the cost of the best solution at the end of the run and the time until convergence.

The results reveal that, according to our assumptions, the known best solution can be easily found. As a result, we assume that the problem is easy. Therefore, GEAs using Prüfer numbers fail due to the low locality of the encoding. GEAs using characteristic vectors perform well for the 6 node problems but fail completely for larger problem instances. They are not able to converge within 200 generations and the solution quality is completely unacceptable. As predicted, the use of NetKeys shows similar results as the use of the LNB encoding with $P_1 = 20$ and $P_2 = 0$. Both encodings result

---

[1] The data sets were not listed in the thesis and are not directly available from Palmer.

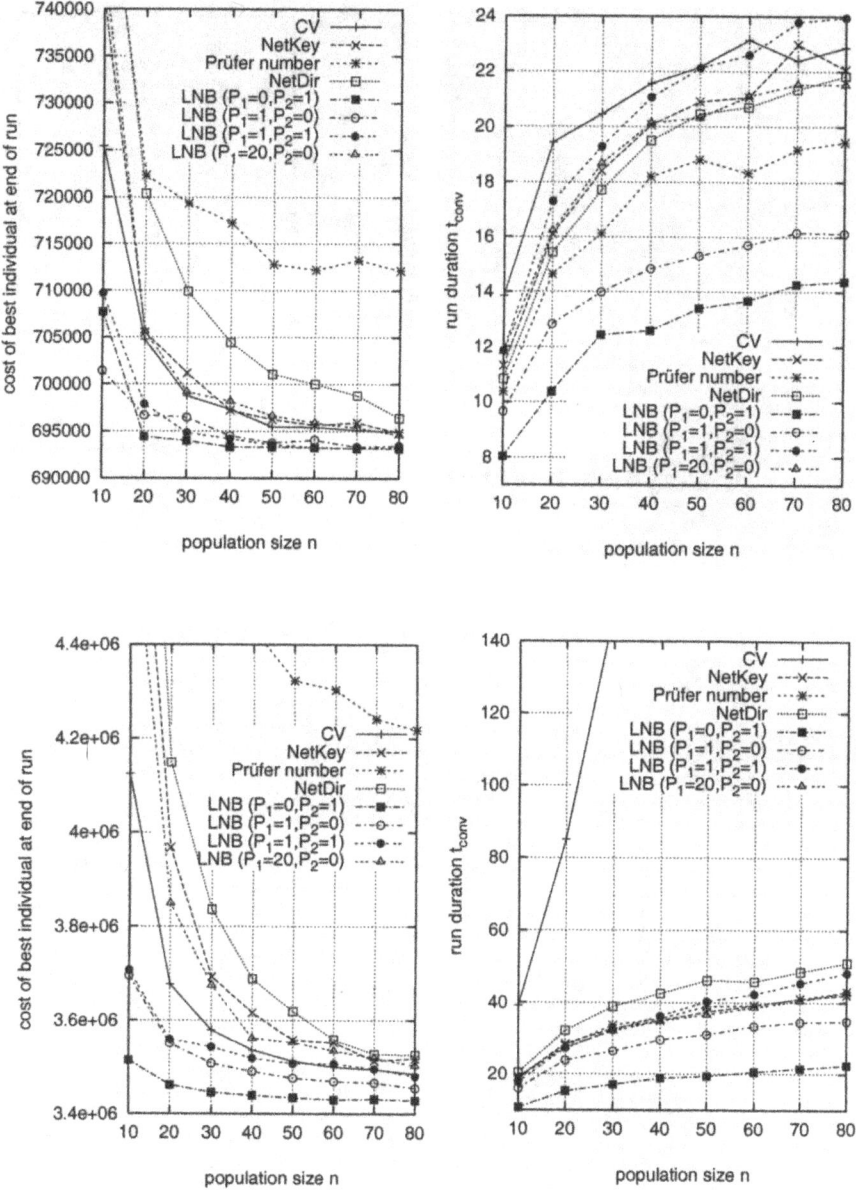

Fig. 8.14. We show the cost of the best individual at the end of the run (left) and the number of generations until convergence (right) over the population size n. The results are for the 6 node (top) and 12 node (bottom) test instance from Palmer (1994). GEAs using LNB encodings with a small bias perform best as the optimal solution is similar to the MST.

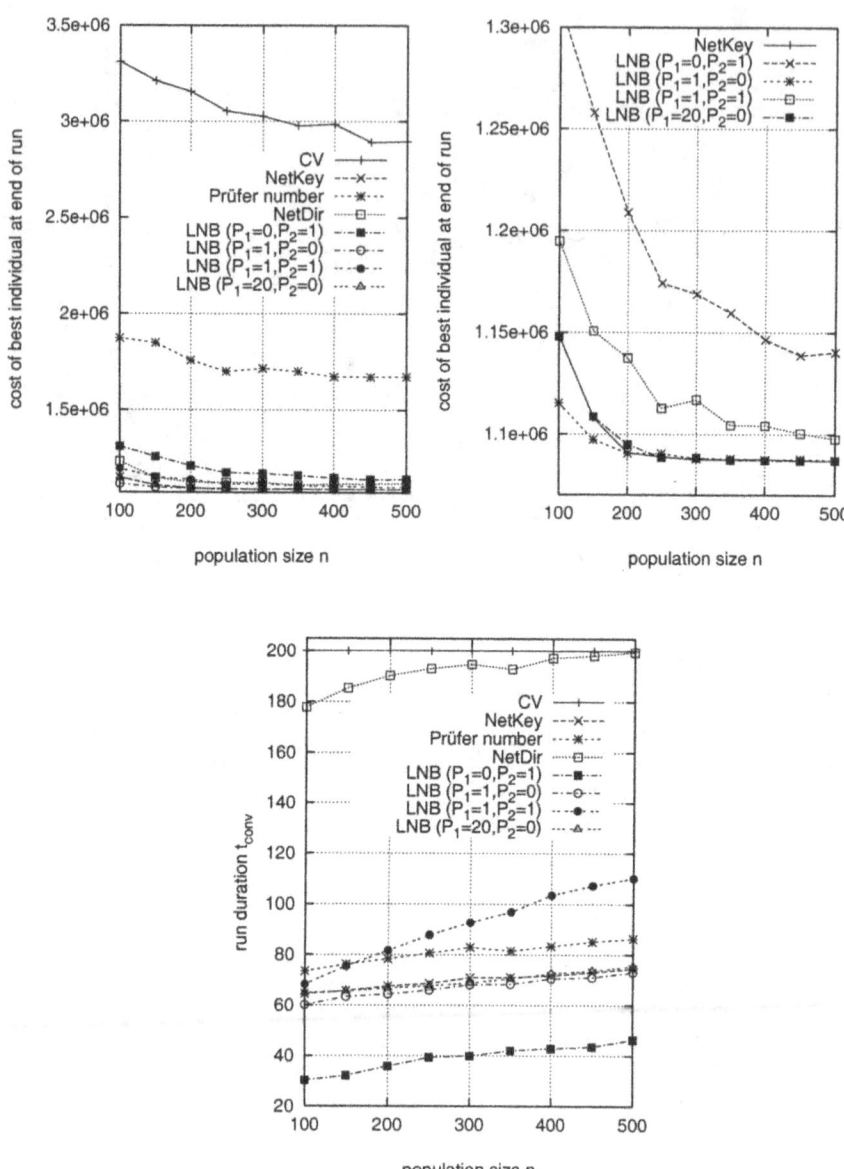

Fig. 8.15. We show the cost of the best individual at the end of the run (top) and the number of generations until convergence (bottom) over the population size n. The results are for the 24 node test instance from Palmer (1994). GEAs using LNB encodings with a small bias perform best as the optimal solution is similar to the MST. GEAs using Prüfer numbers or CVs fail completely.

Table 8.3. We show for a randomly created individual the average minimum phenotypic distance to star networks and the average phenotypic distance to the minimum spanning tree (mean μ and standard deviation σ). The numbers reveal that the Prüfer number, CV, NetKey, NetDir, and LNB ($P_1 = 20, P_2 = 0$) encoding is unbiased.

problem size n		6 node		12 node		24 node	
distance d		μ	σ	μ	σ	μ	σ
Prüfer number	$\min(d_{star})$	2.04	0.61	7.22	0.75	18.50	0.80
	d_{MST}	3.36	0.91	9.17	1.17	21.05	1.30
CV	$\min(d_{star})$	2.04	0.62	7.21	0.76	18.51	0.80
	d_{MST}	3.38	0.91	9.16	1.16	21.04	1.30
NetKey	$\min(d_{star})$	2.12	0.60	7.3	0.71	18.61	0.78
	d_{MST}	3.34	0.94	9.15	1.18	21.07	1.30
NetDir	$\min(d_{star})$	2.05	0.61	7.22	0.75	18.50	0.80
	d_{MST}	3.39	0.90	9.19	1.16	21.05	1.30
LNB ($P_1 = 0, P_2 = 1$)	$\min(d_{star})$	1.00	0.78	5.49	1.33	7.59	4.18
	d_{MST}	2.37	0.80	6.12	1.07	18.68	1.65
LNB ($P_1 = 1, P_2 = 0$)	$\min(d_{star})$	2.08	0.60	7.50	0.63	18.61	0.79
	d_{MST}	2.50	0.96	6.55	1.40	19.45	1.62
LNB ($P_1 = 1, P_2 = 1$)	$\min(d_{star})$	1.59	0.72	6.22	1.08	15.39	1.66
	d_{MST}	2.68	0.87	7.12	1.33	20.21	1.42
LNB ($P_1 = 20, P_2 = 0$)	$\min(d_{star})$	2.13	0.60	7.30	0.71	18.59	0.79
	d_{MST}	3.28	0.92	9.0	1.19	21.01	1.32

Table 8.4. Properties of the best ever found solutions to Palmer's 6, 12, and 24 node test instances

	6 node	12 node	24 node
cost	693 180	3 428 509	1 086 656
$\min(d_{star})$	2	7	17
d_{MST}	1	5	12
$\min(d_{MST,star})$	3	9	20

in high performance and are able to reliably find the best solution after a short time. Even higher performance of GEAs can be obtained when using LNB encoding with either a small link- or a small node-bias. These types of encodings are biased towards the MST and result in very high performance because the best solutions are similar to MSTs. However, the results also show that GEA behavior for different values of P_1 and P_2 is difficult to predict. Sometimes a small link-bias and sometimes a small node-bias result in best GEA performance. Which of the LNB encodings performs best depends on the exact structure of the optimal solution.

8.2.4 Raidl's Test Instances

This subsection presents results for some OCSTP test instances from Raidl (2001).

Problem Description

To compare the performance of GEAs on OCST problems, Raidl (2001) proposed several test instances ranging from 10 to 100 nodes. The distances between the nodes and the traffic demands have been generated randomly. In analogy to Palmer's test instances, the overall cost of a tree is calculated as

$$C = \sum_{l_{i,j}} d_{i,j} * t_{i,j},$$

where $l_{i,j}$ denotes the used links, $d_{i,j}$ the distance between two nodes i and j and $t_{i,j}$ the traffic over this link. The distance matrix and traffic demands are summarized in appendix A.2.

Empirical Results

Due to the large complexity of the problems and the limited available computer power, we restrict our investigations to the 10 and 20 node problem instances.

Table 8.5. We show for a randomly created individual the average minimum distance to star networks and the average distance to the minimum spanning tree (mean μ and standard deviation σ). The numbers reveal that the Prüfer number, CV, NetKey, NetDir, and LNB ($P_1 = 20, P_2 = 0$) encoding is unbiased.

problem size n		10 node		20 node	
distance d		μ	σ	μ	σ
Prüfer number	$\min(d_{star})$	5.42	0.70	14.69	0.77
	d_{MST}	7.20	1.1	17.07	1.27
CV	$\min(d_{star})$	5.41	0.70	14.67	0.79
	d_{MST}	7.23	1.1	17.05	1.27
NetKey	$\min(d_{star})$	5.51	0.66	14.78	0.77
	d_{MST}	7.21	1.1	17.11	1.27
NetDir	$\min(d_{star})$	5.42	0.70	14.68	0.80
	d_{MST}	7.22	1.1	17.04	1.28
LNB ($P_1 = 0, P_2 = 1$)	$\min(d_{star})$	4.12	1.1	10.75	2.24
	d_{MST}	4.24	1.09	11.63	1.3
LNB ($P_1 = 1, P_2 = 0$)	$\min(d_{star})$	5.41	0.69	14.78	0.78
	d_{MST}	4.61	1.27	12.74	1.85
LNB ($P_1 = 1, P_2 = 1$)	$\min(d_{star})$	4.66	0.96	12.48	1.45
	d_{MST}	5.11	1.20	14.02	1.68
LNB ($P_1 = 20, P_2 = 0$)	$\min(d_{star})$	5.51	0.67	14.78	0.78
	d_{MST}	7.02	1.14	16.77	1.34

In analogy to Palmer's test instances, in Table 8.5 we present for a randomly generated individual the average minimum phenotypic distance towards a star, $\min(d_{star})$, and the average phenotypic distance towards the

Table 8.6. Properties of the best ever found solutions for Raidl's 10 and 20 node OCST problems

	10 node	20 node
optimum	53 674	157 570
$\min(d_{star})$	4	14
d_{MST}	3	4
$\min(d_{MST,star})$	6	16

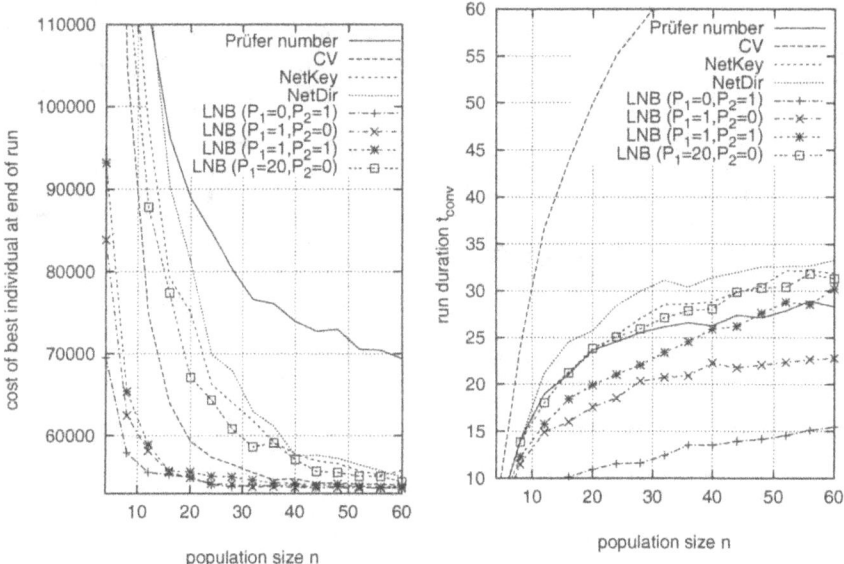

Fig. 8.16. We show the cost of the best individual at the end of the run (left) and the number of generations until convergence (right) over the population size n. The results are for the 10 node test instance from Raidl (2001). GEAs using LNB encodings with a small bias perform best as the optimal solution is similar to the MST.

MST, d_{MST}. As before, we created 10 000 individuals for each representation. For the LNB encodings, the calculation of the distances between the nodes is based on Raidl's test instances.

Table 8.6 shows the properties of the best found solutions. The minimal distance towards star networks, $\min(d_{star})$, is about the same for randomly created unbiased individuals than for the optimal solutions. However, the optimal solutions are very similar to the MST. For unbiased representations, the average distance towards the MST is on average $d_{MST} \approx 7.2$ (10 node) and $d_{MST} \approx 17.1$ (20 node). In contrast, the optimal solution has a minimum distance of $d_{MST} = 3$ (10 node) and a distance of $d_{MST} = 4$ (20 node) only. As a result, we expect GEAs using the LNB encoding with a small link- or node-specific bias to perform very well.

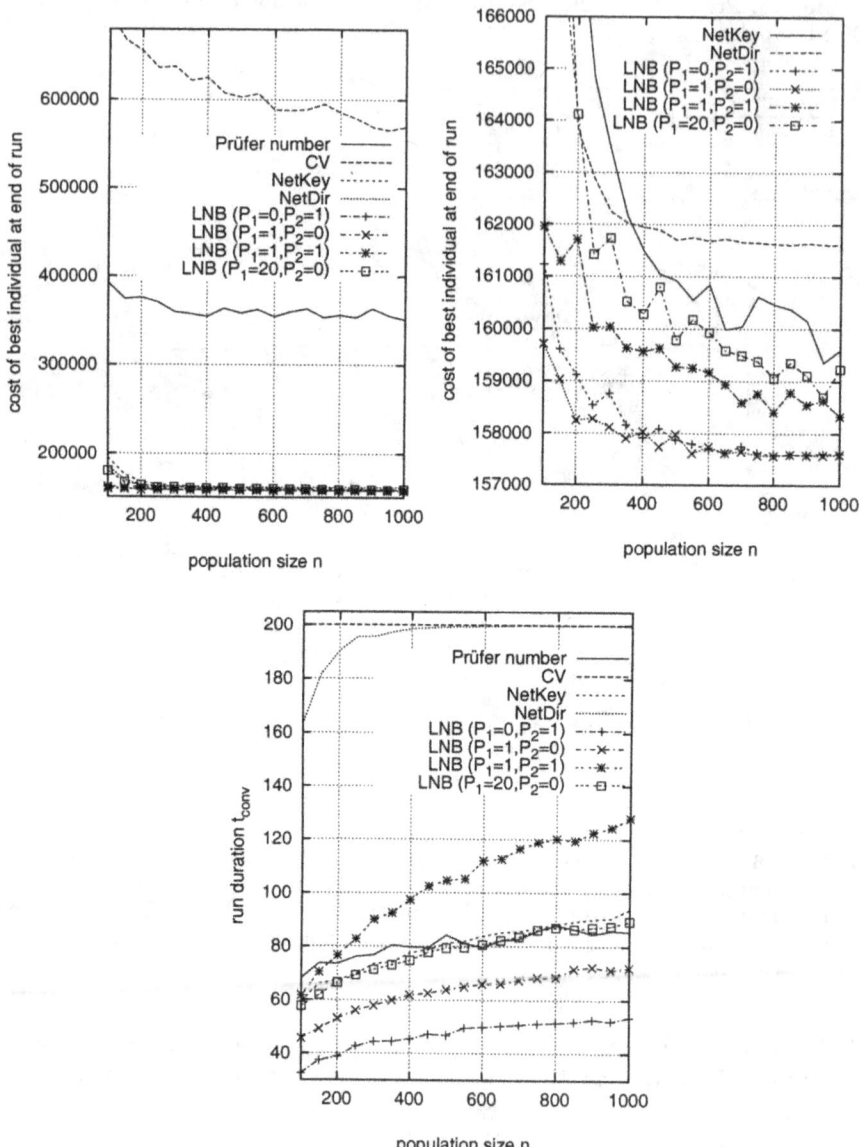

Fig. 8.17. We show the cost of the best individual at the end of the run (top) and the number of generations until convergence (bottom) over the population size n. The results are for the 20 node test instance from Raidl (2001). GEAs using LNB encodings with a small bias perform best as the optimal solution is similar to the MST. GEAs using the Prüfer number representation fail due to the low locality of the representation.

The left plot in Figure 8.16 (10 nodes) and the plots at the top of Figure 8.17 (20 nodes) show the solution quality over the population size. The time to convergence is shown at the right of Figure 8.16 (10 node) and at the bottom of Figure 8.17 (20 node). As before, we use a simple genetic algorithm without mutation, uniform crossover, and tournament selection without replacement of size 2. The runs are stopped after the population is converged or 200 generations are exceeded.

The results indicate the same pattern as for the test instances from Palmer (1994). GEAs using Prüfer numbers fail completely. GEAs using the characteristic vector encoding perform well for very small problems but need a huge number of generations for larger problems and are not able to find their way to the good solutions. Stealth mutation does not allow GEAs to perform well and results in a random search.

As before, NetKeys result in similar performance to the LNB encoding with large link-specific bias. Both encodings allow GEAs to find high-quality solutions in a short time. GEAs perform best when using LNB encodings with a small link- or node specific bias. When using these types of encodings, MST-like structures are overrepresented in the initial population and GEAs find the best solutions after a small number of generations.

8.2.5 Test Instances from Berry, Murtagh, and McMahon (1995)

This subsection presents results for the test instances from Berry, Murtagh, and McMahon (1995).

Problem Description

Berry, Murtagh, and McMahon (1995) presented three different instances of the OCST problem. They proposed a 6 node and two 35 node problems. The distance matrix and the traffic demands are listed in appendix A.3. As before, the cost of a network is calculated as

$$C = \sum_{l_{i,j}} d_{i,j} * t_{i,j},$$

where $l_{i,j}$ denotes the used links, $d_{i,j}$ the distances between the nodes i and j, and $t_{i,j}$ the corresponding traffic over this link.

Both 35 node problems use the same traffic demands, but differ in the distances between the nodes. One problem has uniform distances with $d_{i,j} = 1$, and the other has non-uniform distances. The best solution found by Berry et al. (1995) for the 35 node problem with uniform distances was 16 915 and 30 467 for the problem with non-uniform distances. Li and Bouchebaba (1999) improved these results and found the best solution for the 35 node problem with uniform distances to have cost 16 420 and for the 35 node problem with non-uniform distances to have cost 16 915.

Empirical Results

As for the previous problems, we present in Table 8.7 for a randomly created individual the average minimum phenotypic distance towards a star and the average phenotypic distance towards the MST. We create 10 000 individuals for each representation.

Table 8.7. We show for a randomly created individual the average minimum distance to star networks and the average distance to the minimum spanning tree. The numbers reveal that the Prüfer number, CV, NetKey, NetDir, and LNB ($P_1 = 20, P_2 = 0$) encoding is unbiased.

problem size n		6 node		35 node uniform		35 node non-uniform	
distance d		μ	σ	μ	σ	μ	σ
Prüfer number	$\min(d_{star})$	2.03	0.61	29.19	0.829	29.16	0.83
	d_{MST}	3.51	0.83	n.a.	n.a.	32.05	1.32
CV	$\min(d_{star})$	2.04	0.61	29.17	0.83	29.18	0.83
	d_{MST}	3.54	0.84	n.a.	n.a.	32.04	1.32
NetKey	$\min(d_{star})$	2.11	0.60	29.27	0.80	29.27	0.80
	d_{MST}	3.34	0.90	n.a.	n.a.	32.04	1.33
NetDir	$\min(d_{star})$	2.04	0.61	29.19	0.83	29.17	0.82
	d_{MST}	3.50	0.85	n.a.	n.a.	32.01	1.34
LNB ($P_1 = 0, P_2 = 1$)	$\min(d_{star})$	1.48	0.56	0.0	0.0	21.63	3.54
	d_{MST}	1.78	0.87	n.a.	n.a.	22.10	2.68
LNB ($P_1 = 1, P_2 = 0$)	$\min(d_{star})$	2.15	0.53	29.26	0.80	28.92	0.95
	d_{MST}	1.86	0.89	n.a.	n.a.	23.93	2.44
LNB ($P_1 = 1, P_2 = 1$)	$\min(d_{star})$	1.77	0.63	24.63	1.93	25.39	2.03
	d_{MST}	2.14	0.90	n.a.	n.a.	26.54	2.18
LNB ($P_1 = 20, P_2 = 0$)	$\min(d_{star})$	2.11	0.61	29.27	0.80	29.24	0.81
	d_{MST}	3.25	0.92	n.a.	n.a.	31.41	1.49

Table 8.8. Properties of the best ever found solutions for the test instances from Berry, Murtagh, and Mc Mahon (1995).

	6 node	35 node uniform	35 node non-uniform
optimum	534	16 273	16915
$\min(d_{star})$	2	28	30
d_{MST}	0	-	1
$\min(d_{MST,star})$	2	-	30

Table 8.8 presents the properties of the best ever found solutions. For the 35 node problem with uniform distances between the nodes, the minimum spanning tree is not unique. Therefore, we can not calculate any distances for the 35 node problem with uniform distances. The results for the 6 node

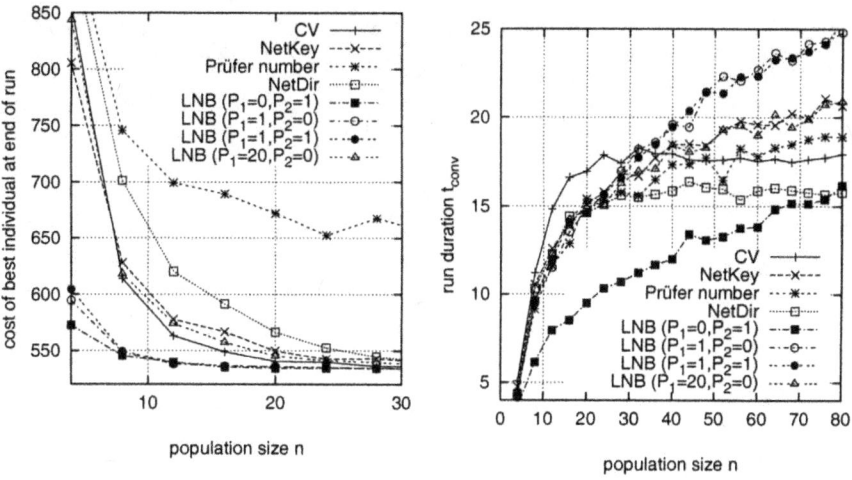

Fig. 8.18. We show the cost of the best individual at the end of the run (left) and the number of generations until convergence (right) over the population size n for the 6 node test instance from Berry, Murtagh, and Mc Mahon (1995). GEAs using LNB encodings with a small bias perform best as the optimal solution is the MST. GEAs using the Prüfer number representation fail due to the low locality of the representation.

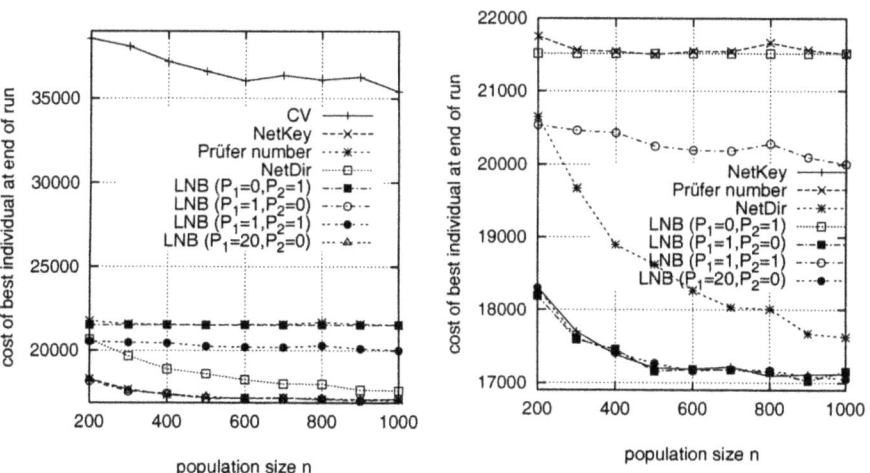

Fig. 8.19. We show the cost of the best individual at the end of the run over the population size n for the 35 node test instance with uniform distances from Berry, Murtagh, and Mc Mahon (1995). GEAs using LNB encodings with a small bias perform best as the optimal solution is similar to the MST. GEAs using either NetKeys or LNBs with no node-specific bias show the same performance because the link-bias is large enough to represent the individuals uniformly.

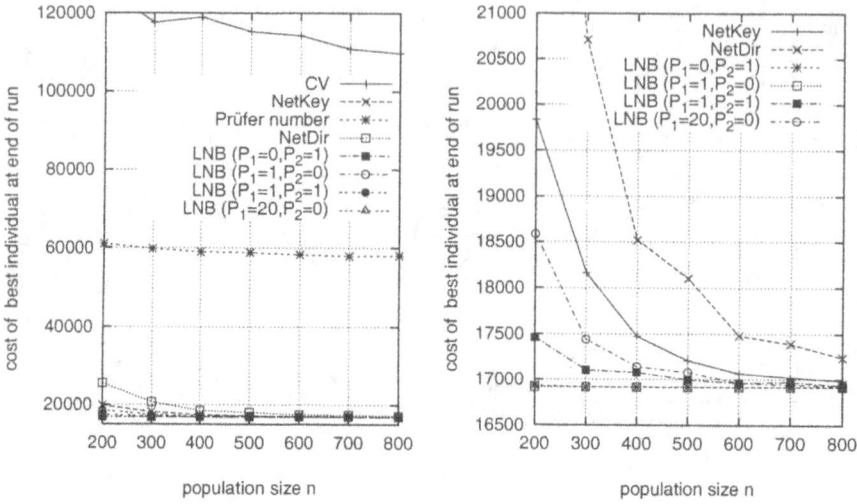

Fig. 8.20. We show the cost of the best individual at the end of the run over the population size n for the 35 node test instance with non-uniform distances from Berry, Murtagh, and Mc Mahon (1995). GEAs using LNB encodings with a small bias perform best as the optimal solution is similar to the MST. GEAs using CVs fail due to problems with stealth mutation.

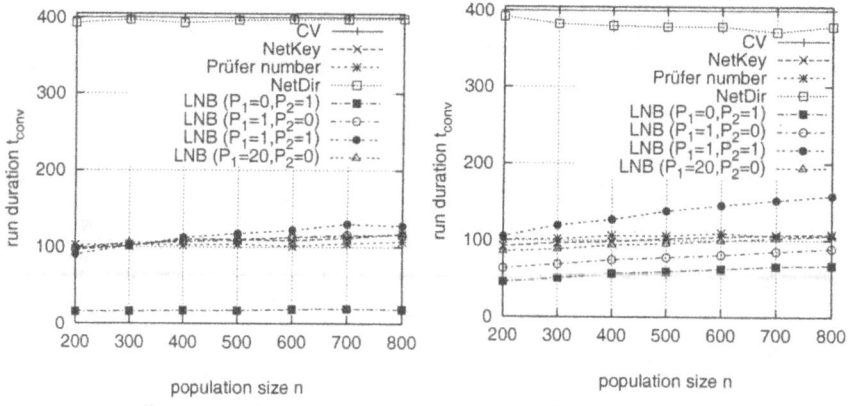

(a) 35 node with uniform distances. GEAs using the LNB encoding with $P_1 = 0/P_2 = 1$ are converged after a few generations to a non-optimal star network and fail.

(b) 35 node with non-uniform distances. GEAs using LNB encodings with a small bias perform best as the optimal solution is similar to the MST.

Fig. 8.21. We show the number of generations until convergence over the population size n for the 35 node test instances from Berry, Murtagh, and Mc Mahon (1995). GEAs using NetDir or CV encoding are not able to converge.

and 35 node (non-uniform) problem show that the best found solutions have about the same distance towards a star as a randomly generated unbiased individual. However, the best solution to the 6 node problem is the MST itself and the best solution for the 35 node problem only has a distance of $d_{MST} = 1$ towards the MST. This means that the MST has on average 33 out of 34 links in common with the optimal solution. Therefore, we expect GEAs using the LNB encoding with small biases to significantly outperform all other representations.

Figure 8.18 (6 node), Figure 8.19 and Figure 8.21(a) (35 node uniform), and Figure 8.20 and Figure 8.21(b) (35 node non-uniform) show the cost of the best individual at the end of the run and the number of generations until convergence for the three problem instances. We use a genetic algorithm with no mutation and uniform crossover. For selection we use tournament selection without replacement of size 2. We performed 100 runs for every population size and the runs were stopped after the population is fully converged or the number of generations exceeded 200.

The results for the 6 node and 35 node problem with non-uniform distances are similar to the results for the test instances from Palmer (1994) or Raidl (2001). GEAs using Prüfer numbers fail. GEAs using CVs perform well for the small problem but fail completely for larger problems. We get the best results when using LNB encodings with small link- or node-bias. Then, the individuals are biased towards the MST and the best solutions which are very similar to the MST can easily be found.

The results for the 35 node non-uniform problem in Figure 8.19 (solution quality) and Figure 8.21(b) (time to convergence) are more interesting. Our investigations into the properties of the LNB encoding revealed that if the distances between the nodes are the same, and only a link-specific bias is used, then the encoding can only encode stars (see subsection 6.3.4). As a result, GEAs using LNB encodings with $P_1 = 0, P_2 = 1$ can only find stars and no other solutions. The cost of the best star is about 21 500. Furthermore, the figures reveal that GEAs using either NetKeys or LNB encodings with no node bias show the same performance and are able to find the best solution fast and reliably. The behavior can be explained because the link-bias of the LNB encodings is large enough in comparison to the distances between the nodes to represent trees uniformly and not to be biased towards the MST.

8.2.6 Selected Real-World Test Instances

This subsection presents results for some real-world test instances proposed in Rothlauf, Goldberg, and Heinzl (2002).

Problem Description

The presented OCST problems are derived from a real-world 26-node problem from a company with locations all over Germany.

For fulfilling the demands between the nodes, different line types with only discrete capacities and cost are available. The cost for installing a link consists of a fixed and length dependent share. Both depend on the capacity of the link. The cost are based on the tariffs of the German Telecom from 1996. For an exact description on how the cost of a link depends on its length and its capacity the reader is referred to appendix A.4.

We present four different problems. In problem 1 we have a 16-node problem with traffic ending only at node 1. In the second problem, one of the nodes is removed and some additional traffic is added. The third problem is similar to problem 1 but uses a modified cost-function for the lines. Finally, we look at a 16-node problem with traffic between all nodes. For all problems, the distances between the nodes are fixed. The distances between the nodes (cities) are calculated using the Euclidean distances. In the following we give a brief description of the four problems:

- **Problem 1: One headquarter and 15 branch offices**
  This problem is the original design problem. All 15 branch offices (node 2 to 16) communicate only with the headquarter (node 1). The cost of renting a line depends on the length and the capacity of the link. Possible line capacities are 64 kBit/s, 512 kBit/s, and 2048 kBit/s. The optimal solution for this problem is shown in Figure 8.22(a). The complexity of the problem is low.
- **Problem 2: One headquarter and only 14 branches**
  If one node is left out and some additional traffic is added, finding the best solution is slightly more involved than in problem 1. The optimal solution is shown in Figure 8.22(b).
- **Problem 3: One headquarter, 15 branches and cheap lines for everybody**
  In the scenario shown in Figure 8.22(c), the fixed cost for installing a high-capacity line is only 10% of the cost in problem 1. Therefore, the cost of a link is mainly determined by its length. Hence, the optimal structure is more like a minimum spanning tree. If the cost of the link would only be determined by the length of the link, and if there was only one possible capacity, the optimal solution would be the minimum spanning tree. Otherwise the problem is exactly like problem 1.
- **Problem 4: 4 headquarters, 12 branches and all are working together**
  In problem 4 depicted in Figure 8.22(d), the demand matrix is completely filled. Between every node i and j some traffic exists. Between the four headquarters (node 1, 2, 3 and 4), the traffic is uniformly distributed between 256 kBit/s and 512 kBit/s. Every other node communicates with the four headquarters and has a uniform demand between 0 and 512 kBit/s. This demand is split into the headquarters at a ratio of 0.4, 0.3, 0.2, and

0.1 for the node 1, 2, 3, and 4^2. Between all 12 branch offices the demand of the traffic is uniformly distributed between 0 and 64 kBit/s. To make the problem more realistic, two additional line types are available. It is possible to use a line of 128 kBit/s and 4096 kBit/s with twice the cost of a 64kBit/s respectively the 2048 kBit/s line.

In the following, we present empirical results for these four real-world problems.

Empirical Results

In analogy to the previous investigations, Table 8.9 presents for a randomly created individual the average minimum distance towards a star, $\min(d_{star})$, and the average distance towards the minimum spanning tree, d_{MST}. We create 10 000 individuals for each representation and each problem. In Table 8.10, we present the properties of the best solutions we ever found for the four problems. Figure 8.22 shows the corresponding structure of the optimal trees. The results show that the best solution has about the same distance towards a star as a randomly generated individual. However, the distance of the optimal solutions to the MST is much smaller than the distance of a randomly created individual towards the MST. Therefore, the best solution is biased towards the MST and we expect GEAs using representations that are biased towards the MST to perform well.

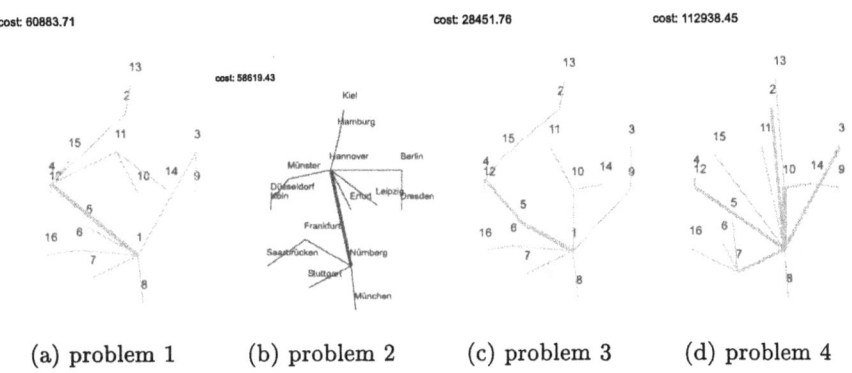

| (a) problem 1 | (b) problem 2 | (c) problem 3 | (d) problem 4 |

Fig. 8.22. Optimal solutions for the four real world problem instances.

Furthermore, we investigated the influence of representations on the performance of GEAs for the 4 different problem instances. Figure 8.23 (problem

---

^{2} Node 1 is the most important node and 40% of the traffic of the branches ends there, in node 2, 30% of the traffic ends, and so on.

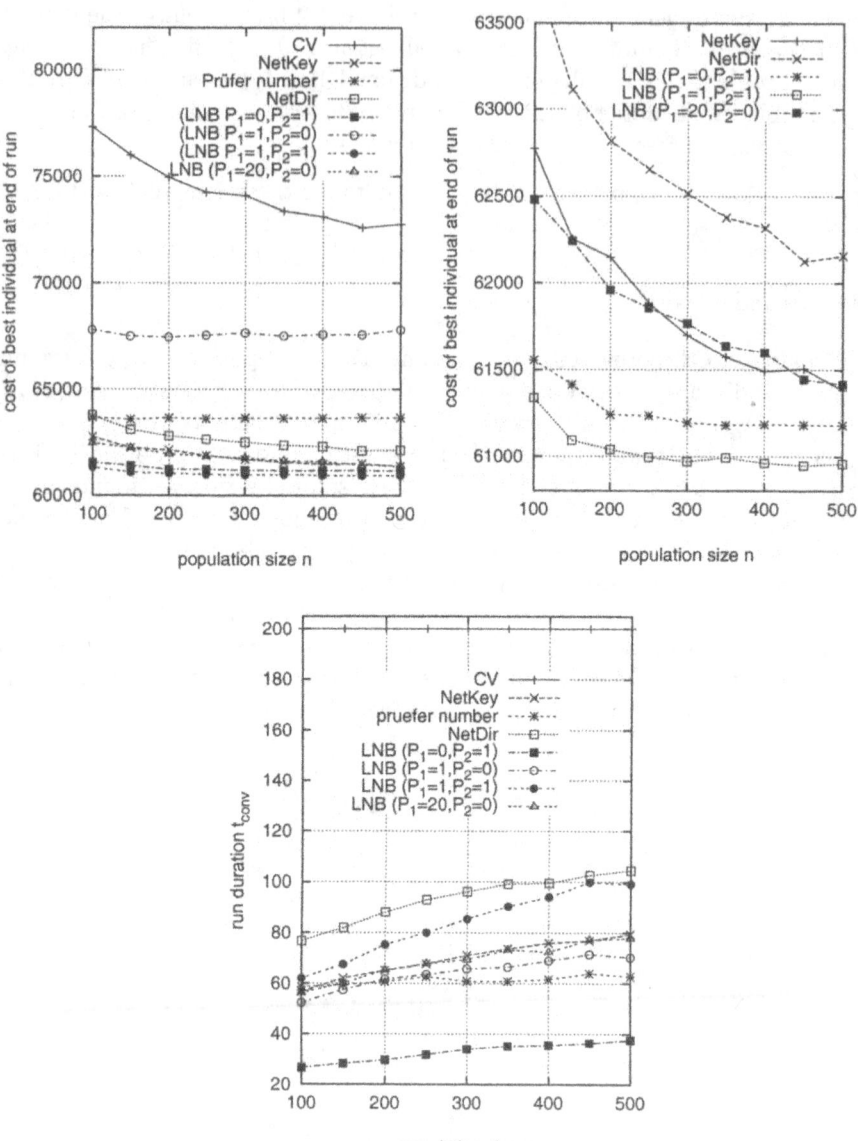

Fig. 8.23. We show the cost of the best individual at the end of the run (top) and the number of generations until convergence (bottom) over the population size n. The results represent the first out of four real-world OCST problems (problem 1). GEAs using LNB encodings with a small bias perform best as the optimal solution is similar to the MST. GEAs using CVs fail due to problems with stealth mutation.

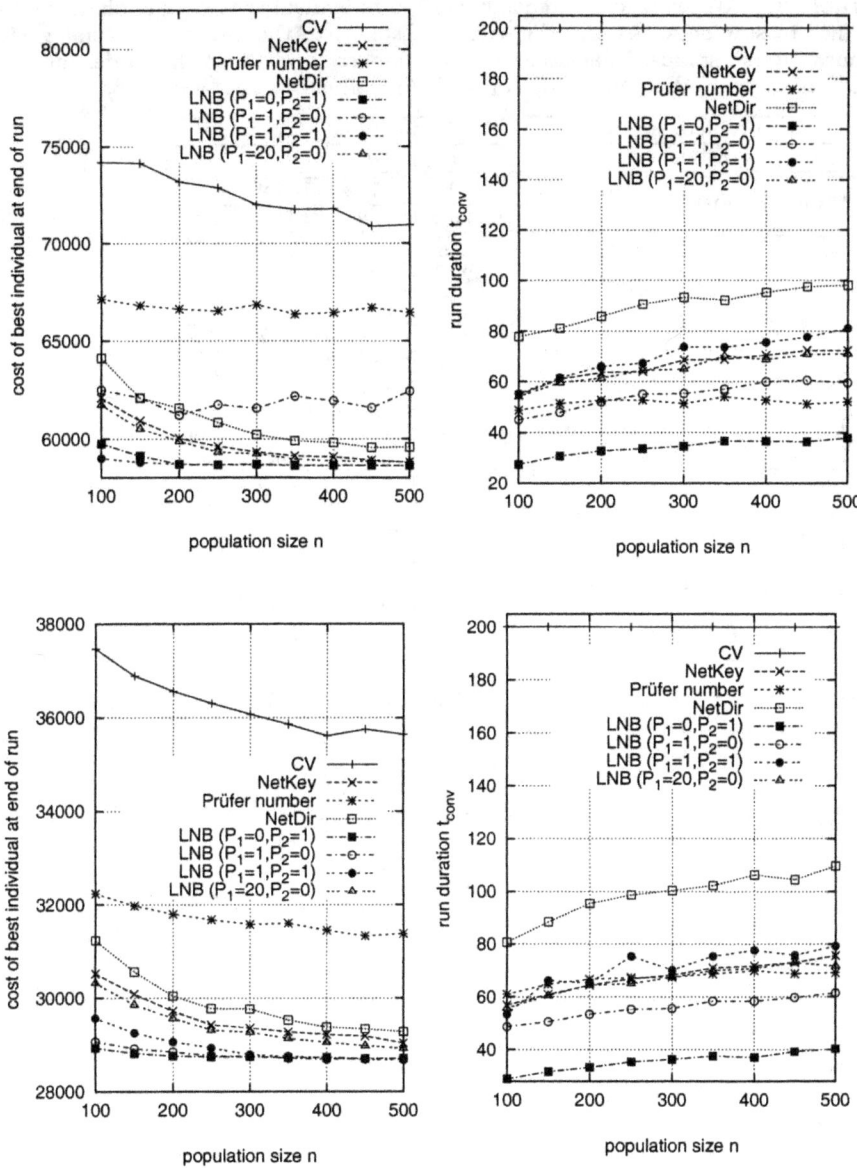

Fig. 8.24. We show for problem 2 (top) and problem 3 (bottom) the cost of the best individual at the end of the run (left) and the number of generations until convergence (right) over the population size n.

Table 8.9. We show for a randomly created individual the average minimum distance to star networks and the average distance to the minimum spanning tree (mean μ and standard deviation σ). The numbers reveal that the Prüfer number, CV, NetKey, NetDir, and LNB ($P_1 = 20, P_2 = 0$) encoding is unbiased.

problem size n		15 node		16 node	
distance d		μ	σ	μ	σ
Prüfer number	$\min(d_{star})$	9.99	0.772	10.89	0.80
	d_{MST}	12.08	1.08	13.07	1.24
CV	$\min(d_{star})$	9.98	0.76	10.91	0.79
	d_{MST}	12.11	1.04	13.10	1.23
NetKey	$\min(d_{star})$	10.07	0.75	11.0	0.76
	d_{MST}	12.13	1.03	13.12	1.23
NetDir	$\min(d_{star})$	9.98	0.778	10.91	0.78
	d_{MST}	12.09	1.06	13.08	1.24
LNB ($P_1 = 0, P_2 = 1$)	$\min(d_{star})$	7.62	1.80	7.91	1.85
	d_{MST}	8.06	1.381	8.66	1.52
LNB ($P_1 = 1, P_2 = 0$)	$\min(d_{star})$	10.05	0.77	11.02	0.76
	d_{MST}	8.85	1.60	9.55	1.66
LNB ($P_1 = 1, P_2 = 1$)	$\min(d_{star})$	8.36	1.29	9.20	1.31
	d_{MST}	9.75	1.43	10.53	1.49
LNB ($P_1 = 20, P_2 = 0$)	$\min(d_{star})$	10.07	0.77	11.0	0.76
	d_{MST}	11.93	1.27	12.89	1.29

Table 8.10. Properties of the optimal solution for the four real-world communication network problems

	problem 1	problem 2	problem 3	problem 4
optimum	60 883	58 619	28 451	112 938
$\min(d_{star})$	9	8	9	7
d_{MST}	7	4	6	9
$\min(d_{MST,star})$	12	11	12	12

1), Figure 8.24 (problem 2 and 3), and Figure 8.25 (problem 4) show the cost of the best individual at the end of the run and the number of generations until the population is completely converged. We use a simple genetic algorithm without mutation for our experiments. Furthermore, we use uniform crossover and tournament selection without replacement of size 2. As before, we stopped the runs after the population is fully converged or the number of generations exceeded 200 generations. We performed 100 runs for every population size.

The GEAs using different types of representations show similar performance as for the other OCST problems. GEAs using Prüfer number fail due to the low locality of the encoding. Due to problems with stealth mutation, GEAs using the characteristic vector encoding are not able to reliably detect the good solutions and are always stopped after 200 generations. The performance of GEAs using NetKeys is similar as when using the LNB encoding

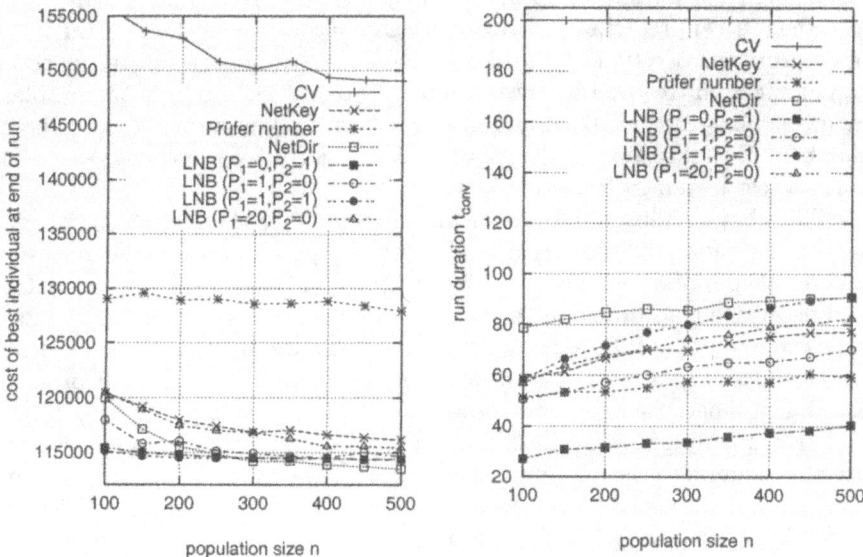

Fig. 8.25. We show the cost of the best individual at the end of the run (left) and the number of generations until convergence (right) over the population size n. The results are for the fourth out of four real-world OCST problems (problem 4).

with a large link-specific bias. Both encodings show good results and find the good solutions after short time. As for the previous problems, GEAs perform mostly at their best when using LNB encodings with a small bias. These encodings are non-uniformly redundant and overrepresent MST-like structures. Because the optimal solutions are similar to the MST, GEAs using these encodings find the best solutions in the shortest time.

However, finding exactly the best setting for the link- and node-specific bias is tricky as the bias of the encoding depends on P_1 and P_2 as well as on the distances between the nodes. If the biases P_1 and P_2 are too small, only a small fraction of MST-like solutions can be represented; if the link-specific bias is too large, GEAs perform the same as when using NetKeys; if the node-specific bias is too large, only stars can be represented and GEA performance is reduced.

8.3 Summary

This chapter predicted and compared GEA performance for different tree problems. Section 8.1 focused on the performance of GEAs using different types of representations for the one-max tree and the deceptive tree problem. As the optimal solution is known a priori, we were able to predict the expected GEA performance, solution quality and time, for both problems well. The theoretical predictions were empirically verified for various instances of both problems. Section 8.2 focused on GEA performance for the optimal communication tree problem. After a brief problem definition of the problem in subsection 8.2.1, we discussed the expected GEA performance in subsection 8.2.2. In subsection 8.2.3, 8.2.4, 8.2.5, and 8.2.6, we presented results for various test instances from the literature.

This chapter illustrated how GEA users can use the framework of representations for predicting and explaining the performance of GEAs using different types of representations. When applying GEAs to the one-max tree and deceptive trap problem, the optimal solutions are known a priori and it is possible to predict the expected GEA performance. In contrast, predictions of GEA performance for the optimal communication tree (OCST) problem are difficult because the good solutions are in general not known a priori. However, many real-world OCST problems are easy and the good solutions are often similar to the minimum spanning tree. Based on this assumption, we can use our knowledge about representations for quite accurate predictions of GEA performance on OCST problems.

The results for the scalable test problems confirmed the theoretical predictions. GEAs using Prüfer numbers fail for the one-max tree problem but perform well for the deceptive tree problem. Due to stealth mutation, GEAs using characteristic vectors are able to find a large proportion of correct BBs if the problem is easy or small but need a huge number of generations. LNB encodings with a small bias are biased towards stars and minimum spanning trees (MST). Therefore, GEAs perform well if the optimal solution is either star-like or MST-like. For other problems, GEAs using LNB encodings with a small bias fail. As predicted, random network keys (NetKeys) show about the same performance as LNB encodings with a large link-specific bias. GEAs using these two encodings performed reliably well on all test instances and were in general the best choice for the one-max tree and deceptive tree problem.

The situation is slightly different when using GEAs for OCST problems. Many instances of these problems are not too difficult and the optimal solutions are often similar to the MST. Therefore, GEAs using LNB encoding with a small bias perform best. NetKeys and LNB encoding with a large link-specific bias perform well, whereas Prüfer numbers, characteristic vectors and LNB encodings with a large node-specific bias fail.

This chapter nicely illustrated the benefits we gain from analyzing the properties of encodings and using the framework about representations for predicting the expected GEA performance. We want to encourage researchers to use the proposed tools for analyzing the properties of representations for other problem domains. Examining redundancy, scaling, and locality of an encoding a priori allows us to predict GEA performance based on the presented framework about representations.

9. Summary, Conclusions and Future Work

The purpose of this work was to understand the influence of representations on the performance of genetic and evolutionary algorithms. This chapter summarizes the work contained in this study, lists its major contributions, and points to some directions of further research.

9.1 Summary

The study started in **chapter 2** by providing the necessary background for examining representations for genetic and evolutionary algorithms. Researchers recognized early that representations are likely to have a large influence on the performance of GEAs. Consequently, after a brief introduction into representations and GEAs, we discussed how the influence of representations on problem difficulty can be measured. The chapter ended with prior guidelines for choosing high-quality representations. Most of them are mainly based on empirical observations and intuition and not on theoretical analysis.

Therefore, we presented in **chapter 3** three aspects of a theory of representations for genetic algorithms and evolutionary algorithms (GEAs). We investigated how redundant encodings, encodings using exponentially scaled BBs, and encodings that change the distances between corresponding genotypes and phenotypes influence GEA performance. The performance of genetic and evolutionary algorithms is determined by the solution quality at the end of a run and the number of generations until the population is converged. Consequently, we presented for redundant and exponentially scaled encodings population sizing models and described how the time to convergence is changed. We were able to demonstrate that encodings which do not change the distances between the individuals do not change the order and length of the BBs. Therefore, the complexity of the problem remains untouched. By using encodings that change the distances between the individuals the difficulty of problems can be completely changed. For all three types of encodings, the theoretical models were verified with empirical results.

In **chapter 4**, we combined the three elements of representation theory from chapter 3 to form a framework for a theory-guided design and analysis of representations. The framework describes how the time to convergence and the solution quality of GEAs depends on the redundancy, BB-scaling, and distance distortion of a representation. Using this framework, we could estimate the performance of GEAs using different types of representations in a theory-guided manner. We recognized that different types of representations can dramatically change the behavior of GEAs. Consequently, we presented some major implications when using different types of representations.

In **chapter 5** and **chapter 6**, we used the framework for an analysis of existing representations in a theory-guided matter. We analyzed the redundancy, BB-scaling, and BB-distortion of different representations and made predictions regarding GEA performance based on the framework. These predictions were verified by empirical results. In chapter 5, we focused on binary representations for integers. We compared the performance of binary, gray, and unary encodings for integer problems. Based on the properties of the encodings, the framework was able to explain existing performance differences. Chapter 6 focused on tree representations. In analogy to binary representations for integers, we analyzed the properties of tree representations, namely Prüfer numbers, characteristic vector encodings, and link and node biased encodings. Based on these investigations, we were able to predict GEA performance using the framework.

The framework about representations can not only be used for analysis of representations, but also for the design of representations. Based on the insights into the insufficiencies of existing tree representations, we combined in **chapter 7** the advantageous elements of the characteristic vector and the link and node biased encoding to form the NetKey encoding. The NetKey encoding is designed to allow GEAs to perform well independently of the structure of the optimal solution and to be able to distinguish between important and unimportant links in a tree. Furthermore, we presented a population sizing model for the NetKey encoding and verified for the one-max tree problem that the time to convergence depends linearly on the problem size. Finally, we finished our theoretical analysis and design of representations with the development of the direct NetDir representation. With direct representations we come from representations to genetic operators. Direct representations are defined directly on phenotypes. Therefore, standard genetic operators can not be used any more. As a result, problem-specific operators which are defined on the phenotypes must be developed and the presented framework about representations can not be used any more. We illustrated that for creating efficient GEAs, it is necessary either to design problem-specific genotype-phenotype mappings (representations) and to use standard operators like one-point or uniform crossover, or to develop problem-specific operators and to use direct representations (phenotypes).

Finally, **chapter 8** illustrated how GEA users can use the provided theory about representations for estimating the influence of different representations on GEA performance. Based on the analysis of redundancy, BB-scaling, and distance distortion of representations, we compared GEAs performance for the one-max tree problem, the deceptive tree problem, and various instances of the optimal communication spanning tree problem. The comparison of performance of the different types of tree representations is based on extensive empirical tests.

9.2 Conclusions

We want to summarize the most important contributions of this work.

Framework for design and analysis of representations for genetic and evolutionary algorithms. The main purpose of this study was to present a framework which describes how genetic representations influence the performance of GEAs. The performance of GEAs is measured by the solution quality at the end of the run and the number of generations until the population is converged. The proposed framework allows us to analyze the influence of existing representations on GEA performance and to develop efficient new representations in a theory-guided way. Therefore, the development of good representations remains not only a matter of intuition and random search but becomes a design issue that can be solved by using the presented framework. Even though more work is needed, we believe that the results presented are sufficiently compelling to recommend increased use of the framework for the analysis and design of representations.

Redundancy, BB-Scaling, and Distance Distortion. These are the three elements of the proposed framework of representations. We demonstrated that these three properties of representations influence GEA performance and presented theoretical models to predict how solution quality and time to convergence changes. By examining the redundancy, BB-scaling, and distance distortion of an encoding, we are able to predict GEA performance based on the framework.

The theoretical analysis shows that the redundancy of an encoding influences the supply of building blocks (BB) in the initial population. With r denoting the number of genotypic BBs that represent the best phenotypic BB, and k_r denoting the order of redundancy, the probability of GEA failure goes with $O(\exp(-r/2^{k_r}))$ for uniformly scaled representations and with $O(\exp(-\sqrt{r/2^{k_r}}))$ for exponentially scaled representations. Therefore, GEA performance increases if the representation overrepresents high-quality BBs. If a representation is uniformly redundant, that means each phenotype is represented by the same number of genotypes, GEA performance remains unchanged.

The analysis of encodings using exponentially scaled BBs reveals that non-uniformly scaled representations modify the dynamics of genetic search. If exponentially scaled representations are used, the alleles are solved serially which increases the overall time until convergence and results in problems with genetic drift but allows rough approximations of the expected optimal solution after a few generations.

We know from previous work that high locality of an encoding is a necessary condition for efficient mutation-based search. Our investigations into the effect of encodings that change the distances between corresponding genotypes and phenotypes revealed that low distance distortion is necessary for efficient crossover-based search. Furthermore, we have seen that high locality is a necessary condition for low distance distortion. An encoding has high locality if neighboring phenotypes correspond to neighboring genotypes and low distance distortion if genotypic distances correspond to phenotypic distances. We showed that the distance distortion determines whether the difficulty of a problem is changed by the representation. Only representations that do not change distances guarantee not to change the difficulty of the problems. Easy phenotypic problems remain easy for GEAs and difficult phenotypic problems remain difficult. If the distances are changed, problem difficulty is also changed and the resulting problem difficulty depends on the specific fitness function and genotype-phenotype mapping.

Population sizing and time to convergence models for redundant encodings, exponentially scaled encodings, and NetKeys. Based on a better understanding of redundancy and scaling, we were able to formulate population sizing models and time to convergence models for redundant and exponentially scaled encodings. The models show that for redundant encodings the population size grows with $O(2^{k_r}/r)$ and the time to convergence goes with $O(\text{const} - r/2^{k_r k - 1})$, where k denotes the order of building blocks. When using exponentially scaled encodings, we have to distinguish whether we want to consider genetic drift or not. When neglecting genetic drift, the population size n is independent of the length of an exponentially scaled BB l_s, but depends only on the number of competing exponentially scaled BBs m ($n = O(\sqrt{m})$). The time to convergence goes with $O(l_s \sqrt{m})$. For considering genetic drift, we developed two different population sizing models based on the model for the non-drift case (stair-case drift model and approximated drift model). Due to genetic drift, the ability of GEAs to decide well between competing schemata decreases with increasing number of generations. Therefore, the probability of GEA failure and the number of individuals n increase with larger l_s.

Instead of using binary encodings, the presented NetKey encoding uses continuous variables for representing trees. We presented for NetKeys a population sizing model and showed for the easy one-max tree problem that the population size n goes with $O(z^{1.5})$, where z denotes the number of nodes. The time to convergence is increasing linearly with $O(z)$.

Analysis of binary representations for integers. The framework about representations can be used for explaining the performance differences of GEAs using different types of binary representations for integer problems. The analysis of binary, gray, and unary encoding has shown that the unary encoding is non-uniformly redundant, that the binary encoding has exponentially scaled BBs, and the gray encoding has high locality but does – like the binary encoding – not preserve the structure of the BBs.

Therefore, the performance of GEAs using unary encoding depends on the structure of the optimal solution. If the optimal solution is underrepresented, GEAs fail; in contrast, if the optimal solution is overrepresented GEAs using the unary encoding perform well. Furthermore, we know that when using the binary encoding that the alleles are solved serially and the time to convergence increases. Some low salient genes are randomly fixed due to genetic drift before they can be reached by the search process. Finally, binary and gray encoding change the distances between corresponding genotypes and phenotypes and therefore change the difficulty of the optimization problem. Thus, the resulting problem difficulty depends not only on the used representation but also on the considered optimization problem. Some problems like the easy integer one-max problem remain easier using binary encoding than using gray encoding but there are other easy problems that will be easier using gray encoding.

Analysis of tree representations. The framework about representations can also be used for analyzing the influence of tree representations on GEA performance. Based on the properties of the Prüfer number, characteristic vector, and link and node biased encoding, the proposed framework allows us to predict GEA performance. The analysis of the Prüfer number encoding revealed that the locality of the encoding is high around stars and low elsewhere. Therefore, GEA performance is low on easy problems if the optimal solution is not a star. On the other hand, the performance is high if the problem is fully difficult and the optimal solution is a non-star.

The link and node biased (LNB) encoding proposed by Palmer (1994) uses a link- and node-specific bias to control the representations influence on the structure of the represented phenotype. The investigation into the properties of the encoding reveals that the encoding is uniformly redundant if a large link-specific bias is used. If the link-specific bias is small the encoding is non-uniformly redundant and biased towards the minimum spanning tree. The use of a node-specific bias results in an additional bias of the encoding towards star-like tree structures. Therefore, only GEAs using LNB encodings with a large link-specific bias perform independently of the structure of the optimal solution. If the link-specific bias is small, GEAs only perform well when searching for optimal minimum spanning tree-like phenotypes. When using an additional node-specific bias, GEAs also perform well when searching for optimal solutions similar to stars.

Analyzing the characteristic vector encoding revealed that the encoding is uniformly redundant but affected by stealth mutation. Stealth mutation is a result of repairing invalid solutions and has the same effects as additional mutation. Therefore, the performance of genetic algorithms using only crossover is increased for small problem instances. The additional mutation helps GEAs to find the good solutions. However, for larger problem instances stealth mutation reduces GEA performance and strongly increases the number of generations. Because the repair operator must repair many links in an offspring to represent a valid tree, guided search is no longer possible and GEAs behave like random search.

Development of new representations for trees. The framework about representations can not only be used for the analysis of existing representations but rather for the development of new representations. Based on the insights into representation theory, we proposed the NetKey encoding. The NetKey encoding is based on the characteristic vector encoding but uses continuous variables for encoding a tree. The encoding is uniformly redundant, has uniformly scaled BBs, and does not change problem difficulty. In contrast to other encodings, GEAs using NetKeys are able to distinguish between important and unimportant links. The high performance of GEAs using NetKeys was verified by empirical results and proposes an increasing use of that encoding.

Furthermore, we presented the NetDir representation as an example of a direct encoding. Direct encodings do not use an explicit representation but apply genetic operators directly to the phenotype. Therefore, problem-specific operators are necessary and the framework about representations can not be used any more. Direct encodings are at the intersection between representations and operators and illustrate that when using no explicit representation, the design of proper operators becomes the critical design issue.

Scalable test problems for trees. Last but not least, we presented the fully easy one-max tree problem and the deceptive tree problem as examples for scalable test problems for trees. For these types of problems the optimal solution is determined a priori and the distance of an individual towards the optimal solution determines the fitness of the individual. Both problems, the one-max tree problem and the deceptive tree problem, are easy to implement and can be advantageously used for comparing the performance of GEAs on tree problems. By providing scalable test instances the performance of different types of representations or new, more effective search methods can be compared more easily.

9.3 Future Work

With respect to future work, we would like to address some open problems.

- The theoretical model for redundant encodings presented in section 3.1 focused on linear redundancy on bit strings only and neglected the effect of improper mixing and other effects. Although the proposed models can explain the behavior of GEAs using redundant encodings well, we believe that more aspects of redundancy must be considered to get a better understanding of redundant encodings.

- When modeling the influence of non-uniformly scaled encodings on the performance of GEAs, we assumed a strictly serial solving of the alleles. Therefore, the size of the convergence window was assumed to be one. However, in reality the solving process is not strictly serial but more than one allele is solved at the same time. The more uniformly scaled the BBs are, the more alleles which are solved in parallel. At the extreme where the alleles are uniformly scaled, the size of the convergence window is the same as the length of the string and all alleles are solved in parallel. Therefore, for a better understanding of scaled encodings, it is necessary to develop a model that describes how the size of the convergence window depends on the scaling of the encoding. Then we could substitute the two separate models for uniformly scaled and non-uniformly scaled encodings by one model that describes in general how GEA performance depends on the scaling of an encoding.
- The analysis of locality and distance distortion of an encoding reveals whether the problem difficulty changes, or not. Predictions about how exactly the difficulty of the problem changes are difficult as they depend on the exact structure of the problem and the genotype-phenotype mapping. However, the benefits we can get from estimating the influence of locality and distance distortion on the difficulty of a problem are tremendous. We would be able to design in a theory guided manner representations that not only preserve problem difficulty but also reduce problem difficulty. Although this problem is difficult to solve in general, even small steps in that direction would allow us to use GEAs much more efficiently.
- In this study, we analyzed the influence of redundant encodings, exponentially scaled encodings, and encodings that change distances between individuals on the performance of GEAs. As a result, the framework about representations consists currently of three parts: redundancy, scaling, and distance distortion. However, the presented framework is only one step towards a more general theory of representations. More steps are necessary. The next steps are the identification of further elements of the framework and a better understanding of the already existing three elements. Furthermore, we have to investigate whether there are interdependencies between the different elements of the framework. When performing these tasks, we believe that the black magic of finding good representations becomes a well understood engineering design task. Then, with a general theory of representations at hand, we can unleash the full power of genetic search and solve even difficult problems fast and reliably.

A. Optimal Communication Spanning Tree Test Instances

Searching the literature for standard test problems for the optimal communication spanning tree (OCST) problem reveals that many researchers use their private test problems which are mostly not published. As a result, the comparison of different search algorithms or representations is a difficult and time consuming task. It is not possible to check quickly if a new search method is better than the existing ones. Furthermore, applicants hesitate to use new and efficient search methods or representations if they can not be tested on a variety of different test problems and solve these problems well and reliably. Therefore, the building up of a collection of test instances for the OCST problem is necessary.

The purpose of this appendix is to go one step in this direction and to present a collection of different test instances for the OCST problem. It gives exact details concerning the properties of the problems we used in section 8.2 for the comparison of different types of representations. Based on the summarized test instances, a fair and standardized comparison of new search techniques and representations is possible.

For each test problem we present the best known solution, and the demands and distances between all nodes. For all demands and distances the upper right corner of the matrices specifies the demands and distances between nodes. In section A.1 we summarize the properties of the test instances from Palmer (1994). We are not able to present data for the 47 and 98 node problems because these are no longer available. In section A.2 we present the details for the 10, 20, and 50 node OCST problem from Raidl (2001). Subsequently, A.3 specifies the 6 node, 35 node with uniform distances, and 35 node with non-uniform distances OCST problems presented by Berry, Murtagh, and McMahon (1995). Finally, A.4 summarizes the exact specifications of four real-world test problems from Rothlauf, Goldberg, and Heinzl (2002).

A.1 Palmer's Test Instances

We present the details for the 6, 12, and 24 node test instances presented by Palmer (1994).

Table A.1 gives an overview of the best known solutions for the three test problems. The cost of a link is calculated as the traffic over a link multiplied by the length of the link. For the 6 node problem the demand matrix is shown in Table A.2 and the corresponding distances between the links are presented in Table A.3. The 12 node problem from Palmer (1994) is specified in Table A.4 (demands) and Table A.5 (distances). Finally, the 24 node problem is described by Table A.6 (demands) and Table A.7 (distances).

Table A.1. Cost and structure of the best solutions to the test instances from Palmer (1994)

problem	cost	links
6 node	693 180	DET-CHI, PHIL-DET, PHIL-NY, HOU-LA, HOU-DET
12 node	3 428 509	SD-PHNX, SD-LA, SF-SD, DET-CHI, PHIL-BALT, PHIL-DET, PHIL-NY, DAL-SD, DAL-DET, HOU-DAL, SANAN-DAL
24 node	1 086 656	LA-PHNX, SD-LA, SF-LA, SJ-LA, CHI-LA, CHI-DEN, DET-CHI, DET-IND, NY-BOS, CLEVE-DET, COL-DET, PHIL-WDC, PHIL-JACK, PHIL-BALT, PHIL-DET, PHIL-NY, HOU-NO, HOU-DET, HOU-MEMP, HOU-DAL, HOU-ELPAS, SANAN-HOU, MILW-CHI

Table A.2. Demand matrix for the 6 node problem from Palmer (1994)

	LA	CHI	DET	NY	PHIL	HOU
LA	0	1	1	1	1	2
CHI	-	0	10	3	4	3
DET	-	-	0	5	6	2
NY	-	-	-	0	31	2
PHIL	-	-	-	-	0	2
HOU	-	-	-	-	-	0

Table A.3. Distance matrix for the 6 node problem from Palmer (1994)

	LA	CHI	DET	NY	PHIL	HOU
LA	0	16661	18083	21561	21099	13461
CHI	-	0	5658	9194	8797	10440
DET	-	-	0	7230	6899	11340
NY	-	-	-	0	4300	13730
PHIL	-	-	-	-	0	13130
HOU	-	-	-	-	-	0

Table A.4. Demand matrix for the 12 node problem from Palmer (1994)

	PHNX	LA	SD	SF	CHI	BALT	DET	NY	PHIL	DAL	HOU	SANAN
PHNX	0	7	8	4	2	1	2	1	1	3	3	3
LA	-	0	25	7	1	1	1	1	1	2	2	2
SD	-	-	0	6	1	1	1	1	1	2	2	2
SF	-	-	-	0	1	1	1	1	1	2	2	2
CHI	-	-	-	-	0	4	11	4	4	3	3	2
BALT	-	-	-	-	-	0	6	15	29	2	2	2
DET	-	-	-	-	-	-	0	5	6	3	2	2
NY	-	-	-	-	-	-	-	0	33	2	2	2
PHIL	-	-	-	-	-	-	-	-	0	2	2	2
DAL	-	-	-	-	-	-	-	-	-	0	11	10
HOU	-	-	-	-	-	-	-	-	-	-	0	14
SANAN	-	-	-	-	-	-	-	-	-	-	-	0

Table A.5. Distance matrix for the 12 node problem from Palmer (1994)

	PHNX	LA	SD	SF	CHI	BALT	DET	NY	PHIL	DAL	HOU	SANAN
PHNX	0	6490	5903	8484	14561	18359	15976	19360	18867	10090	10883	9665
LA	-	0	4523	6256	16661	20618	18083	21561	21099	12639	13461	12236
SD	-	-	0	6908	16414	20292	17829	21263	20787	12073	12802	11540
SF	-	-	-	0	17328	21452	18714	22286	21874	14234	15259	14136
CHI	-	-	-	-	0	8425	5658	9194	8797	9603	10440	11237
BALT	-	-	-	-	0	0	6621	5067	4439	12385	12526	13722
DET	-	-	-	-	-	-	0	7230	6899	10720	11340	12297
NY	-	-	-	-	-	-	-	0	4300	13531	13730	14912
PHIL	-	-	-	-	-	-	-	-	0	12967	13130	14319
DAL	-	-	-	-	-	-	-	-	-	0	4888	5076
HOU	-	-	-	-	-	-	-	-	-	-	0	4478
SANAN	-	-	-	-	-	-	-	-	-	-	-	0

Table A.6. Demand matrix for the 24 node problem from Palmer (1994)

	PHNX	LA	SD	SF	SJ	DEN	WDC	JACK	CHI	IND	NO	BOS	BALT	DET	NY	CLEVE	COL	PHIL	MEMP	DAL	ELPAS	HOU	SANAN	MILW
PHNX	0	100	0	0	0	0	0	0	0	0	0	0	0	0	0	0	0	0	0	0	0	0	0	0
LA	-	0	100	100	0	0	0	2	0	0	0	0	0	0	0	0	0	0	0	0	0	0	0	0
SD	-	-	0	0	0	0	0	0	0	0	0	0	0	0	0	0	0	0	0	0	0	0	0	0
SF	-	-	-	0	0	0	0	0	0	0	0	0	0	0	0	0	0	0	0	0	0	0	0	0
SJ	-	-	-	-	0	100	0	0	0	0	0	0	0	0	0	0	0	0	0	0	0	0	0	0
DEN	-	-	-	-	-	0	0	100	0	0	0	0	0	100	0	0	0	0	0	0	0	0	0	0
WDC	-	-	-	-	-	-	0	0	0	0	0	0	0	0	0	0	0	100	0	0	0	0	0	0
JACK	-	-	-	-	-	-	-	0	0	0	0	0	0	0	0	0	0	100	0	0	0	0	0	0
CHI	-	-	-	-	-	-	-	-	0	2	0	0	0	0	0	0	0	0	0	0	0	0	0	100
IND	-	-	-	-	-	-	-	-	-	0	100	0	0	0	100	0	0	0	0	0	0	0	0	0
NO	-	-	-	-	-	-	-	-	-	-	0	100	0	0	100	0	0	0	0	0	0	0	0	0
BOS	-	-	-	-	-	-	-	-	-	-	-	0	0	0	100	0	0	0	0	0	0	100	0	0
BALT	-	-	-	-	-	-	-	-	-	-	-	-	0	0	0	100	100	0	0	0	0	0	0	0
DET	-	-	-	-	-	-	-	-	-	-	-	-	-	0	0	0	100	2	0	0	0	2	0	0
NY	-	-	-	-	-	-	-	-	-	-	-	-	-	-	0	0	0	2	0	0	0	0	0	0
CLEVE	-	-	-	-	-	-	-	-	-	-	-	-	-	-	-	0	0	0	0	0	0	0	0	0
COL	-	-	-	-	-	-	-	-	-	-	-	-	-	-	-	-	0	0	0	0	0	0	0	0
PHIL	-	-	-	-	-	-	-	-	-	-	-	-	-	-	-	-	-	0	0	0	0	0	0	0
MEMP	-	-	-	-	-	-	-	-	-	-	-	-	-	-	-	-	-	-	0	0	0	100	0	0
DAL	-	-	-	-	-	-	-	-	-	-	-	-	-	-	-	-	-	-	-	0	0	100	0	0
ELPAS	-	-	-	-	-	-	-	-	-	-	-	-	-	-	-	-	-	-	-	-	0	100	0	0
HOU	-	-	-	-	-	-	-	-	-	-	-	-	-	-	-	-	-	-	-	-	-	0	100	0
SANAN	-	-	-	-	-	-	-	-	-	-	-	-	-	-	-	-	-	-	-	-	-	-	0	0
MILW	-	-	-	-	-	-	-	-	-	-	-	-	-	-	-	-	-	-	-	-	-	-	-	0

Table A.7. Distance matrix for the 24 node problem from Palmer (1994)

	PHNX	LA	SD	SF	SJ	DEN	WDC	JACK	CHI	IND	NO	BOS	BALT	DET	NY	CLEVE	COL	PHIL	MEMP	DAL	ELPAS	HOU	SANAN	MILW
PHNX	0	649	590	848	824	808	1810	1700	1456	1489	1336	2059	1835	1597	1936	1674	1588	1886	1289	1009	602	1088	966	1461
LA	-	0	453	626	599	980	2034	1955	1661	1717	1595	2275	2068	1803	2151	1890	1813	2109	1537	1269	852	1341	1226	1664
SD	-	-	0	698	664	967	2003	1897	1644	1688	1539	2241	2022	1789	2123	1864	1784	2077	1483	1203	791	1282	1150	1644
SF	-	-	-	0	282	1056	2126	2106	1738	1795	1758	2339	2142	1874	2226	1963	1892	2184	1663	1424	1051	1529	1416	1715
SJ	-	-	-	-	0	1046	2114	2088	1720	1782	1742	2323	2136	1860	2228	1955	1887	2176	1643	1404	1031	1508	1390	1719
DEN	-	-	-	-	-	0	1451	1456	1068	1129	1154	1674	1471	1203	1561	1295	1223	1515	1005	833	741	975	920	1065
WDC	-	-	-	-	-	-	0	855	829	754	1064	675	387	657	524	615	605	463	914	1206	1582	1225	1348	853
JACK	-	-	-	-	-	-	-	0	1036	929	752	1144	888	983	1005	979	870	942	817	1037	1424	954	1092	1097
CHI	-	-	-	-	-	-	-	-	0	549	995	1030	845	568	914	651	600	877	736	963	1264	1040	1127	475
IND	-	-	-	-	-	-	-	-	-	0	919	993	771	567	875	618	528	821	667	932	1280	995	1089	592
NO	-	-	-	-	-	-	-	-	-	-	0	1373	1100	1047	1226	1074	958	1161	621	675	1058	577	714	1056
BOS	-	-	-	-	-	-	-	-	-	-	-	0	655	823	530	813	857	585	1203	1494	1851	1511	1631	1038
BALT	-	-	-	-	-	-	-	-	-	-	-	-	0	661	507	633	627	449	940	1235	1617	1256	1372	862
DET	-	-	-	-	-	-	-	-	-	-	-	-	-	0	720	463	491	689	812	1070	1404	1130	1227	576
NY	-	-	-	-	-	-	-	-	-	-	-	-	-	-	0	702	722	430	1068	1351	1727	1370	1492	934
CLEVE	-	-	-	-	-	-	-	-	-	-	-	-	-	-	-	0	493	660	858	1123	1472	1170	1279	663
COL	-	-	-	-	-	-	-	-	-	-	-	-	-	-	-	-	0	671	739	1019	1377	1060	1162	635
PHIL	-	-	-	-	-	-	-	-	-	-	-	-	-	-	-	-	-	0	1000	1297	1663	1310	1439	896
MEMP	-	-	-	-	-	-	-	-	-	-	-	-	-	-	-	-	-	-	0	652	1044	680	795	792
DAL	-	-	-	-	-	-	-	-	-	-	-	-	-	-	-	-	-	-	-	0	744	488	506	990
ELPAS	-	-	-	-	-	-	-	-	-	-	-	-	-	-	-	-	-	-	-	-	0	809	674	1280
HOU	-	-	-	-	-	-	-	-	-	-	-	-	-	-	-	-	-	-	-	-	-	0	448	1084
SANAN	-	-	-	-	-	-	-	-	-	-	-	-	-	-	-	-	-	-	-	-	-	-	0	1166
MILW	-	-	-	-	-	-	-	-	-	-	-	-	-	-	-	-	-	-	-	-	-	-	-	0

A.2 Raidl's Test Instances

This section presents the details for the 10, 20, and 50 node test instances from Raidl (2001). We do not list the 75 and 100 nodes test problems herein because they are too extensive to be published. However, the details of the test instances are available and can be directly obtained from Günther Raidl[1].

All demands and distances between the nodes were generated randomly and uniformly distributed. As before, the cost of a link is calculated as the amount of traffic over a link multiplied by its length. The nodes are labeled with numbers starting from zero.

In Table A.8 we present the properties of the best known solutions to the different problem instances. The demands for the various test problems are presented in Table A.9 (10 nodes), Table A.11 (20 nodes), and Table A.13 (50 nodes). The distances are shown in Table A.10 (10 nodes), Table A.12 (20 nodes), and Table A.14 (50 nodes).

Table A.8. Cost and structure of the best solutions to the test instances from Raidl (2001)

problem	cost	links
10 node	53 674	1-0, 2-0, 3-0, 4-1, 5-0, 6-0, 7-3, 8-1, 9-1
20 node	157 570	2-0, 7-5, 9-6, 9-7, 10-0, 11-0, 12-4, 13-0, 13-1, 13-3, 13-4, 14-10, 16-2, 17-0, 17-15, 18-8, 18-9, 18-10, 19-10

Table A.9. Demand matrix for the 10 node problem from Raidl (2001)

	0	1	2	3	4	5	6	7	8	9
0	0	34	97	50	93	100	89	24	89	3
1	-	0	79	65	78	81	82	66	98	72
2	-	-	0	11	36	87	23	78	97	81
3	-	-	-	0	23	88	40	91	83	84
4	-	-	-	-	0	80	16	47	96	9
5	-	-	-	-	-	0	46	84	100	0
6	-	-	-	-	-	-	0	53	78	66
7	-	-	-	-	-	-	-	0	98	58
8	-	-	-	-	-	-	-	-	0	13
9	-	-	-	-	-	-	-	-	-	0

---

[1] Address: Günther Raidl, Institute of Computer Graphics, Vienna University of Technology, Favoritenstrae 9-11/1861, 1040 Vienna, Austria. E-Mail: raidl@apm.tuwien.ac.at

Table A.10. Distance matrix for the 10 node problem from Raidl (2001)

	0	1	2	3	4	5	6	7	8	9
0	0	8	17	1	41	12	7	16	90	47
1	-	0	47	31	17	87	59	14	5	9
2	-	-	0	53	36	29	47	14	18	84
3	-	-	-	0	53	83	72	6	79	36
4	-	-	-	-	0	64	39	52	16	31
5	-	-	-	-	-	0	63	75	47	5
6	-	-	-	-	-	-	0	21	45	87
7	-	-	-	-	-	-	-	0	89	31
8	-	-	-	-	-	-	-	-	0	45
9	-	-	-	-	-	-	-	-	-	0

Table A.11. Demand matrix for the 20 node problem from Raidl (2001)

	0	1	2	3	4	5	6	7	8	9	10	11	12	13	14	15	16	17	18	19
0	0	19	7	97	99	22	65	82	53	76	2	44	7	40	67	100	7	40	94	90
1	-	0	37	18	98	75	99	90	42	51	4	91	76	91	10	49	53	75	72	17
2	-	-	0	56	91	59	24	34	33	30	0	32	38	6	25	94	43	9	57	18
3	-	-	-	0	8	13	26	25	17	16	67	74	93	16	26	33	54	10	90	44
4	-	-	-	-	0	69	80	44	1	10	10	100	14	16	92	7	26	0	30	44
5	-	-	-	-	-	0	75	43	36	66	26	18	33	100	11	15	26	44	69	2
6	-	-	-	-	-	-	0	100	79	37	80	22	39	56	32	4	70	48	96	77
7	-	-	-	-	-	-	-	0	74	63	73	84	3	16	86	70	8	4	2	8
8	-	-	-	-	-	-	-	-	0	82	84	0	92	52	2	58	30	39	3	18
9	-	-	-	-	-	-	-	-	-	0	44	59	50	15	28	64	77	71	4	5
10	-	-	-	-	-	-	-	-	-	-	0	43	88	9	25	40	79	34	44	47
11	-	-	-	-	-	-	-	-	-	-	-	0	8	92	30	8	83	82	77	40
12	-	-	-	-	-	-	-	-	-	-	-	-	0	78	82	43	96	93	68	11
13	-	-	-	-	-	-	-	-	-	-	-	-	-	0	7	96	75	84	66	79
14	-	-	-	-	-	-	-	-	-	-	-	-	-	-	0	90	76	33	99	0
15	-	-	-	-	-	-	-	-	-	-	-	-	-	-	-	0	73	43	0	83
16	-	-	-	-	-	-	-	-	-	-	-	-	-	-	-	-	0	90	8	74
17	-	-	-	-	-	-	-	-	-	-	-	-	-	-	-	-	-	0	86	83
18	-	-	-	-	-	-	-	-	-	-	-	-	-	-	-	-	-	-	0	22
19	-	-	-	-	-	-	-	-	-	-	-	-	-	-	-	-	-	-	-	0

Table A.12. Distance matrix for the 20 node problem from Raidl (2001)

	0	1	2	3	4	5	6	7	8	9	10	11	12	13	14	15	16	17	18	19
0	0	30	6	15	67	100	99	34	85	56	3	13	23	2	72	24	53	12	94	57
1	-	0	98	30	53	35	38	59	85	82	85	78	16	3	59	73	17	77	73	15
2	-	-	0	62	9	70	65	21	44	18	44	68	71	56	13	79	5	43	83	39
3	-	-	-	0	80	93	23	61	78	52	28	80	62	1	48	38	25	62	100	33
4	-	-	-	-	0	84	16	6	27	85	49	46	7	4	59	37	8	53	19	98
5	-	-	-	-	-	0	32	12	20	92	41	71	20	72	32	72	19	22	96	80
6	-	-	-	-	-	-	0	73	80	15	88	85	94	72	34	39	79	89	49	15
7	-	-	-	-	-	-	-	0	49	1	86	46	32	97	66	76	37	88	47	8
8	-	-	-	-	-	-	-	-	0	72	18	78	93	65	57	65	44	24	4	29
9	-	-	-	-	-	-	-	-	-	0	17	75	14	55	5	54	56	72	2	56
10	-	-	-	-	-	-	-	-	-	-	0	57	100	40	5	17	67	93	4	13
11	-	-	-	-	-	-	-	-	-	-	-	0	52	75	82	29	19	46	37	83
12	-	-	-	-	-	-	-	-	-	-	-	-	0	41	60	38	21	76	13	86
13	-	-	-	-	-	-	-	-	-	-	-	-	-	0	83	69	40	90	40	93
14	-	-	-	-	-	-	-	-	-	-	-	-	-	-	0	97	48	92	36	52
15	-	-	-	-	-	-	-	-	-	-	-	-	-	-	-	0	8	2	44	12
16	-	-	-	-	-	-	-	-	-	-	-	-	-	-	-	-	0	66	95	38
17	-	-	-	-	-	-	-	-	-	-	-	-	-	-	-	-	-	0	99	23
18	-	-	-	-	-	-	-	-	-	-	-	-	-	-	-	-	-	-	0	57
19	-	-	-	-	-	-	-	-	-	-	-	-	-	-	-	-	-	-	-	0

Table A.13. Demand matrix for the 50 node problem from Raidl (2001)

	0	1	2	3	4	5	6	7	8	9	10	11	12	13	14	15	16	17	18	19	20	21	22	23	24	25	26	27	28	29	30	31	32	33	34	35	36	37	38	39	40	41	42	43	44	45	46	47	48	49
0	0	62	56	24	24	29	17	9	97	47	42	5	25	22	11	59	16	97	87	26	65	7	38	9	66	74	94	2	23	14	65	42	42	87	32	66	16	49	75	79	97	83	50	21	71	62	47	87	24	100
1	-	0	79	55	6	16	64	73	57	23	41	80	37	72	88	45	58	19	78	40	69	18	86	31	1	35	18	38	63	65	25	64	70	8	37	87	39	9	43	62	50	22	66	66	51	10	10	45	96	88
2	-	-	0	86	30	73	37	61	40	72	46	78	1	10	2	55	41	73	29	78	25	68	53	68	97	2	56	62	90	32	38	34	95	93	86	35	0	91	52	4	27	98	82	28	74	51	49	14	90	44
3	-	-	-	0	58	14	12	10	48	75	79	70	36	68	69	40	69	63	32	20	53	97	9	4	67	36	1	49	30	42	66	79	22	55	22	81	35	0	91	83	41	36	18	77	3	53	84	38	15	15
4	-	-	-	-	0	25	69	79	0	73	45	36	41	60	32	49	92	77	71	46	65	51	81	66	8	44	14	50	13	67	33	18	49	14	43	17	59	43	56	71	45	56	63	97	11	89	53	37		

Table A.14. Distance matrix for the 50 node problem from Raidl (2001)

	0	1	2	3	4	5	6	7	8	9	10	11	12	13	14	15	16	17	18	19	20	21	22	23	24	25	26	27	28	29	30	31	32	33	34	35	36	37	38	39	40	41	42	43	44	45	46	47	48	49
0	0	71	20	56	94	50	76	57	57	45	65	60	80	19	64	100	4	73	91	32	27	93	37	83	85	8	3	82	90	45	14	92	15	85	47	60	86	23	69	95	19	33	6	98	51	69	50	7	41	92
1		0	90	19	36	26	53	20	34	8	53	23	52	1	66	19	55	64	30	41	38	98	87	56	82	92	6	85	61	7	91	35	98	80	24	33	6	28	4	91	35	8	65	39	78	30	57	33	93	86
2			0	25	31	36	11	38	17	54	8	45	37	51	1	91	19	50	23	94	28	34	21	89	70	56	97	34	94	26	63	2	58	7	87	34	89	22	96	27	91	50	73	1	15	74	34	85	75	78
3				0	9	35	51	97	4	6	45	37	95	12	21	51	52	80	9	90	66	98	64	61	76	62	48	96	29	73	29	13	48	84	43	8	70	45	4	73	2	61	52	71	11	55	2	20	96	58
4					0	67	69	11	80	96	16	41	95	12	21	68	80	92	34	66	98	64	74	97	29	70	73	29	30	76	51	93	82	53	64	77	19	84	88	98	79	55	91	73	18	11	98	20	96	75
5						0	37	69	11	80	96	41	95	6	70	44	88	97	95	80	78	99	95	17	78	46	67	8	100	9	80	18	71	72	79	67	81	15	41	16	51	39	31	57	60	74	44	9	20	75
6							0	86	18	69	44	87	6	70	44	88	5	13	36	58	83	59	36	1	92	67	8	95	52	84	88	95	62	51	3	82	25	41	99	45	43	8	52	61	7	16	92	3	36	34
7								0	70	88	92	5	40	83	6	33	41	57	68	29	17	63	43	67	65	24	44	57	74	88	41	11	82	65	51	79	77	56	64	46	43	8	2	34	42	7	67	83	15	97
8									0	86	18	69	44	87	49	64	21	6	84	61	16	65	25	18	96	53	25	59	98	19	66	51	53	8	9	19	42	24	4	56	6	52	61	7	16	92	3	36	92
9										0	63	31	48	5	50	13	80	45	21	6	84	61	16	65	24	44	57	26	9	54	82	67	5	94	25	72	85	28	8	34	14	56	49	16	80	82	14	57	26	92
10											0	87	49	64	3	73	34	98	26	58	9	75	77	26	29	33	34	99	26	72	24	38	57	37	45	5	5	77	39	70	33	30	43	68	49	41	28	48	24	6
11												0	74	84	68	99	61	48	84	46	98	9	7	51	42	73	8	35	97	79	38	45	10	10	43	8	68	39	74	90	20	82	74	77	2	10	19	5	25	30
12													0	69	37	90	52	64	60	91	79	26	86	99	68	20	39	42	34	22	36	49	45	89	61	63	24	22	62	47	99	6	15	79	62	44	84	1
13														0	95	4	55	32	54	22	51	44	15	85	65	3	85	9	64	52	23	10	38	61	63	99	68	74	20	90	20	79	67	24	8	13	79	62	44	84
14															0	35	95	27	1	31	92	55	15	70	60	55	93	76	5	83	56	52	92	93	12	54	92	42	20	67	49	84	45	62	28	28	48	74	6	1
15																0	4	49	7	70	1	77	81	7	21	1	90	76	5	33	69	68	39	12	10	78	10	94	94	67	84	21	1	71	94	58	23	49	7	30
16																	0	59	6	50	65	26	3	6	17	39	21	74	77	84	35	86	61	45	79	75	34	68	99	86	90	44	44	13	92	2	42	13	60
17																		0	47	63	24	24	65	30	76	23	20	49	96	32	83	33	45	73	89	98	70	9	7	16	87	6	11	51	49	12	92	62	71	90
18																			0	76	46	65	40	27	41	14	98	41	61	89	8	78	12	3	80	75	43	29	45	92	45	5	99	71	63	6	60	65	74	71
19																				0	44	57	57	35	36	49	48	33	50	99	5	9	76	16	11	30	95	38	72	75	34	27	59	59	38	13	41	52	18	17
20																					0	9	17	79	17	11	41	13	66	57	14	64	61	75	91	76	37	20	97	45	85	43	49	71	88	23	31	60	100	48
21																						0	22	25	11	1	43	22	3	14	85	32	9	51	40	75	66	100	13	57	85	28	1	11	23	50	17	80	69
22																							0	92	64	34	33	40	45	85	32	9	51	40	77	84	66	24	58	88	8	85	89	28	3	64	66	11	3
23																								0	65	1	28	12	44	39	27	78	72	18	74	56	49	37	24	41	62	31	27	96	28	23	31	19	12
24																									0	33	15	71	64	35	35	17	62	46	60	1	24	37	24	41	62	89	49	80	83	53	43	71	84	27
25																										0	45	47	83	3	57	48	5	42	14	75	48	90	39	82	89	3	27	74	2	88	46	14	94	37
26																											0	78	59	76	95	14	99	77	84	62	6	42	4	37	67	12	31	86	2	26	66	91	78	100
27																												0	27	76	95	32	78	72	18	72	6	90	37	85	49	39	24	80	39	88	23	31	66	34
28																													0	51	42	80	88	43	57	28	53	86	86	39	82	89	49	83	53	43	71	87	84	50
29																														0	86	14	38	99	70	28	25	30	12	6	85	97	62	47	10	1	42	67	54	27
30																															0	52	44	60	93	92	91	75	97	85	97	83	68	82	96	58	80	65	37
31																																0	56	47	100	62	77	13	60	87	14	1	5	67	79	56	10	39	100
32																																	0	1	81	27	49	66	23	31	33	5	79	69	60	22	47	67	83	67
33																																		0	78	57	18	91	16	56	56	35	67	66	67	23	44	61	11	39
34																																			0	54	31	68	62	100	7	12	55	14	4	39	27	61	32	30	6
35																																				0	33	20	84	35	62	81	64	26	14	80	75	6	35	77
36																																					0	54	31	36	33	0	63	26	14	90	58	40	85	76	87
37																																						0	33	20	84	51	7	53	95	90	87	13	7	35	92
38																																							0	36	51	66	98	93	56	3	23	23	64	32	87
39																																								0	33	92	81	49	89	47	87	84	31	91	49
40																																									0	0	63	26	14	80	58	40	85	76	38
41																																										0	0	81	64	20	27	5	58	39	67
42																																											0	86	21	99	49	92	12
43																																												0	0	0	65	64	93	6
44																																													0	0	0	30	44	75
45																																														0	25	54	50	58
46																																															0	65	64	15
47																																																0	30	77
48																																																	0	44
49																																																		0

A.3 Berry's Test Instances

This section presents the details for the 6 node, 35 node with uniform distances, and 35 node with non-uniform distances problem instances proposed by Berry, Murtagh, and McMahon (1995).[2]

In analogy to the previous section in Table A.15 we present the properties of the best known solutions to the different problem instances. The demands for the test problems are presented in Table A.16 (6 node) and Table A.18 (35 node). The distances for the 6 node problem are shown in Table A.17. Table A.19 illustrates the distances for the 35 node problem with non-uniform distances. For the 35 node problem with uniform distances the distances between all nodes are 1. The demands are the same for the 35 node problem with uniform and non-uniform distances.

Table A.15. Cost and structure of the best solutions for the test instances from Berry, Murtagh, and McMahon (1995)

problem	cost	links
6 node	534	1-0, 3-1, 5-2, 5-3, 5-4
35 node (uniform)	16 273	1-0, 8-2, 11-4, 12-9, 12-10, 13-8, 15-9, 16-2, 17-3, 18-6, 19-15, 20-9, 21-8, 25-1, 25-3, 25-19, 25-24, 26-22, 27-15, 28-9, 29-8, 29-11, 29-25, 30-5, 30-14, 30-21, 30-22, 31-7, 31-12, 31-23, 32-18, 32-25, 33-25, 34-29
35 node (non-uniform)	16 915	1-0, 8-2, 11-4, 12-9, 12-10, 13-8, 16-2, 17-3, 18-6, 19-15, 20-9, 20-15, 21-8, 24-1, 24-17, 24-18, 25-3, 25-8, 25-19, 26-22, 27-15, 28-9, 29-11, 29-25, 30-5, 30-14, 30-21, 30-22, 31-7, 31-12, 31-23, 32-18, 33-24, 34-29

Table A.16. Demand matrix for the 6 node problem from Berry, Murtagh, and McMahon (1995)

	0	1	2	3	4	5
0	0	5	13	12	8	9
1	-	0	7	4	2	6
2	-	-	0	3	10	15
3	-	-	-	0	11	7
4	-	-	-	-	0	12
5	-	-	-	-	-	0

---

[2] The exact specification of the test instances can also be found at http://www.cse.rmit.edu.au/~rdslw/research.html

Table A.17. Distance matrix for the 6 node problem from Berry, Murtagh, and McMahon (1995)

	0	1	2	3	4	5
0	0	3	6	5	9	7
1	-	0	3	2	4	8
2	-	-	0	3	7	2
3	-	-	-	0	9	2
4	-	-	-	-	0	1
5	-	-	-	-	-	0

Table A.18. Demand matrix for the 35 node problems from Berry, Murtagh, and McMahon (1995)

	0	1	2	3	4	5	6	7	8	9	10	11	12	13	14	15	16	17	18	19	20	21	22	23	24	25	26	27	28	29	30	31	32	33	34
0	0	639	0	0	0	0	0	0	0	0	0	0	0	0	0	0	0	0	0	0	0	0	0	0	93	0	0	0	0	0	0	0	129	0	0
1	-	0	0	0	0	0	0	0	0	0	0	0	0	0	0	0	0	0	0	0	0	147	0	0	0	0	0	0	0	0	0	0	83	0	0
2	-	-	0	0	0	0	0	189	0	0	0	0	0	0	0	0	99	0	0	0	0	0	0	0	0	0	0	0	0	0	0	0	43	0	0
3	-	-	-	0	0	0	0	0	0	0	0	0	0	0	531	0	0	0	0	0	0	0	0	623	0	0	0	0	0	0	0	0	0	0	0
4	-	-	-	-	0	0	0	0	53	0	0	0	0	0	0	0	0	39	0	0	0	0	0	0	0	0	0	0	0	0	0	0	0	0	0
5	-	-	-	-	-	0	43	0	0	0	0	0	0	0	0	119	0	0	0	0	329	0	0	0	0	0	0	0	651	0	0	0	0	0	0
6	-	-	-	-	-	-	0	0	0	0	0	0	0	0	371	0	0	0	0	0	0	0	0	0	0	0	0	0	0	0	171	0	0	0	0
7	-	-	-	-	-	-	-	0	0	0	0	0	0	0	23	0	0	0	0	0	0	9	0	0	0	0	0	0	0	0	0	0	0	0	0
8	-	-	-	-	-	-	-	-	0	0	0	41	0	0	0	0	0	189	0	0	0	0	0	0	0	0	0	0	123	0	0	0	0	0	0
9	-	-	-	-	-	-	-	-	-	0	0	0	61	0	0	351	0	0	0	0	0	0	0	0	0	0	0	0	217	0	0	0	0	0	0
10	-	-	-	-	-	-	-	-	-	-	0	81	0	0	0	0	11	0	0	0	0	0	0	0	0	0	0	0	133	0	0	0	0	0	0
11	-	-	-	-	-	-	-	-	-	-	-	0	0	0	0	0	0	0	0	0	0	0	0	0	0	0	0	0	0	0	0	0	0	0	0
12	-	-	-	-	-	-	-	-	-	-	-	-	0	0	27	0	0	0	0	0	0	0	161	0	0	0	0	0	0	0	0	261	0	0	0
13	-	-	-	-	-	-	-	-	-	-	-	-	-	0	0	0	0	0	0	0	0	0	0	0	0	0	0	0	0	0	0	0	0	0	0
14	-	-	-	-	-	-	-	-	-	-	-	-	-	-	0	0	0	0	0	0	0	261	0	0	0	0	0	0	0	639	0	0	0	0	0
15	-	-	-	-	-	-	-	-	-	-	-	-	-	-	-	0	0	147	0	0	0	0	0	0	0	423	0	0	0	0	0	0	0	0	0
16	-	-	-	-	-	-	-	-	-	-	-	-	-	-	-	-	0	0	0	0	0	0	0	0	69	0	0	0	0	0	0	0	0	0	0
17	-	-	-	-	-	-	-	-	-	-	-	-	-	-	-	-	-	0	0	0	0	0	351	0	0	0	0	0	0	0	0	117	0	0	0
18	-	-	-	-	-	-	-	-	-	-	-	-	-	-	-	-	-	-	0	0	0	0	243	0	0	0	0	0	0	0	0	873	0	0	0
19	-	-	-	-	-	-	-	-	-	-	-	-	-	-	-	-	-	-	-	0	0	0	639	0	0	0	0	0	0	0	0	0	0	119	0
20	-	-	-	-	-	-	-	-	-	-	-	-	-	-	-	-	-	-	-	-	0	91	0	0	0	0	57	0	0	387	0	0	0	0	0
21	-	-	-	-	-	-	-	-	-	-	-	-	-	-	-	-	-	-	-	-	-	0	0	0	89	0	0	0	0	0	0	0	0	0	0
22	-	-	-	-	-	-	-	-	-	-	-	-	-	-	-	-	-	-	-	-	-	-	0	0	0	0	0	0	0	651	0	0	0	0	0
23	-	-	-	-	-	-	-	-	-	-	-	-	-	-	-	-	-	-	-	-	-	-	-	0	0	0	0	0	133	0	0	0	0	0	0
24	-	-	-	-	-	-	-	-	-	-	-	-	-	-	-	-	-	-	-	-	-	-	-	-	0	0	21	0	0	0	0	0	0	0	0
25	-	-	-	-	-	-	-	-	-	-	-	-	-	-	-	-	-	-	-	-	-	-	-	-	-	0	0	0	0	0	0	0	0	0	0
26	-	-	-	-	-	-	-	-	-	-	-	-	-	-	-	-	-	-	-	-	-	-	-	-	-	-	0	0	111	0	0	0	0	0	0
27	-	-	-	-	-	-	-	-	-	-	-	-	-	-	-	-	-	-	-	-	-	-	-	-	-	-	-	0	0	0	0	0	0	0	0
28	-	-	-	-	-	-	-	-	-	-	-	-	-	-	-	-	-	-	-	-	-	-	-	-	-	-	-	-	0	0	0	0	0	0	63
29	-	-	-	-	-	-	-	-	-	-	-	-	-	-	-	-	-	-	-	-	-	-	-	-	-	-	-	-	-	0	0	0	0	0	0
30	-	-	-	-	-	-	-	-	-	-	-	-	-	-	-	-	-	-	-	-	-	-	-	-	-	-	-	-	-	-	0	0	0	0	0
31	-	-	-	-	-	-	-	-	-	-	-	-	-	-	-	-	-	-	-	-	-	-	-	-	-	-	-	-	-	-	-	0	0	0	0
32	-	-	-	-	-	-	-	-	-	-	-	-	-	-	-	-	-	-	-	-	-	-	-	-	-	-	-	-	-	-	-	-	0	71	0
33	-	-	-	-	-	-	-	-	-	-	-	-	-	-	-	-	-	-	-	-	-	-	-	-	-	-	-	-	-	-	-	-	-	0	0
34	-	-	-	-	-	-	-	-	-	-	-	-	-	-	-	-	-	-	-	-	-	-	-	-	-	-	-	-	-	-	-	-	-	-	0

Table A.19. Distance matrix for the 35 node problem with non-uniform distances from Berry, Murtagh, and McMahon (1995)

	0	1	2	3	4	5	6	7	8	9	10	11	12	13	14	15	16	17	18	19	20	21	22	23	24	25	26	27	28	29	30	31	32	33	34
0	0	1	7	4	8	9	4	12	6	9	11	7	10	7	9	7	8	3	3	6	8	7	10	12	2	5	9	8	10	6	8	11	4	3	7
1	-	0	6	3	7	8	3	11	5	8	10	6	9	6	8	6	7	2	2	5	7	6	9	11	1	4	8	7	9	5	7	10	3	2	6
2	-	-	0	3	5	4	7	9	1	6	8	4	7	2	4	4	1	4	6	3	5	2	5	9	5	2	4	5	7	3	3	8	7	6	4
3	-	-	-	0	4	5	4	8	2	5	7	3	6	3	5	3	4	1	3	2	4	3	6	8	2	1	5	4	6	2	4	7	4	3	3
4	-	-	-	-	0	7	8	10	4	7	9	1	8	5	7	5	6	5	7	4	6	5	8	10	6	3	7	6	8	2	6	9	8	7	3
5	-	-	-	-	-	0	9	11	3	8	10	6	9	4	2	6	5	6	8	5	7	2	3	11	7	4	2	7	9	5	1	10	9	8	6
6	-	-	-	-	-	-	0	12	6	9	11	7	10	7	9	7	8	3	1	6	8	7	10	12	2	5	9	8	10	6	8	11	2	3	7
7	-	-	-	-	-	-	-	0	8	3	3	9	2	9	11	5	10	9	11	6	4	9	12	2	10	7	11	6	4	8	10	1	12	11	9
8	-	-	-	-	-	-	-	-	0	5	7	3	6	1	3	3	2	3	5	2	4	1	4	8	4	1	3	4	6	2	2	7	6	5	3
9	-	-	-	-	-	-	-	-	-	0	2	6	1	6	8	2	7	6	8	3	1	6	9	3	7	4	8	3	1	5	7	2	9	8	6
10	-	-	-	-	-	-	-	-	-	-	0	8	1	8	10	4	9	8	10	5	3	8	11	3	9	6	10	5	3	7	9	2	11	10	8
11	-	-	-	-	-	-	-	-	-	-	-	0	7	4	6	4	5	4	6	3	5	4	7	9	5	2	6	5	7	1	5	8	7	6	2
12	-	-	-	-	-	-	-	-	-	-	-	-	0	7	9	3	8	7	9	4	2	7	10	2	8	5	9	4	2	6	8	1	10	9	7
13	-	-	-	-	-	-	-	-	-	-	-	-	-	0	4	4	3	4	6	3	5	2	5	9	5	2	4	5	7	3	3	8	7	6	4
14	-	-	-	-	-	-	-	-	-	-	-	-	-	-	0	6	5	6	8	5	7	2	3	11	7	4	2	7	9	5	1	10	9	8	6
15	-	-	-	-	-	-	-	-	-	-	-	-	-	-	-	0	5	4	6	1	1	4	7	5	5	2	6	1	3	3	5	4	7	6	4
16	-	-	-	-	-	-	-	-	-	-	-	-	-	-	-	-	0	5	7	4	6	3	6	10	6	3	5	6	8	4	4	9	8	7	5
17	-	-	-	-	-	-	-	-	-	-	-	-	-	-	-	-	-	0	2	3	5	4	7	9	1	2	6	5	7	3	5	8	3	2	4
18	-	-	-	-	-	-	-	-	-	-	-	-	-	-	-	-	-	-	0	5	7	6	9	11	1	4	8	7	9	5	7	10	1	2	6
19	-	-	-	-	-	-	-	-	-	-	-	-	-	-	-	-	-	-	-	0	2	3	6	6	4	1	5	2	4	2	4	5	6	5	3
20	-	-	-	-	-	-	-	-	-	-	-	-	-	-	-	-	-	-	-	-	0	5	8	4	6	3	7	2	2	4	6	3	8	7	5
21	-	-	-	-	-	-	-	-	-	-	-	-	-	-	-	-	-	-	-	-	-	0	3	9	5	2	2	5	7	3	1	8	7	6	4
22	-	-	-	-	-	-	-	-	-	-	-	-	-	-	-	-	-	-	-	-	-	-	0	12	8	5	1	8	10	6	2	11	10	9	7
23	-	-	-	-	-	-	-	-	-	-	-	-	-	-	-	-	-	-	-	-	-	-	-	0	10	7	11	6	4	8	10	1	12	11	9
24	-	-	-	-	-	-	-	-	-	-	-	-	-	-	-	-	-	-	-	-	-	-	-	-	0	3	7	6	8	4	6	9	2	1	5
25	-	-	-	-	-	-	-	-	-	-	-	-	-	-	-	-	-	-	-	-	-	-	-	-	-	0	4	3	5	1	3	6	5	4	2
26	-	-	-	-	-	-	-	-	-	-	-	-	-	-	-	-	-	-	-	-	-	-	-	-	-	-	0	7	9	5	1	10	9	8	6
27	-	-	-	-	-	-	-	-	-	-	-	-	-	-	-	-	-	-	-	-	-	-	-	-	-	-	-	0	4	4	6	5	8	7	5
28	-	-	-	-	-	-	-	-	-	-	-	-	-	-	-	-	-	-	-	-	-	-	-	-	-	-	-	-	0	6	8	3	10	9	7
29	-	-	-	-	-	-	-	-	-	-	-	-	-	-	-	-	-	-	-	-	-	-	-	-	-	-	-	-	-	0	4	7	6	5	1
30	-	-	-	-	-	-	-	-	-	-	-	-	-	-	-	-	-	-	-	-	-	-	-	-	-	-	-	-	-	-	0	9	8	7	5
31	-	-	-	-	-	-	-	-	-	-	-	-	-	-	-	-	-	-	-	-	-	-	-	-	-	-	-	-	-	-	-	0	11	10	8
32	-	-	-	-	-	-	-	-	-	-	-	-	-	-	-	-	-	-	-	-	-	-	-	-	-	-	-	-	-	-	-	-	0	3	7
33	-	-	-	-	-	-	-	-	-	-	-	-	-	-	-	-	-	-	-	-	-	-	-	-	-	-	-	-	-	-	-	-	-	0	6
34	-	-	-	-	-	-	-	-	-	-	-	-	-	-	-	-	-	-	-	-	-	-	-	-	-	-	-	-	-	-	-	-	-	-	0

A.4 Real World Problems

This section presents the properties of four selected real-world problems.

For fulfilling the demands between the nodes, different line types with only discrete capacities and cost are available. The cost of installing a link consists of a fixed and length dependent share. Both depend on the capacity of the link. The cost are based on the tariffs of the German Telecom from 1996 and represent the amount of money (in German Marks) a company has to pay for a telecommunication line of a specific length and capacity per month. For a detailed description of the four different problems the reader is referred to subsection 8.2.6.

In particular, the overall cost of a communication network is calculated as

$$C = \sum_i f(d_i, b_i),$$

where i denotes the used links, d_i the length of a link, and b_i the capacity of a link. The length of a link connecting the nodes x and y is calculated according to the Euclidean distance between the nodes x and y. The capacity of a link must be selected at a higher level than the overall traffic over the link. Therefore,

$$t_i \leq b_i,$$

where t_i denotes the overall traffic over the link i. This means that to every link a line is assigned with the next higher available capacity.

We illustrate this with a brief example. If there are three line types available with capacity 64 kBit/s, 512 kBit/s, and 2048 kBit/s, a line with capacity 64 kBit/s is assigned to all links with less than 64 kBit/s of traffic. If the traffic over a link is between 64 kBit/s and 512 kBit/s the 512 kBit/s line is chosen. If the traffic over a link exceeds 512 kBit/s the 2048 kBit/s line must be chosen.

In analogy to the previous sections in Table A.20 we present the properties of the best known solutions to the four problems. Table A.21 (problem 1 and 2), Table A.22 (problem 3), and Table A.23 (problem 4) illustrate how the cost of a link depends on the capacity b_i and length l_i of the used line. The largest available line type is 2048 kBit/s. If the traffic over a link exceeds 2048 kBit/s a large penalty is used.

In Table A.24 (problem 1 and 3), Table A.25 (problem 2), and Table A.26 (problem 4) we present the demands between the 15 respective 16 cities. Table A.27 lists the coordinates of the cities. The distances $d_{i,j}$ between two cities i and j are calculated as

$$d_{i,j} = \sqrt{(x_i - y_i)^2 + (y_i - y_j)},$$

where x and y denote the coordinates of the cities. To get the distances and coordinates in kilometer, the distances and coordinates must be multiplied by 14.87. The factor 14.87 results from the used "Gebührenfeldverfahren" of the German Telekom.

Table A.20. Cost and structure of the best solutions to selected real-world test instances

problem	cost	used links
1	60 883	2-0, 3-0, 3-1, 4-0, 5-0, 6-0, 7-0, 8-2, 10-3, 10-9, 11-3, 12-1, 13-10, 14-3, 15-5
2	58 619	4-0, 5-0, 6-0, 7-2, 9-0, 9-1, 9-2, 9-8, 10-3, 11-1, 12-9, 13-3, 13-9, 14-4
3	28 451	4-0, 4-3, 5-0, 6-0, 7-0, 8-0, 8-2, 9-0, 10-9, 11-3, 12-1, 13-9, 14-1, 14-3, 15-5
4	112 938	1-0, 2-0, 6-0, 6-4, 6-5, 7-0, 9-0, 10-0, 11-0, 11-3, 12-9, 13-8, 13-9, 14-0, 15-6

Table A.21. Cost structure for real-world problems 1 and 2

capacity b	distance d	cost $c = f(d, b)$
64 kBit/s	[0.00; 1.00]	$334.58l + 385.00$
	]1.00; 3.00]	$148.70l + 572.00$
	]3.00; 10.00]	$29.74l + 972.50$
	]10.00; ∞]	$22.31l + 1047.00$
512 kBit/s	[0.00; 1.00]	$1107.00l + 975.00$
	]1.00; 3.00]	$520.00l + 1567.00$
	]3.00; 10.00]	$178.00l + 2717.00$
	]10.00; ∞]	$111.53l + 3392.00$
2048 kBit/s	[0.00; 1.00]	$2215.00l + 1950.00$
	]1.00; 3.00]	$1040.90l + 3135.00$
	]3.00; 10.00]	$356.88l + 5435.00$
	]10.00; ∞]	$223.05l + 6785.00$
> 2048 kBit/s	[0.00; ∞]	$500000l + 50000$

Table A.22. Cost structure for real-world problem 3

capacity b	distance d	cost $c = f(d, b)$
64 kBit/s	[0.00; 1.00]	$334.58l + 385.00$
	]1.00; 3.00]	$148.70l + 572.00$
	]3.00; 10.00]	$29.74l + 972.50$
	]10.00; ∞]	$22.31l + 1047.00$
512 kBit/s	[0.00; 1.00]	$1107.00l + 97.50$
	]1.00; 3.00]	$520.00l + 156.70$
	]3.00; 10.00]	$178.00l + 271.70$
	]10.00; ∞]	$111.53l + 339.20$
2048 kBit/s	[0.00; 1.00]	$2215.00l + 195.00$
	]1.00; 3.00]	$1040.90l + 313.50$
	]3.00; 10.00]	$356.88l + 543.50$
	]10.00; ∞]	$223.05l + 678.50$
> 2048 kBit/s	[0.00; ∞]	$500000l + 50000$

Table A.23. Cost structure for real-world problem 4

capacity b	distance d	cost $c = f(d, b)$
64 kBit/s	[0.00; 1.00]	$334.58l + 385.00$
	]1.00; 3.00]	$148.70l + 572.00$
	]3.00; 10.00]	$29.74l + 972.50$
	]10.00; ∞]	$22.31l + 1047.00$
128 kBit/s	[0.00; 1.00]	$669.16l + 770.00$
	]1.00; 3.00]	$297.40l + 1144.00$
	]3.00; 10.00]	$59.48l + 1945.00$
	]10.00; ∞]	$44.62l + 2094.00$
512 kBit/s	[0.00; 1.00]	$1107.00l + 975.00$
	]1.00; 3.00]	$520.00l + 1567.00$
	]3.00; 10.00]	$178.00l + 2717.00$
	]10.00; ∞]	$111.53l + 3392.00$
2048 kBit/s	[0.00; 1.00]	$2215.00l + 1950.00$
	]1.00; 3.00]	$1040.90l + 3135.00$
	]3.00; 10.00]	$356.88l + 5435.00$
	]10.00; ∞]	$223.05l + 6785.00$
4096 kBit/s	[0.00; 1.00]	$4430.00l + 3900.00$
	]1.00; 3.00]	$2081.80l + 6270.00$
	]3.00; 10.00]	$713.76l + 10870.00$
	]10.00; ∞]	$446.10l + 13570.00$
> 2048 kBit/s	[0.00; ∞]	$500000l + 50000$

Table A.24. Demand matrix for real-world problem 1 and 3

	0	1	2	3	4	5	6	7	8	9	10	11	12	13	14	15
0	0	424	458	727	468	414	440	521	50	48	381	34	28	48	34	28
1	-	0	0	0	0	0	0	0	0	0	0	0	0	0	0	0
2	-	-	0	0	0	0	0	0	0	0	0	0	0	0	0	0
3	-	-	-	0	0	0	0	0	0	0	0	0	0	0	0	0
4	-	-	-	-	0	0	0	0	0	0	0	0	0	0	0	0
5	-	-	-	-	-	0	0	0	0	0	0	0	0	0	0	0
6	-	-	-	-	-	-	0	0	0	0	0	0	0	0	0	0
7	-	-	-	-	-	-	-	0	0	0	0	0	0	0	0	0
8	-	-	-	-	-	-	-	-	0	0	0	0	0	0	0	0
9	-	-	-	-	-	-	-	-	-	0	0	0	0	0	0	0
10	-	-	-	-	-	-	-	-	-	-	0	0	0	0	0	0
11	-	-	-	-	-	-	-	-	-	-	-	0	0	0	0	0
12	-	-	-	-	-	-	-	-	-	-	-	-	0	0	0	0
13	-	-	-	-	-	-	-	-	-	-	-	-	-	0	0	0
14	-	-	-	-	-	-	-	-	-	-	-	-	-	-	0	0
15	-	-	-	-	-	-	-	-	-	-	-	-	-	-	-	0

Table A.25. Demand matrix for real-world problem 2

	0	1	2	3	4	5	6	7	8	9	10	11	12	13	14
0	0	424	458	200	468	440	521	50	48	600	34	28	48	34	28
1	-	0	0	0	0	0	0	0	0	0	0	0	0	40	0
2	-	-	0	0	0	0	0	0	0	0	0	0	0	0	0
3	-	-	-	0	0	0	0	0	0	0	0	0	0	0	0
4	-	-	-	-	0	0	0	0	0	0	0	0	0	0	100
5	-	-	-	-	-	0	0	0	0	0	0	0	0	0	0
6	-	-	-	-	-	-	0	0	0	0	0	0	0	0	0
7	-	-	-	-	-	-	-	0	0	0	0	0	0	0	0
8	-	-	-	-	-	-	-	-	0	0	0	0	0	0	0
9	-	-	-	-	-	-	-	-	-	0	0	0	0	0	0
10	-	-	-	-	-	-	-	-	-	-	0	0	0	0	0
11	-	-	-	-	-	-	-	-	-	-	-	0	0	0	0
12	-	-	-	-	-	-	-	-	-	-	-	-	0	0	0
13	-	-	-	-	-	-	-	-	-	-	-	-	-	0	0
14	-	-	-	-	-	-	-	-	-	-	-	-	-	-	0

Table A.26. Demand matrix for real-world problem 4

	0	1	2	3	4	5	6	7	8	9	10	11	12	13	14	15
0	0	308	491	364	36	51	195	72	114	111	14	150	78	136	33	44
1	-	0	503	323	27	38	146	54	86	83	11	112	59	102	24	33
2	-	-	0	272	18	25	97	36	57	55	7	75	39	68	16	22
3	-	-	-	0	9	12	48	18	28	27	3	37	19	34	8	11
4	-	-	-	-	0	51	17	1	34	40	54	36	47	45	25	11
5	-	-	-	-	-	0	15	63	22	16	31	42	28	54	33	7
6	-	-	-	-	-	-	0	5	26	62	54	45	39	12	16	18
7	-	-	-	-	-	-	-	0	32	13	40	22	20	34	61	38
8	-	-	-	-	-	-	-	-	0	35	16	54	13	38	49	17
9	-	-	-	-	-	-	-	-	-	0	10	12	47	4	5	49
10	-	-	-	-	-	-	-	-	-	-	0	49	10	55	28	39
11	-	-	-	-	-	-	-	-	-	-	-	0	10	4	48	37
12	-	-	-	-	-	-	-	-	-	-	-	-	0	19	41	38
13	-	-	-	-	-	-	-	-	-	-	-	-	-	0	17	34
14	-	-	-	-	-	-	-	-	-	-	-	-	-	-	0	36
15	-	-	-	-	-	-	-	-	-	-	-	-	-	-	-	0

Table A.27. Position of the nodes for the four selected real-world problems

problem	node	x	y	node	x	y	node	x	y
1, 3, 4	1	29	12	2	26	42	3	41	34
	4	10	27	5	18	18	6	16	13
	7	19	7	8	30	2	9	41	25
	10	29	25	11	24	34	12	10	25
	13	27	48	14	35	26	15	14	32
	16	9	12						
2	1	29	12	2	26	42	3	41	34
	4	10	27	5	18	18	6	19	7
	7	30	2	8	41	25	9	29	25
	10	24	34	11	10	25	12	27	48
	13	35	26	14	14	32	15	9	12

Table A.28. Distance matrix for real-world problems 1, 3, and 4

0	1	2	3	4	5	6	7	8	9	10	11	12	13	14	15
0	0	30.15	25.06	24.21	12.53	13.04	11.18	10.05	17.69	13.00	22.56	23.02	36.06	15.23	25.00	20.00
1	-	0	17.00	21.93	25.30	30.68	35.69	40.20	22.67	17.26	8.25	23.35	6.08	18.36	15.62	34.48
2	-	-	0	31.78	28.02	32.65	34.83	33.84	9.00	15.00	17.00	32.28	19.80	10.00	27.07	38.83
3	-	-	-	0	12.04	15.23	21.93	32.02	31.06	19.10	15.65	2.00	27.02	25.02	6.40	15.03
4	-	-	-	-	0	5.39	11.05	20.00	24.04	13.04	17.09	10.63	31.32	18.79	14.56	10.82
5	-	-	-	-	-	0	6.71	17.80	27.73	17.69	22.47	13.42	36.69	23.02	19.10	7.07
6	-	-	-	-	-	-	0	12.08	28.43	20.59	27.46	20.12	41.77	24.84	25.50	11.18
7	-	-	-	-	-	-	-	0	25.50	23.02	32.56	30.48	46.10	24.52	34.00	23.26
8	-	-	-	-	-	-	-	-	0	12.00	19.24	31.00	26.93	6.08	27.89	34.54
9	-	-	-	-	-	-	-	-	-	0	10.30	19.00	23.09	6.08	16.55	23.85
10	-	-	-	-	-	-	-	-	-	-	0	16.64	14.32	13.60	10.20	26.63
11	-	-	-	-	-	-	-	-	-	-	-	0	28.60	25.02	8.06	13.04
12	-	-	-	-	-	-	-	-	-	-	-	-	0	23.41	20.62	40.25
13	-	-	-	-	-	-	-	-	-	-	-	-	-	0	21.84	29.53
14	-	-	-	-	-	-	-	-	-	-	-	-	-	-	0	20.62
15	-	-	-	-	-	-	-	-	-	-	-	-	-	-	-	0

Table A.29. Distance matrix for real-world problem 2

0	1	2	3	4	5	6	7	8	9	10	11	12	13	14
0	0	30.15	25.06	24.21	12.53	11.18	10.05	17.69	13.00	22.56	23.02	36.06	15.23	25.00	20.00
1	-	0	17.00	21.93	25.30	35.69	40.20	22.67	17.26	8.25	23.35	6.08	18.36	15.62	34.48
2	-	-	0	31.78	28.02	34.83	33.84	9.00	15.00	17.00	32.28	19.80	10.00	27.07	38.83
3	-	-	-	0	12.04	21.93	32.02	31.06	19.10	15.65	2.00	27.02	25.02	6.40	15.03
4	-	-	-	-	0	11.05	20.00	24.04	13.04	17.09	10.63	31.32	18.79	14.56	10.82
5	-	-	-	-	-	0	12.08	28.43	20.59	27.46	20.12	41.77	24.84	25.50	11.18
6	-	-	-	-	-	-	0	25.50	23.02	32.56	30.48	46.10	24.52	34.00	23.26
7	-	-	-	-	-	-	-	0	12.00	19.24	31.00	26.93	6.08	27.89	34.54
8	-	-	-	-	-	-	-	-	0	10.30	19.00	23.09	6.08	16.55	23.85
9	-	-	-	-	-	-	-	-	-	0	16.64	14.32	13.60	10.20	26.63
10	-	-	-	-	-	-	-	-	-	-	0	28.60	25.02	8.06	13.04
11	-	-	-	-	-	-	-	-	-	-	-	0	23.41	20.62	40.25
12	-	-	-	-	-	-	-	-	-	-	-	-	0	21.84	29.53
13	-	-	-	-	-	-	-	-	-	-	-	-	-	0	20.62
14	-	-	-	-	-	-	-	-	-	-	-	-	-	-	0

References

Abramowitz, M., & Stegun, I. A. (1972). *Handbook of mathematical functions*. New York: Dover Publications.

Abuali, F. N., Schoenefeld, D. A., & Wainwright, R. L. (1994). Designing telecommunications networks using genetic algorithms and probabilistic minimum spanning trees. In Deaton, E., Oppenheim, D., Urban, J., & Berghel, H. (Eds.), *Proceedings of the 1994 ACM Symposium on Applied Computing* (pp. 242–246). ACM Press.

Abuali, F. N., Wainwright, R. L., & Schoenefeld, D. A. (1995). Determinant factorization: A new encoding scheme for spanning trees applied to the probabilistic minimum spanning tree problem. See Eschelman (1995), pp. 470–477.

Ackley, D. H. (1987). *A connectionist machine for genetic hill climbing*. Boston: Kluwer Academic.

Albuquerque, P., Chopard, B., Mazza, C., & Tomassini, M. (2000). On the impact of the representation on fitness landscapes. In Poli, R., Banzhaf, W., Langdon, W. B., Miller, J., Nordin, P., & Fogarty, T. C. (Eds.), *Genetic Programming: Third European Conference* (pp. 1–15). Berlin: Springer-Verlag.

Altenberg, L. (1997). Fitness distance correlation analysis: An instructive counterexample. In Bäck, T. (Ed.), *Proceedings of the Seventh International Conference on Genetic Algorithms* (pp. 57–64). San Francisco: Morgan Kaufmann.

Angeline, P. J., Michalewicz, Z., Schoenauer, M., Yao, X., Zalzala, A., & Porto, W. (Eds.) (1999). *Proceedings of the 1999 IEEE congress on evolutionary computation*. IEEE Press.

Asoh, H., & Mühlenbein, H. (1994). On the mean convergence time of evolutionary algorithms without selection and mutation. See Davidor, Schwefel, and Männer (1994), pp. 88–97.

Bäck, T., Fogel, D. B., & Michalewicz, Z. (Eds.) (1997). *Handbook of Evolutionary Computation*. Bristol and New York: Institute of Physics Publishing and Oxford University Press.

Bäck, T., & Schwefel, H.-P. (1995). Evolution strategies I: Variants and their computational implementation. In Winter, G., Périaux, J., Galán, M., & Cuesta, P. (Eds.), *Genetic Algorithms in Engineering and Com-*

puter Science (Chapter 6, pp. 111–126). Chichester: John Wiley and Sons.

Bagley, J. D. (1967). *The behavior of adaptive systems which employ genetic and correlation algorithms*. Doctoral dissertation, University of Michigan. (University Microfilms No. 68-7556).

Banzhaf, W., Daida, J., Eiben, A. E., Garzon, M. H., Honavar, V., Jakiela, M., & Smith, R. E. (Eds.) (1999). *Proceedings of the Genetic and Evolutionary Computation Conference: Volume 1*. San Francisco, CA: Morgan Kaufmann Publishers.

Bean, J. C. (1992, June). *Genetics and random keys for sequencing and optimization* (Technical Report 92-43). Ann Arbor, MI: Department of Industrial and Operations Engineering, University of Michigan.

Bean, J. C. (1994). Genetic algorithms and random keys for sequencing and optimization. *ORSA Journal on Computing, 6*(2), 154–160.

Beasley, D., Bull, D. R., & Martin, R. R. (1993). Reducing epitasis in combinatorial problems by expansive coding. See Forrest (1993), pp. 400–407.

Belew, R. K., & Booker, L. B. (Eds.) (1991). *Proceedings of the Fourth International Conference on Genetic Algorithms*. San Mateo, CA: Morgan Kaufmann.

Berry, L. T. M., Murtagh, B. A., & McMahon, G. (1995). Applications of a genetic-based algorithm for optimal design of tree-structured communication networks. In *Proceedings of the Regional Teletraffic Engineering Conference of the International Teletraffic Congress* (pp. 361–370). Pretoria, South Africa.

Berry, L. T. M., Murtagh, B. A., McMahon, G., & Sugden, S. (1997). Optimization models for communication network design. In *Proceedings of the Fourth International Meeting Decision Sciences Institute* (pp. 67–70). Sydney, Australia.

Berry, L. T. M., Murtagh, B. A., McMahon, G., Sugden, S., & Welling, L. (1999). An integrated GA–LP approach to communication network design. *Telecommunication Systems, 12*(2), 265–280.

Berry, L. T. M., Murtagh, B. A., & Sugden, S. J. (1994). A genetic-based approach to tree network synthesis with cost constraints. In Zimmermann, H. J. (Ed.), *Second European Congress on Intelligent Techniques and Soft Computing - EUFIT'94*, Volume 2 (pp. 626–629). Promenade 9, D-52076 Aachen: Verlag der Augustinus Buchhandlung.

Bethke, A. D. (1981). *Genetic algorithms as function optimizers*. Doctoral dissertation, University of Michigan. (University Microfilms No. 8106101).

Brittain, D. (1999). *Optimisation of the telecommunications access network*. Unpublished doctoral dissertation, University of Bristol, Bristol.

Brittain, D., Williams, J. S., & McMahon, C. (1997). A genetic algorithm approach to planning the telecommunications access network. In Bäck,

T. (Ed.), *Proceedings of the Seventh International Conference on Genetic Algorithms* (pp. 623–628). San Francisco: Morgan Kaufmann.

Cahn, R. S. (1998). *Wide area network design, concepts and tools for optimization*. San Francisco: Morgan Kaufmann Publishers.

Caruana, R. A., & Schaffer, J. D. (1988). Representation and hidden bias: Gray vs. binary coding for genetic algorithms. In Laird, L. (Ed.), *Proceedings of the Fifth International Workshop on Machine Learning* (pp. 153–161). San Mateo, CA: Morgan Kaufmann.

Caruana, R. A., Schaffer, J. D., & Eshelman, L. J. (1989). Using multiple representations to improve inductive bias: Gray and binary coding for genetic algorithms. In Spatz, B. (Ed.), *Proceedings of the Sixth International Workshop on Machine Learning* (pp. 375–378). San Mateo, CA: Morgan Kaufmann.

Cavicchio, Jr., D. J. (1970). *Adaptive search using simulated evolution*. Unpublished doctoral dissertation, University of Michigan, Ann Arbor, MI. (University Microfilms No. 25-0199).

Cayley, A. (1889). A theorem on trees. *Quarterly Journal of Mathematics, 23*, 376–378.

Celli, G., Costamagna, E., & Fanni, A. (1995, October). Genetic algorithm for telecommunication network optimization. In *IEEE Int. Conf. on Systems, Man and Cybernetics*, Volume 2 (pp. 1227–1232). IEEE Systems, Man and Cybernetics Society, Vancouver.

Chu, C.-H., Premkumar, G., Chou, C., & Sun, J. (1999). Dynamic degree constrained network design: A genetic algorithm approach. See Banzhaf, Daida, Eiben, Garzon, Honavar, Jakiela, and Smith (1999), pp. 141–148.

Chu, C. H.and Chou, H., & Premkumar, G. (1999). *Digital data networks design using genetic algorithms* (Technical Report 07-1999). Penn State, USA.

Cohoon, J. P., Hegde, S. U., Martin, W. N., & Richards, D. (1988). Floorplan design using distributed genetic algorithms. In *IEEE International Conference on Computer Aided-Design* (pp. 452–455). IEEE.

Coli, M., & Palazzari, P. (1995a). Searching for the optimal coding in genetic algorithms. In *1995 IEEE International Conference on Evolutionary Computation*, Volume 1 (pp. 92–96). Piscataway, NJ: IEEE Service Center.

Coli, M., & Palazzari, P. (1995b). Searching for the optimal coding in genetic algorithms. *1995 IEEE International Conference on Evolutionary Computation, 1*, 92–96.

Darwin, C. (1859). *On the origin of species*. London: John Murray.

Davidor, Y. (1989). *Epistasis variance – Suitability of a representation to genetic algorithms* (Tech. Rep. No. CS89-25). Rehovot, Israel: Department of Applied Mathematics and Computer Science, The Weizmann Institute of Science.

Davidor, Y. (1991). Epistasis variance: A viewpoint on GA-hardness. See Rawlins (1991), pp. 23–35.

Davidor, Y., Schwefel, H.-P., & Männer, R. (Eds.) (1994). *Parallel Problem Solving from Nature- PPSN III*. Berlin: Springer-Verlag.

Davis, L. (1987). *Genetic algorithms and simulated annealing*. San Mateo, CA: Morgan Kaufmann.

Davis, L. (1989). Adapting operator probabilities in genetic algorithms. See Schaffer (1989), pp. 61–69.

Davis, L., Orvosh, D., Cox, A., & Qiu, Y. (1993). A genetic algorithm for survivable network design. See Forrest (1993), pp. 408–415.

De Jong, K. A. (1975). *An analysis of the behavior of a class of genetic adaptive systems*. Doctoral dissertation, University of Michigan, Ann Arbor. (University Microfilms No. 76-9381).

Deb, K., Altenberg, L., Manderick, B., Bäck, T., Michalewicz, Z., Mitchell, M., & Forrest, S. (1997). Theoretical foundations and properties of evolutionary computations: fitness landscapes. See Bäck, Fogel, and Michalewicz (1997) (pp. B2.7:1–B2.7:25).

Deb, K., & Goldberg, D. E. (1993). Analyzing deception in trap functions. See Whitley (1993), pp. 93–108.

Deb, K., & Goldberg, D. E. (1994). Sufficient conditions for deceptive and easy binary functions. *Annals of Mathematics and Artificial Intelligence*, *10*, 385–408.

Dengiz, B., Altiparmak, F., & Smith, A. E. (1997a). Efficient optimization of all-terminal reliable networks, using an evolutionary approach. *IEEE Transactions on Reliability*, *46*(1), 18–26.

Dengiz, B., Altiparmak, F., & Smith, A. E. (1997b). Local search genetic algorithm for optimal design of reliable networks. *IEEE Transactions on Evolutionary Computation*, *1*(3), 179–188.

Dengiz, B., Altiparmak, F., & Smith, A. E. (1997c). Local search genetic algorithm for optimization of highly reliable communications networks. In Bäck, T. (Ed.), *Proceedings of the Seventh International Conference on Genetic Algorithms* (pp. 650–657). San Francisco: Morgan Kaufmann.

Ebner, M., Langguth, P., Albert, J., Shackleton, M., & Shipman, R. (2001, 27-30 May). On neutral networks and evolvability. In *Proceedings of the 2001 Congress on Evolutionary Computation CEC2001* (pp. 1–8). COEX, World Trade Center, 159 Samseong-dong, Gangnam-gu, Seoul, Korea: IEEE Press.

Edelson, W., & Gargano, M. L. (2000). Feasible encodings for GA solutions of constrained minimal spanning tree problems. See Whitley, Goldberg, Cantú-Paz, Spector, Parmee, and Beyer (2000), pp. 754.

Edelson, W., & Gargano, M. L. (2001, 7 July). Leaf constrained minimal spanning trees solved by a GA with feasible encodings. In Wu, A. S. (Ed.), *Proceedings of the 2001 Genetic and Evolutionary Computaton*

Conference Workshop Program (pp. 268–271). San Francisco, California, USA.

Elbaum, R., & Sidi, M. (1996). Topological design of local-area networks using genetic algorithms. *IEEE/ACM Transactions on Networking*, *4*(5), 766–778.

Eschelman, L. (Ed.) (1995). *Proceedings of the Sixth International Conference on Genetic Algorithms*. San Francisco, CA: Morgan Kaufmann.

Eshelman, L. J., & Schaffer, J. D. (1991). Preventing premature convergence in genetic algorithms by preventing incest. See Belew and Booker (1991), pp. 115–122.

Even, S. (1973). *Algorithmic combinatorics*. New York: The Macmillan Company.

Feller, W. (1957). *An introduction to probability theory and its applications* (1st ed.), Volume 1. New York: John Wiley & Sons.

Forrest, S. (Ed.) (1993). *Proceedings of the Fifth International Conference on Genetic Algorithms*. San Mateo, CA: Morgan Kaufmann.

Fox, B. R., & McMahon, M. B. (1991). Genetic operators for sequencing problems. See Rawlins (1991), pp. 284–300.

Gale, J. S. (1990). *Theoretical population genetics*. London: Unwin Hyman.

Garey, M. R., & Johnson, D. S. (1979). *Computers and intractability: A guide to the theory of NP-completeness*. New York: W. H. Freeman.

Gargano, M. L., Edelson, W., & Koval, O. (1998). A genetic algorithm with feasible search space for minimal spanning trees with time-dependent edge costs. In Koza, J. R., Banzhaf, W., Chellapilla, K., Deb, K., Dorigo, M., Fogel, D. B., Garzon, M. H., Goldberg, D. E., Iba, H., & Riolo, R. L. (Eds.), *Genetic Programming 98* (pp. 495). San Francisco: Morgan Kaufmann Publishers.

Gaube, T. (2000, Februar). *Optimierung baumförmiger Netzwerkstrukturen mit Hilfe des Link and Node Biased Encodings*. Master's thesis, Universität Bayreuth, Lehrstuhl für Wirtschaftsinformatik.

Gaube, T., & Rothlauf, F. (2001). The link and node biased encoding revisited: Bias and adjustment of parameters. In Boers, E. J. W., Cagnoni, S., Gottlieb, J., Hart, E., Lanzi, P. L., Raidl, G. R., Smith, R. E., & Tijink, H. (Eds.), *Applications of evolutionary Computing: Proc. EvoWorkshops 2001* (pp. 1–10). Berlin: Springer.

Gen, M., Ida, K., & Kim, J. (1998). A spanning tree-based genetic algorithm for bicriteria topological network design. See Institute of Electrical and Electronics Engineers (1998), pp. 15–20.

Gen, M., & Li, Y. (1999). Spanning tree-based genetic algorithms for the bicriteria fixed charge transportation problem. See Angeline, Michalewicz, Schoenauer, Yao, Zalzala, and Porto (1999), pp. 2265–2271.

Gen, M., Zhou, G., & Takayama, M. (1998). A comparative study of tree encodings on spanning tree problems. See Institute of Electrical and Electronics Engineers (1998), pp. 33–38.

Gerrits, M., & Hogeweg, P. (1991). Redundant coding of an NP-complete problem allows effective Genetic Algorithm search. In Schwefel, H.-P., & Männer, R. (Eds.), *Parallel Problem Solving from Nature* (pp. 70–74). Berlin: Springer-Verlag.

Gerstacker, J. (1999, Februar). *Netzwerkplanung durch Einsatz naturanaloger Verfahren*. Master's thesis, Universität Bayreuth, Lehrstuhl für Wirtschaftsinformatik.

Goldberg, D. E. (1987). Simple genetic algorithms and the minimal, deceptive problem. See Davis (1987) (Chapter 6, pp. 74–88).

Goldberg, D. E. (1989a). Genetic algorithms and Walsh functions: Part I, a gentle introduction. *Complex Systems*, *3*(2), 129–152.

Goldberg, D. E. (1989b). Genetic algorithms and Walsh functions: Part II, deception and its analysis. *Complex Systems*, *3*, 153–171.

Goldberg, D. E. (1989c). *Genetic algorithms in search, optimization, and machine learning*. Reading, MA: Addison-Wesley.

Goldberg, D. E. (1990a). A note on Boltzmann tournament selection for genetic algorithms and population-oriented simulated annealing. *Complex Systems*, *4*(4), 445–460. (Also IlliGAL Report No. 90003).

Goldberg, D. E. (1990b, September). *Real-coded genetic algorithms, virtual alphabets, and blocking* (IlliGAL Report No. 90001). Urbana, IL: University of Illinois at Urbana-Champaign.

Goldberg, D. E. (1991a). *Genetic algorithm theory*. Fourth International Conference conference on Genetic Algorithms Tutorial, unpublished manuscript.

Goldberg, D. E. (1991b). Real-coded genetic algorithms, virtual alphabets, and blocking. *Complex Systems*, *5*(2), 139–167. (Also IlliGAL Report 90001).

Goldberg, D. E. (1992). Construction of high-order deceptive functions using low-order Walsh coefficients. *Annals of Mathematics and Artificial Intelligence*, *5*, 35–48.

Goldberg, D. E. (1998). *The race, the hurdle, and the sweet spot: Lessons from the genetic algorithms for the automation of design innovation and creativity* (IlliGAL Report No. 98007). Urbana, IL: University of Illinois at Urbana-Champaign.

Goldberg, D. E. (1999). The race, the hurdle, and the sweet spot. In Bentley, P. J. (Ed.), *Evolutionary Design by Computers* (pp. 105–118). San Francisco, CA: Morgan Kaufmann.

Goldberg, D. E. (2002). *The design of innovation*. Series on Genetic Algorithms and Evolutionary Computation. Dordrecht, The Netherlands: Kluwer. to appear soon.

Goldberg, D. E., Deb, K., & Clark, J. H. (1991). *Genetic algorithms, noise, and the sizing of populations* (IlliGAL Report No. 91010). Urbana, IL: University of Illinois at Urbana-Champaign.

Goldberg, D. E., Deb, K., & Clark, J. H. (1992). Genetic algorithms, noise, and the sizing of populations. *Complex Systems*, *6*, 333–362.

Goldberg, D. E., Deb, K., Kargupta, H., & Harik, G. (1993). Rapid, accurate optimization of difficult problems using fast messy genetic algorithms. See Forrest (1993), pp. 56–64.

Goldberg, D. E., Deb, K., & Thierens, D. (1993). Toward a better understanding of mixing in genetic algorithms. *Journal of the Society of Instrument and Control Engineers*, *32*(1), 10–16.

Goldberg, D. E., Korb, B., & Deb, K. (1989). Messy genetic algorithms: Motivation, analysis, and first results. *Complex Systems*, *3*(5), 493–530.

Goldberg, D. E., & Segrest, P. (1987). Finite Markov chain analysis of genetic algorithms. In Grefenstette, J. J. (Ed.), *Proceedings of the Second International Conference on Genetic Algorithms* (pp. 1–8). Hillsdale, NJ: Lawrence Erlbaum Associates.

Gottlieb, J., & Eckert, C. (2000). A comparison of two representations for the fixed charge transportation problem. See Schoenauer, Deb, Rudolph, Yao, Lutton, Merelo, and Schwefel (2000), pp. 345–354.

Gottlieb, J., Julstrom, B. A., Raidl, G. R., & Rothlauf, F. (2001). Prüfer numbers: A poor representation of spanning trees for evolutionary search. In Spector, L., E., G., Wu, A., B., L. W., Voigt, H.-M., Gen, M., Sen, S., Dorigo, M., Pezeshk, S., Garzon, M., & Burke, E. (Eds.), *Proceedings of the Genetic and Evolutionary Computation Conference 2001* (pp. 343–350). San Francisco, CA: Morgan Kaufmann Publishers.

Gottlieb, J., & Raidl, G. R. (2000). The effects of locality on the dynamics of decoder-based evolutionary search. See Whitley, Goldberg, Cantú-Paz, Spector, Parmee, and Beyer (2000), pp. 283–290.

Grasser, C. (2000, Mai). *Multiperiodenplanung von Kommunikationsnetzwerken mit naturanalogen Verfahren*. Master's thesis, Universität Bayreuth, Lehrstuhl für Wirtschaftsinformatik.

Güls, D. (1996, August). *Optimierung der Netzkonfiguration eines Corporate network am Beispiel des DATEV-Genossenschaftsnetzes*. Master's thesis, Universität Bayreuth, Lehrstuhl für Wirtschaftsinformatik.

Hamming, R. (1980). *Coding and information theory*. Prentice-Hall.

Harik, G. (1999). *Linkage learning via probabilistic modeling in the ECGA* (IlliGAL Report No. 99010). Urbana, IL: University of Illinois at Urbana-Champaign.

Harik, G., Cantú-Paz, E., Goldberg, D. E., & Miller, B. L. (1999). The gambler's ruin problem, genetic algorithms, and the sizing of populations. *Evolutionary Computation*, *7*(3), 231–253.

Harik, G. R., Cantú-Paz, E., Goldberg, D. E., & Miller, B. L. (1997). The gambler's ruin problem, genetic algorithms, and the sizing of populations. In Bäck, T. (Ed.), *Proceedings of the Forth International Conference on Evolutionary Computation* (pp. 7–12). New York: IEEE Press.

Harik, G. R., & Goldberg, D. E. (1996). Learning linkage. In Belew, R. K., & Vose, M. D. (Eds.), *Foundations of Genetic Algorithms 4* (pp. 247–262). San Francisco, CA: Morgan Kaufmann.

Hartl, D. L., & Clark, A. G. (1997). *Principles of population genetics* (3 ed.). Sunderland, Massachusetts: Sinauer Associates.

Hinterding, R. (2000, 6-9 July). Representation, mutation and crossover issues in evolutionary computation. In *Proceedings of the 2000 Congress on Evolutionary Computation CEC00* (pp. 916–923). La Jolla Marriott Hotel La Jolla, California, USA: IEEE Press.

Holland, J. H. (1975). *Adaptation in natural and artificial systems*. Ann Arbor, MI: University of Michigan Press.

Horn, J. (1995). *Genetic algorithms, problem difficulty, and the modality of fitness landscapes* (IlliGAL Report No. 95004). Urbana, IL: University of Illinois at Urbana-Champaign.

Hu, T. C. (1974, September). Optimum communication spanning trees. *SIAM Journal on Computing, 3*(3), 188–195.

Huynen, M., Stadler, P., & Fontana, W. (1996). Smoothness within ruggedness: The role of neutrality in adaptation. In *Proc. Natl. Acad. Sci. USA, 93* (pp. 397–401).

Igel, C. (1998). Causality of hierarchical variable length representations. See Institute of Electrical and Electronics Engineers (1998), pp. 324–329.

Institute of Electrical and Electronics Engineers (Ed.) (1998). *Proceedings of 1998 IEEE International Conference on Evolutionary Computation*. Piscataway, NJ: IEEE Service Center.

Jones, T. (1995). *Evolutionary algorithms and heuristic search*. Unpublished doctoral dissertation, University of New Mexico, Alberquerque, NM.

Jones, T., & Forrest, S. (1995). Fitness distance correlation as a measure of problem difficulty for genetic algorithms. See Eschelman (1995), pp. 184–192.

Julstrom, B. A. (1993). A genetic algorithm for the rectilinear steiner problem. See Forrest (1993), pp. 474–480.

Julstrom, B. A. (1999). Redundant genetic encodings may not be harmful. See Banzhaf, Daida, Eiben, Garzon, Honavar, Jakiela, and Smith (1999), pp. 791.

Julstrom, B. A. (2000). Comparing lists of edges with two other genetic codings of rectilinear steiner trees. In *Late Breaking Papers at the 2000 Genetic And Evolutionary Computation Conference* (pp. 155–161). Madison, WI: Omni Press.

Julstrom, B. A. (2001, 7 July). The blob code: A better string coding of spanning trees for evolutionary search. In Wu, A. S. (Ed.), *Proceedings of the 2001 Genetic and Evolutionary Computaton Conference Workshop Program* (pp. 256–261). San Francisco, California, USA.

Kargupta, H. (2000a). The genetic code and the genome representation. In Annie, S. W. (Ed.), *Proceedings of the 2000 Genetic and Evolutionary Computaton Conference Workshop Program* (pp. 179–184). Las Vegas Nevada: University of Central Florida.

Kargupta, H. (2000b). The genetic code-like transformations and their effect on learning functions. See Schoenauer, Deb, Rudolph, Yao, Lutton, Merelo, and Schwefel (2000), pp. 99–108.

Kargupta, H., Deb, K., & Goldberg, D. E. (1992). Ordering genetic algorithms and deception. In Männer, R., & Manderick, B. (Eds.), *Parallel Problem Solving from Nature- PPSN II* (pp. 47–56). Amsterdam: Elsevier Science.

Kershenbaum, A. (1993). *Telecommunications network design algorithms.* New York: McGraw Hill.

Kim, J. R., & Gen, M. (1999). Genetic algorithm for solving bicriteria network topology design problem. See Angeline, Michalewicz, Schoenauer, Yao, Zalzala, and Porto (1999), pp. 2272–2279.

Kimura, M. (1962). On the probability of fixation of mutant genes in a population population.. *Genetics, 47*, 713–719.

Kimura, M. (1964). Diffusion models in population genetics. *J. Appl. Prob., 1*, 177–232.

Knjazew, D. (2000). *Application of the fast messy genetic algorithm to permutation and scheduling problems* (IlliGAL Report No. 2000022). Urbana, IL: University of Illinois at Urbana-Champaign.

Knjazew, D., & Goldberg, D. E. (2000). *Large-scale permutation optimization with the ordering messy genetic algorithm* (IlliGAL Report No. 2000013). Urbana, IL: University of Illinois at Urbana-Champaign.

Knowles, J., Corne, D., & Oates, M. (1999). A new evolutionary approach to the degree constrained minimum spanning tree problem. See Banzhaf, Daida, Eiben, Garzon, Honavar, Jakiela, and Smith (1999), pp. 794.

Ko, K.-T., Tang, K.-S., Chan, C.-Y., Man, K.-F., & Kwong, S. (1997, August). Using genetic algorithms to design mesh networks. *Computer, 30*(8), 56–61.

Krishnamoorthy, M., Ernst, A. T., & Sharaiha, Y. M. (1999). *Comparison of algorithms for the degree constrained minimum spanning tree* (Tech. Rep.). Clayton, Australia: CSIRO Mathematical and Information Sciences.

Lewontin, R. C. (1974). *The genetic basis of evolutionary change.* Number 25 in Columbia biological series. New York: Columbia University Press.

Li, Y. (2001). An effective implementation of a ga using a direct tree representation for constrained minimum spanning tree problems. In Boers,

E. J. W., Cagnoni, S., Gottlieb, J., Hart, E., Lanzi, P. L., Raidl, G. R., Smith, R. E., & Tijink, H. (Eds.), *Applications of evolutionary Computing: Proc. EvoWorkshops 2001* (pp. 11–19). Berlin: Springer.

Li, Y., & Bouchebaba, Y. (1999). A new genetic algorithm for the optimal communication spanning tree problem. In Fonlupt, C., Hao, J.-K., Lutton, E., Ronald, E., & Schoenauer, M. (Eds.), *Proceedings of Artificial Evolution: Fifth European Conference* (pp. 162–173). Berlin: Springer.

Li, Y., Gen, M., & Ida, K. (1998). Fixed charge transportation problem by spanning tree-based genetic algorithm. *Beijing Mathematics*, *4*(2), 239–249.

Liepins, G. E., & Vose, M. D. (1990). Representational issues in genetic optimization. *Journal of Experimental and Theoretical Artificial Intelligence*, *2*, 101–115.

Liepins, G. E., & Vose, M. D. (1991). Polynomials, basis sets, and deceptiveness in genetic algorithms. *Complex Systems*, *5*(1), 45–61.

Lobo, F. G., Goldberg, D. E., & Pelikan, M. (2000). Time complexity of genetic algorithms on exponentially scaled problems. See Whitley, Goldberg, Cantú-Paz, Spector, Parmee, and Beyer (2000), pp. 151–158.

Mahfoud, S. W., & Goldberg, D. E. (1995). Parallel recombinative simulated annealing: A genetic algorithm. In *Parallel Computing*, Volume 21 (pp. 1–28). Amsterdam, The Netherlands: Elsevier Science.

Manderick, B., de Weger, M., & Spiessens, P. (1991). The genetic algorithm and the structure of the fitness landscape. See Belew and Booker (1991), pp. 143–150.

Mason, A. (1995). A non-linearity measure of a problem's crossover suitability. In *1995 IEEE International Conference on Evolutionary Computation*, Volume 1 (pp. 68–73). Piscataway, NJ: IEEE Service Center.

Mendel, G. (1866). Versuche über Pflanzen-Hybriden. In *Verhandlungen des naturforschenden Vereins*, Volume 4 (pp. 3–47). Brünn: Naturforschender Verein zu Brünn.

Miller, B. L. (1997). *Noise, sampling, and efficient genetic algorithms*. doctoral dissertation, University of Illinois at Urbana-Champaign, Urbana. Also IlliGAL Report No. 97001.

Miller, B. L., & Goldberg, D. E. (1996a). Genetic algorithms, selection schemes, and the varying effects of noise. *Evolutionary Computation*, *4*(2), 113–131.

Miller, B. L., & Goldberg, D. E. (1996b). *Optimal sampling for genetic algorithms* (IlliGAL Report No. 96005). Urbana, IL: University of Illinois at Urbana-Champaign.

Minoux, M. (1987). Network synthesis and dynamic network optimization. *Ann. Discrete Math.*, *31*, 283–323.

Mühlenbein, H., & Schlierkamp-Voosen, D. (1993). Predictive models for the breeder genetic algorithm: I. Continuous parameter optimization. *Evolutionary Computation*, *1*(1), 25–49.

Nagylaki, T. (1992). *Introduction to theoretical population genetics*, Volume 21 of *Biomathematics*. Berlin, Germany / Heidelberg, Germany / London, UK / etc.: Springer-Verlag.

Naudts, B., Suys, D., & Verschoren, A. (1997). Epistasis as a basic concept in formal landscape analysis. In Bäck, T. (Ed.), *Proceedings of the Seventh International Conference on Genetic Algorithms* (pp. 65–72). San Francisco: Morgan Kaufmann.

Norman, B. A. (1995). *Scheduling using the random keys genetic algorithm.* unpublished PhD thesis, University of Michigan, Ann Arbor, Michigan.

Norman, B. A., & Bean, J. C. (1994). *Random keys genetic algorithm for job shop scheduling* (Tech. Rep. No. 94-5). Ann Arbor, MI: The University of Michigan.

Norman, B. A., & Bean, J. C. (1997). Operation sequencing and tool assignment for multiple spindle CNC machines. In *Proceedings of the Forth International Conference on Evolutionary Computation* (pp. 425–430). Piscataway, NJ: IEEE.

Norman, B. A., & Bean, J. C. (2000). Scheduling operations on parallel machines. *IIE Transactions, 32*(5), 449–459.

Norman, B. A., & Smith, A. E. (1997). Random keys genetic algorithm with adaptive penalty function for optimization of constrained facility layout problems. In *Proceedings of the Forth International Conference on Evolutionary Computation* (pp. 407–411). Piscataway, NJ: IEEE.

Norman, B. A., Smith, A. E., & Arapoglu, R. A. (1998). Integrated facility design using an evolutionary approach with a subordinate network algorithm. In Eiben, A. E., Bäck, T., Schoenauer, M., & Schwefel, H.-P. (Eds.), *Parallel Problem Solving from Nature, PPSN V* (pp. 937–946). Berlin: Springer-Verlag.

Oei, C. K. (1992). *Walsh function analysis of genetic algorithms of non-binary strings.* Master's thesis, University of Illinois at Urbana-Champaign, Department of Computer Science, Urbana.

Orvosh, D., & Davis, L. (1993). Shall we repair? Genetic algorithms, combinatorial optimization, and feasibility constraints. See Forrest (1993), pp. 650.

Palmer, C. C. (1994). *An approach to a problem in network design using genetic algorithms.* unpublished PhD thesis, Polytechnic University, Troy, NY.

Palmer, C. C., & Kershenbaum, A. (1994a). Representing trees in genetic algorithms. In *Proceedings of the First IEEE Conference on Evolutionary Computation*, Volume 1 (pp. 379–384). Piscataway, NJ: IEEE Service Center.

Palmer, C. C., & Kershenbaum, A. (1994b). *Two algorithms for finding optimal communication spanning trees.* IBM research report RC-19394.

Pelikan, M., Goldberg, D. E., & Cantú-Paz, E. (1999). *BOA: The Bayesian optimization algorithm* (IlliGAL Report No. 99003). Urbana, IL: University of Illinois at Urbana-Champaign.

Pelikan, M., Goldberg, D. E., & Lobo, F. (1999). *A survey of optimization by building and using probabilistic models* (IlliGAL Report No. 99018). Urbana, IL: University of Illinois at Urbana-Champaign.

Picciotto, S. (1999). *How to encode a tree*. Doctoral dissertation, University of California, San Diego, USA.

Piggott, P., & Suraweera, F. (1993). Encoding graphs for genetic algorithms: An investigation using the minimum spanning tree problem. In Yao, X. (Ed.), *Preprints of the AI'93 Workshop on Evolutionary Computation* (pp. 37–48). Canberra, Australia: University of New South Wales, Australian Defense Force Academy.

Premkumar, G., Chu, C., & Chou, H. (2001). Telecommunications network design decision - a genetic algorithm approach. Forthcoming in Decision Sciences.

Prim, R. (1957). Shortest connection networks and some generalizations. *Bell System Technical Journal, 36*, 1389–1401.

Prüfer, H. (1918). Neuer Beweis eines Satzes über Permutationen. *Archiv für Mathematik und Physik, 27*, 742–744.

Radcliffe, N. J. (1993). Genetic set recombination. See Whitley (1993), pp. 203–219.

Radcliffe, N. J. (1997). Theoretical foundations and properties of evolutionary computations: schema processing. See Bäck, Fogel, and Michalewicz (1997) (pp. B2.5:1–B2.5:10).

Raidl, G. R. (2000). An efficient evolutionary algorithm for the degree-constrained minimum spanning tree problem. In *Proceedings of 2000 IEEE International Conference on Evolutionary Computation* (pp. 43–48). Piscataway, NJ: IEEE.

Raidl, G. R. (2001, February). Various instances of optimal communication spanning tree problems. personal communciation.

Raidl, G. R., & Drexel, C. (2000, 8 July). A predecessor coding in an EA for the capacitated minimum spanning tree problem. In Whitley, D. (Ed.), *Late Breaking Papers at the 2000 Genetic and Evolutionary Computation Conference* (pp. 309–316). Las Vegas, Nevada, USA.

Raidl, G. R., & Julstrom, B. A. (2000). A weighted coding in a genetic algorithm for the degree-constrained minimum spanning tree problem. In Carroll, J., Damiani, E., Haddad, H., & Oppenheim, D. (Eds.), *Proceedings of the 2000 ACM Symposium on Applied Computing* (pp. 440–445). ACM Press.

Rana, S., & Whitley, W. (1998). Search, binary representations, and counting optima. In *Proceeding of a workshop on Evolutionary Algorithms. Sponsored by the Institute for Mathematics and its Applications* (pp. 1–11). Colorado State University. in press.

Rana, S. B., & Whitley, L. D. (1997). Bit representations with a twist. In Bäck, T. (Ed.), *Proceedings of the Seventh International Conference on Genetic Algorithms* (pp. 188–195). San Francisco: Morgan Kaufmann.

Rawlins, G. J. E. (Ed.) (1991). *Foundations of Genetic Algorithms*. San Mateo, CA: Morgan Kaufmann.

Rechenberg, I. (1973). *Evolutionsstrategie: Optimierung technischer Systeme nach Prinzipien der biologischen Evolution*. Stuttgart-Bad Cannstatt: Friedrich Frommann Verlag.

Reeves, C., & Wright, C. (1994). An experimental design perspective on genetic algorithms. In Whitley, L. D., & Vose, M. D. (Eds.), *Foundations of Genetic Algorithms 3* (pp. 7–22). San Francisco, California: Morgan Kaufmann Publishers, Inc.

Reidys, C. M., & Stadler, P. F. (2001, January). Neutrality in fitness landscapes. *Applied Mathematics and Computation, 117*(2–3), 321–350.

Ronald, S. (1995). *Genetic algorithms and permutation-encoded problems: Diversity preservation and a study of multimodality*. Unpublished doctoral dissertation, The University of South Australia.

Ronald, S. (1997). Robust encodings in genetic algorithms: A survey of encoding issues. In *Proceedings of the Forth International Conference on Evolutionary Computation* (pp. 43–48). Piscataway, NJ: IEEE.

Ronald, S., Asenstorfer, J., & Vincent, M. (1995). Representational redundancy in evolutionary algorithms. In *1995 IEEE International Conference on Evolutionary Computation*, Volume 2 (pp. 631–636). Piscataway, NJ: IEEE Service Center.

Rosenberg, R. S. (1967). *Simulation of genetic populations with biochemical properties*. Doctoral dissertation, The University of Michigan. (University Microfilms No. 67-17,836).

Rothlauf, F., & Goldberg, D. E. (1999). Tree network design with genetic algorithms - an investigation in the locality of the prüfernumber encoding. In Brave, S., & Wu, A. S. (Eds.), *Late Breaking Papers at the Genetic and Evolutionary Computation Conference 1999* (pp. 238–244). Orlando, Florida, USA: Omni Press.

Rothlauf, F., & Goldberg, D. E. (2000). Prüfernumbers and genetic algorithms: A lesson on how the low locality of an encoding can harm the performance of GAs. See Schoenauer, Deb, Rudolph, Yao, Lutton, Merelo, and Schwefel (2000), pp. 395–404.

Rothlauf, F., Goldberg, D. E., & Heinzl, A. (2000). Bad codings and the utility of well-designed genetic algorithms. See Whitley, Goldberg, Cantú-Paz, Spector, Parmee, and Beyer (2000), pp. 355–362.

Rothlauf, F., Goldberg, D. E., & Heinzl, A. (2001, 7 July). On the debate concerning evolutionary search using Prüfer numbers. In Wu, A. S. (Ed.), *Proceedings of the 2001 Genetic and Evolutionary Computaton Conference Workshop Program* (pp. 262–267). San Francisco, California, USA.

Rothlauf, F., Goldberg, D. E., & Heinzl, A. (2002). Network random keys – A tree network representation scheme for genetic and evolutionary algorithms. *Evolutionary Computation, 10*(1), 75–97.

Rudnick, W. M. (1992). *Genetic algorithms and fitness variance with an application to the automated design of artificial neural networks.* Unpublished doctoral dissertation, Oregon Graduate Institute of Science & Technology, Beaverton, OR.

Sastry, K., & Goldberg (2001). *Modeling tournament with replacement using apparent added noise* (IlliGAL Report No. 2001014). Urbana, IL: University of Illinois at Urbana-Champaign.

Schaffer, J. D. (Ed.) (1989). *Proceedings of the Third International Conference on Genetic Algorithms.* San Mateo, CA: Morgan Kaufmann.

Schaffer, J. D., Caruana, R. A., Eshelman, L. J., & Das, R. (1989). A study of control parameters affecting online performance of genetic algorithms for function optimization. See Schaffer (1989), pp. 51–60.

Schnier, T., & Yao, X. (2000, 6-9 July). Using multiple representations in evolutionary algorithms. In *Proceedings of the 2000 Congress on Evolutionary Computation CEC00* (pp. 479–486). La Jolla Marriott Hotel La Jolla, California, USA: IEEE Press.

Schoenauer, M., Deb, K., Rudolph, G., Yao, X., Lutton, E., Merelo, J. J., & Schwefel, H.-P. (Eds.) (2000). *Parallel Problem Solving from Nature, PPSN VI.* Berlin: Springer-Verlag.

Schuster, P. (1997). Genotypes with phenotypes: Adventures in an RNA toy world. *Biophys. Chem., 66*, 75–110.

Schwefel, H.-P. (1975). *Evolutionsstrategie und numerische Optimierung.* Doctoral dissertation, Technical University of Berlin.

Schwefel, H.-P. (1981). *Numerical optimization of computer models.* Chichester: John Wiley & Sons.

Schwefel, H.-P. (1995). *Evolution and optimum seeking.* New York: Wisley & Sons.

Sendhoff, B., Kreutz, M., & von Seelen, W. (1997a). *Causality and the analysis of local search in evolutionary algorithms* (Technical Report). Institut für Neuroinformatik, Ruhr-Universität Bochum.

Sendhoff, B., Kreutz, M., & von Seelen, W. (1997b). A condition for the genotype-phenotype mapping: Causality. In Bäck, T. (Ed.), *Proceedings of the Seventh International Conference on Genetic Algorithms* (pp. 73–80). San Francisco: Morgan Kaufmann.

Shackleton, M., Shipman, R., & Ebner, M. (2000, 6-9 July). An investigation of redundant genotype-phenotype mappings and their role in evolutionary search. In *Proceedings of the 2000 Congress on Evolutionary Computation CEC00* (pp. 493–500). La Jolla Marriott Hotel La Jolla, California, USA: IEEE Press.

Shannon, C. E. (1948). A mathematical theory of communication. *Bell Syst. Technical Jrnl., 27*, 379–423, 623–656.

Shannon, C. E., & Weaver, W. (1949). *The mathematical theory of communication*. Urbana, Illinois: University of Illinois Press.

Shipman, R. (1999). Genetic redundancy: Desirable or problematic for evolutionary adaptation? In *Proceedings of the 4th International Conference on Artificial Neural Networks and Genetic Algorithms (ICAN-NGA)* (pp. 1–11). Springer Verlag.

Shipman, R., Shackleton, M., Ebner, M., & Watson, R. (2000). Neutral search spaces for artificial evolution: A lesson from life. In Bedau, M., McCaskill, J., Packard, N., & Rasmussen, S. (Eds.), *Proceedings of Artificial Life VII* (pp. section III (Evolutionary and Adaptive Dynamics)). MIT Press.

Shipman, R., Shackleton, M., & Harvey, I. (2000). The use of neutral genotype-phenotype mappings for improved evolutionary search. *BT Technology Journal, 18*(4), 103–111.

Sinclair, M. C. (1995). Minimum cost topology optimisation of the COST 239 European optical network. In Pearson, D. W., Steele, N. C., & Albrecht, R. F. (Eds.), *Proceedings of the 1995 International Conference on Artificial Neural Nets and Genetic Algorithms* (pp. 26–29). New York: Springer-Verlag.

Smith, T., Husbands, P., & M., O. (2001a). *Evolvability, neutrality and search space* (Technical Report 535). School of Cognitive and Computing Sciences, University of Sussex.

Smith, T., Husbands, P., & M., O. (2001b). Neutral networks and evolvability with comple genotype-phenotype mapping. to be published in Proceedings of the European Converence on Artificial Life: ECAL2001.

Smith, T., Husbands, P., & M., O. (2001c). Neutral networks in an evolutionary robotics search space. In of Electrical, I., & Engineers, E. (Eds.), *Proceedings of 2001 IEEE International Conference on Evolutionary Computation* (pp. 136–145). Piscataway, NJ: IEEE Service Center.

Streng, C. (1997, Juli). *Optimierung eines bundesweiten Corporate Network am Beispiel der Firma DATEV*. Master's thesis, Universität Erlangen-Nürnberg, Institut für Angewandte Mathematik.

Syswerda, G. (1989). Uniform crossover in genetic algorithms. See Schaffer (1989), pp. 2–9.

Tang, K. S., Man, K. F., & Ko, K. T. (1997). Wireless LAN desing using hierarchical genetic algorithm. In Bäck, T. (Ed.), *Proceedings of the Seventh International Conference on Genetic Algorithms* (pp. 629–635). San Francisco: Morgan Kaufmann.

Thierens, D. (1992). *A hamming space analysis of recombination with uniform crossover*. Draft.

Thierens, D. (1995). *Analysis and design of genetic algorithms*. Leuven, Belgium: Katholieke Universiteit Leuven.

Thierens, D., & Goldberg, D. E. (1993). Mixing in genetic algorithms. See Forrest (1993), pp. 38–45.

Thierens, D., & Goldberg, D. E. (1994). Convergence models of genetic algorithm selection schemes. See Davidor, Schwefel, and Männer (1994), pp. 119–129.

Thierens, D., Goldberg, D. E., & Pereira, Â. G. (1998). Domino convergence, drift, and the temporal-salience structure of problems. See Institute of Electrical and Electronics Engineers (1998), pp. 535–540.

Vose, M. D. (1993). Modeling simple genetic algorithms. See Whitley (1993), pp. 63–73.

Vose, M. D., & Wright, A. H. (1998a). The simple genetic algorithm and the Walsh transform: Part I, theory. *Evolutionary Computation, 6*(3), 253–273.

Vose, M. D., & Wright, A. H. (1998b). The simple genetic algorithm and the Walsh transform: Part II, the inverse. *Evolutionary Computation, 6*(3), 275–289.

Weinberger, E. (1990). Correlated and uncorrelated fitness landscapes and how to tell the difference. *Biological Cybernetics, 63*, 325–336.

Whitley, D. (1999). A free lunch proof for gray versus binary encodings. See Banzhaf, Daida, Eiben, Garzon, Honavar, Jakiela, and Smith (1999), pp. 726–733.

Whitley, D. (2000a). Functions as permutations: Implications for no free lunch, walsh analysis and statistics. See Schoenauer, Deb, Rudolph, Yao, Lutton, Merelo, and Schwefel (2000), pp. 169–178.

Whitley, D. (2000b). Local search and high precision gray codes: Convergence results and neighborhoods. In Martin, W., & Spears, W. (Eds.), *Foundations of Genetic Algorithms 6* (pp. unknown). San Francisco, California: Morgan Kaufmann Publishers, Inc. in press.

Whitley, D., Goldberg, D. E., Cantú-Paz, E., Spector, L., Parmee, L., & Beyer, H.-G. (Eds.) (2000). *Proceedings of the Genetic and Evolutionary Computation Conference 2000.* San Francisco, CA: Morgan Kaufmann Publishers.

Whitley, D., & Rana, S. (1997). Representation, search, and genetic algorithms. In *Proceedings of the 14th National Conference on Artificial Intelligence (AAAI-97)* (pp. 497–502). AAAI Press/MIT Press.

Whitley, D., Rana, S., & Heckendorn, R. (1997). Representation issues in neighborhood search and evolutionary algorithms. In *Genetic Algorithms and Evolution Strategy in Engineering and Computer Science* (Chapter 3, pp. 39–58). West Sussex, England: John Wiley & Sons Ltd.

Whitley, L. D. (Ed.) (1993). *Foundations of Genetic Algorithms 2.* San Mateo, CA: Morgan Kaufmann.

Wolpert, D. H., & Macready, W. G. (1995). *No free lunch theorems for search* (Tech. Rep. No. SFI-TR-95-02-010). Santa Fe, NM: Santa Fe Institute.

Zhou, G., & Gen, M. (1997). Approach to degree-constrained minimum spanning tree problem using genetic algorithm. *Engineering Design & Automation, 3*(2), 157–165.

List of Symbols

α	probability of GEA failure
$\bar{\alpha}$	average percentage of incorrect alleles
α_i	ith coefficients of the polynomial decomposition of x
$\boldsymbol{b}$	biased chromosome
b_i	capacity of a link i in a network
c	crossover-point in string
$\boldsymbol{c}$	characteristic vector
C	overall cost of a communication network
d	signal difference / distance between nodes / distance between individuals
d_c	distance distortion of an encoding
d^h	Hamming distance
d_i	length of a link i
d_m	locality of an encoding
d_{MST}	distance of an individual towards the minimum spanning tree
$d_{i,j}$	distance (cost) between two nodes
$d_{\boldsymbol{x},\boldsymbol{y}}$	distance between individual $\boldsymbol{x}$ and $\boldsymbol{y}$
$\delta(\boldsymbol{h})$	defining length of a schema $\boldsymbol{h}$
D	distance matrix
E	set of links
$E(x)$	mean of x
f_g	genotype-phenotype mapping
f_p	phenotype-fitness mapping
$f(\boldsymbol{x})$	fitness of individual $\boldsymbol{x}$
$f(\boldsymbol{h}, t)$	fitness of schema $\boldsymbol{h}$ at time t
$\bar{f}(t)$	average fitness of population at time t
G	graph
γ_{avg}	average number of genotypes
$\boldsymbol{h}$	schema (ternary string of length l, where $h \in \{0, 1, *\}$

h_i	ith allele of a schema h
I	selection intensity
k	size or oder of BB
k_g	genotypic size of BB
k_p	phenotypic size of BB
k_r	order of redundancy
l	length of a string
l_g	length of a genotypic bitstring
l_p	length of a phenotypic bitstring
l_s	length of an exponentially scaled BB
$l_{i,j}$	link between node i and j
λ	dividing line between converged and unconverged alleles (domino convergence model) / order of schema (distance distortion)
λ_c	size of the convergence window
m	number of BBs
m'	m-1
$m(h, t)$	number of instances of schema h at time t
n	population size (binary and integer problems) / number of nodes (tree problems)
n_{drift}	population size necessary for GAs not to be affected by genetic drift
n_f	number of fitness calls
n_p	number of different phenotypes
N	population size
$\mathbb{N}$	integer numbers
$\mathbb{N}()$	normal distribution with mean μ and standard deviation σ
$o(h)$	order or size of a schema h
p	probability of making the right choice between a single sample of each BB / node-specific bias
p_c	probability of crossover
p_m	probability of mutation
P_1	link-specific bias
P_2	node-specific bias
P_n	probability of GEA success
$\psi_j(x)$	jth Walsh function for x
Φ_g	genotypic search space
Φ_p	phenotypic search space
q	probability of making the wrong decision when deciding between a single sample of each BB

r	number of genotypic BBs that represent the optimal phenotypic BB / order of a binary encoded problem
r	random key vector
r_s	permutation vector
$\mathbb{R}$	real numbers
s	order of scaling / size of tournament / number of possibilities
$s(t)$	probability that an allele is fully converged
σ	standard deviation
σ_{BB}	standard deviation of a BB
σ_f	overall variance of function f
σ_N	standard deviation of additional noise
t	time, number of generations
t_{conv}	convergence time
t_{drift}	drift time
t_i	overall traffic over a link i
T	temperature
u	number of ones
$u(x)$	number of ones in x
V	set of nodes
w	Walsh coefficients
χ	cardinality of alphabet
x_0	expected number of copies of the best BB in a randomly initialized population
x_i	ith allele of an individual x
x	vector of decision variables / individual
x_g	genotype of an individual
x_p	phenotype of an individual
x_i^c	contribution of the ith most salient allele to the fitness of the individual
z	number of nodes
$*$	don't care symbol

List of Acronyms

BB	building block
CV	characteristic vector
GEA	genetic and evolutionary algorithm (meaning GEAs using crossover and mutation)
GA	genetic algorithm (meaning selectorecombinative GEAs)
LNB	link and node biased
MST	minimum spanning tree
NB	node biased
NetDir	direct tree
NetKey	network random key
OCST	optimal communication spanning tree
OCSTP	optimal communication spanning tree problem
RK	random key

Index

GPSR Compliance
The European Union's (EU) General Product Safety Regulation (GPSR) is a set
of rules that requires consumer products to be safe and our obligations to
ensure this.

If you have any concerns about our products, you can contact us on

ProductSafety@springernature.com

In case Publisher is established outside the EU, the EU authorized
representative is:

Springer Nature Customer Service Center GmbH
Europaplatz 3
69115 Heidelberg, Germany